Directed Enzyme Evolution: Advances and Applications

Miguel Alcalde
Editor

Directed Enzyme Evolution: Advances and Applications

Springer

Editor
Miguel Alcalde
Department of Biocatalysis
Institute of Catalysis and Petrochemistry, CSIC
Madrid
Spain

ISBN 978-3-319-50411-7 ISBN 978-3-319-50413-1 (eBook)
DOI 10.1007/978-3-319-50413-1

Library of Congress Control Number: 2017932555

Printed on acid-free paper

This Springer imprint is published by Springer Nature
The registered company is Springer International Publishing AG
The registered company address is: Gewerbestrasse 11, 6330 Cham, Switzerland

Preface

The last couple of decades have witnessed a true revolution in protein engineering that very few could have foreseen, a revolution delivered by the hand of directed evolution. This reliable, robust and effective methodology has enabled us to design enzymes with improved or even novel properties, a mere pipedream only a few years ago. Through consecutive rounds of random mutation, recombination and screening/selection, enzymes tailored *a la carte* are meeting the needs of different industrial processes, addressing the challenges of working at high temperature or extreme pHs, in non-natural environments or in the presence of strong inhibitors. More significantly, through laboratory evolution we can now harness the catalytic promiscuity of many enzymes to undertake non-natural chemistry for a range of applications that lie within the biotechnology rainbow. Conversely, the rise of directed evolution has demonstrated that the tailoring of enzymes, metabolic pathways or even whole microorganisms is now no longer beyond our reach.

Although mimicking the natural process of evolution on a laboratory scale might make us feel ashamed of our immense ignorance regarding protein function (for example through the identification of substitutions of interest that we could have never predicted by rational analysis), dozens of enzymes are being rapidly engineered in a manner that is establishing solid links between the structure–function relationship of proteins and their biotechnological applications. The advent of computational protein engineering, the constant expansion of protein and gene databases, together with the birth of more sophisticated (ultra)high-throughput screening protocols and new library creation methods, is paving the way to expand directed enzyme evolution beyond the limits of nature. As such, the number of enzymes undergoing laboratory evolution is becoming overwhelming, covering many aspects of biotechnology, and contributing to the rapid development of systems and synthetic biology.

As we reach the 25th anniversary of the discovery of directed enzyme evolution, this book presents some case studies of evolved enzymes, while updating concepts and methods within this vigorous research ground. The outstanding panel of contributors in *Directed Enzyme Evolution: Advances and Applications* has ensured the compilation of some remarkable examples of enzymes improved by evolution in just a single volume, as well as providing an interesting collection of the cutting-edge approaches and methods currently available or soon to be added to the directed evolution toolbox. The first part of the book (Chaps. 1, 2, 3, 4, 5, 6 and 7) focuses on several evolutionary success stories, involving tryptophan synthases, therapeutic and stereoselective enzymes, CO_2-fixing enzymes, unspecific peroxygenases, phytases as well as the directed evolution of whole cells. From Chaps. 8 to 10, library creation methods, in silico/computational enzyme design and ancestral enzyme resurrection are described in depth, while offering clues to their applications and prospects.

I truly hope that *Directed Enzyme Evolution: Advances and Applications* will become a valuable benchside book for professors, researchers and students working and/or lecturing in the field of protein engineering and biotechnology, complementing other practical texts and volumes on this fascinating field of research.

Madrid, Spain Miguel Alcalde

Contents

1 **Directed Evolution of an Allosteric Tryptophan Synthase to Create a Platform for Synthesis of Noncanonical Amino Acids** 1
Javier Murciano-Calles, Andrew R. Buller, and Frances H. Arnold

2 **Engineering Therapeutic Enzymes** 17
Stefan Lutz, Elsie Williams, and Pravin Muthu

3 **Recent Advances in Directed Evolution of Stereoselective Enzymes** 69
Manfred T. Reetz

4 **Improving CO_2 Fixation by Enhancing Rubisco Performance** 101
Robert H. Wilson and Spencer M. Whitney

5 **Directed Evolution of Unspecific Peroxygenase** 127
Patricia Molina-Espeja, Patricia Gómez de Santos, and Miguel Alcalde

6 **Recent Advances in Directed Phytase Evolution and Rational Phytase Engineering** 145
Amol V. Shivange and Ulrich Schwaneberg

7 **Strain Development by Whole-Cell Directed Evolution** 173
Tong Si, Jiazhang Lian, and Huimin Zhao

8 **Back to Basics: Creating Genetic Diversity** 201
Kang Lan Tee and Tuck Seng Wong

9 **Resurrected Ancestral Proteins as Scaffolds for Protein Engineering** 229
Valeria A. Risso and Jose M. Sanchez-Ruiz

10 **Molecular Modeling in Enzyme Design, Toward In Silico Guided Directed Evolution** 257
Emanuele Monza, Sandra Acebes, M. Fátima Lucas, and Victor Guallar

Directed Evolution of an Allosteric Tryptophan Synthase to Create a Platform for Synthesis of Noncanonical Amino Acids

1

Javier Murciano-Calles, Andrew R. Buller, and Frances H. Arnold

Abstract

Tryptophan and its derivatives are important natural products and have many biochemical and synthetic applications. However, the more elaborate these derivatives are, the more complex the synthesis becomes. In this chapter, we summarize the development of an engineered enzymatic platform for synthesis of diverse tryptophan analogs. This endeavor utilizes the tryptophan synthase (TrpS) enzyme, an $\alpha_2\beta_2$ heterodimeric protein complex that catalyzes the last two steps in the biosynthetic pathway of tryptophan. Although the synthetically useful reaction (indole + Ser = Trp) takes place in the β-subunit (TrpB), the exquisite allosteric regulation of this enzyme impedes the use of isolated TrpB due to its dramatically decreased activity in the absence of the α-subunit (TrpA). This chapter discusses our efforts to engineer TrpB to serve as a general platform for the synthesis of noncanonical amino acids. We used directed evolution to enhance the activity of TrpB from *Pyrococcus furiosus* (*Pf*TrpB), so that it can act as a stand-alone biocatalyst. Remarkably, we found that mutational activation mimics the allosteric activation induced by binding of TrpA. Toward our goal of expanding the substrate scope of this reaction, we activated other homologs with the same mutations discovered for *Pf*TrpB. We found improved catalysts for the synthesis of 5-substituted tryptophans, an important biological motif. Finally, we performed directed evolution of TrpB for synthesis of β-branched amino acids, a group of products whose chemical syntheses are particularly challenging.

J. Murciano-Calles • A.R. Buller • F.H. Arnold (✉)
Division of Chemistry and Chemical Engineering, California Institute of Technology, Pasadena, CA, USA
e-mail: frances@cheme.caltech.edu

M. Alcalde (ed.), *Directed Enzyme Evolution: Advances and Applications*,
DOI 10.1007/978-3-319-50413-1_1

1.1 Introduction

In addition to being one of the standard 20 proteinogenic α-amino acids, tryptophan (Trp) is the immediate precursor of important biomolecules such as the neurotransmitter serotonin [1], vitamin B3 [2], and auxin phytohormones [3, 4]. In biosynthetic pathways of more complex natural products, modified tryptophan is frequently the core of the final biomolecule [5–9]. Tryptophan derivatives have also been used in chemical biology for a variety of applications [10]. These noncanonical amino acids (ncAAs) also serve as intermediates in the production of pharmaceuticals by chemical synthesis [11]. It is therefore important to develop efficient routes to preparing these compounds.

Tryptophan synthase (TrpS) has been used to make a wide variety of Trp analogs. TrpS catalyzes the last steps in the de novo pathway for Trp in all three domains of life. This enzyme synthesizes Trp from 3-indole-D-glycerol phosphate (IGP) and L-serine (Ser) in two steps that take place in two separate subunits of the heterodimeric enzyme [12]. TrpS has been used for the synthesis of many modified tryptophans by reaction of Ser and an appropriate nucleophile, typically a substituted indole [13–20]. However, these reactions frequently proceed with low yields (below 50%). Researchers have tried to expand the substrate scope to nitrogen nucleophiles through protein engineering, but those efforts concomitantly boosted an abortive side deamination reaction that prevented efficient use of the catalyst [21].

We believed that TrpS would be a good starting point to develop a platform for synthesizing a wider variety of Trp analogs than had been reported. In particular, we sought to generate catalysts for C-C, C-N, and C-S bond-forming reactions to make ncAAs from inexpensive starting materials using low catalysts loadings. Our approach was to first simplify the heterodimeric enzyme complex and engineer TrpB to function as a single, stand-alone enzyme. To help the reader understand this strategy, we describe the sophisticated mechanism of allosteric regulation that governs native TrpS activity.

1.1.1 TrpS

In the first steps of the native catalytic cycle, IGP binds in the α-subunit (TrpA), and Ser binds in the β-subunit (TrpB), where it forms a covalent Schiff base linkage to the cofactor, pyridoxal 5'-phosphate (PLP). Once bound, Ser undergoes dehydration to form an electrophilic amino acrylate intermediate. TrpA then catalyzes the retro-aldol cleavage of IGP, releasing indole, which diffuses through a tunnel connecting the two subunits and enters the TrpB active site. There it reacts with the amino acrylate to yield L-tryptophan (Fig. 1.1).

The efficient functioning of TrpS requires that all of the mechanistic steps be carefully synchronized [22]. In essence, both subunits have an open conformation that permits the entry of substrates, but exhibits a slow rate of catalysis, and a closed conformation that accelerates intermediate chemical steps, but cannot bind substrate or release product. Each subunit's equilibrium between open and closed states

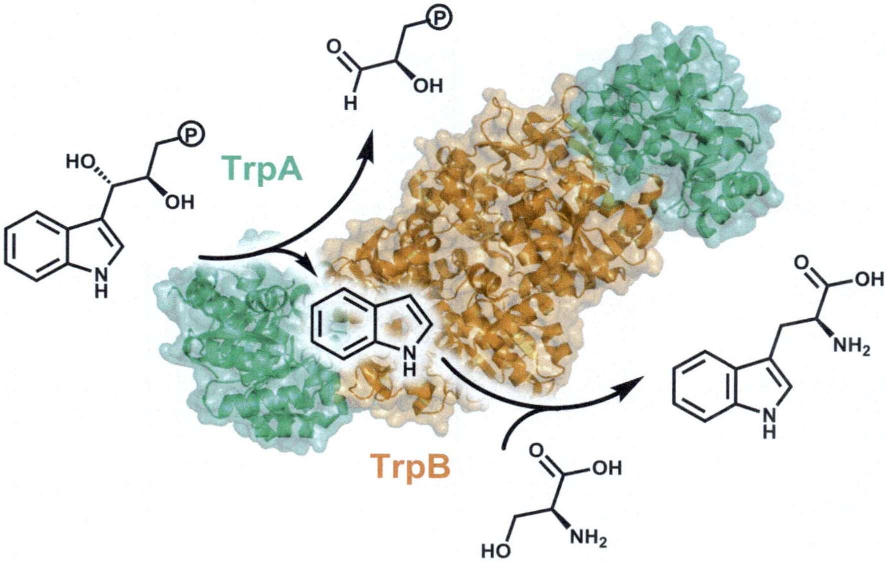

Fig. 1.1 Overall reaction catalyzed by TrpS. The α-subunit (*green*) cleaves IGP in a retro-aldol reaction that releases indole, which diffuses through a molecular tunnel into the β-subunit (*orange*). There, indole reacts with Ser in a PLP-dependent β-substitution reaction to yield Trp

is allosterically modulated by the other subunit, thus ensuring that intermediates are not released prematurely. In particular, the indole must be available to TrpB immediately upon formation of the amino acrylate intermediate, which could otherwise decompose through hydrolysis. However, if indole were released before the subunits are in a fully closed state, it would leak into the cellular medium, whereupon it would diffuse through the membrane and be lost. To accomplish the necessary synchronicity, each subunit acts as an allosteric effector to the other [12, 22].

The molecular basis for this synchronization comprises structural transitions in both subunits [23–25]. In TrpA, the majority of residues in the αL6 loop switch from disordered to well ordered, forming a closed state (Fig. 1.2). In TrpB, the structural change is more significant; around 20% of the residues change conformation. The so-called COMM domain, which refers to the region of TrpB that mediates the communication between subunits, undergoes a rigid-body motion and rotates between open, partially closed, and fully closed conformations (Fig. 1.2). Closure in the COMM domain impedes access to the active site from solution and stabilizes the closed conformation in TrpA. Concurrently with the COMM domain motion, the conformation of TrpB in the interface between the two subunits is also altered. These conformational changes combine to form a 25-Å tunnel between both subunits, allowing indole to diffuse from TrpA to the TrpB active site.

TrpS thus exhibits a sophisticated allosteric control mechanism in which both subunits play a fundamental role. However, only the β-subunit performs catalysis in the β-substitution reaction that is useful to make ncAAs. Unfortunately, the

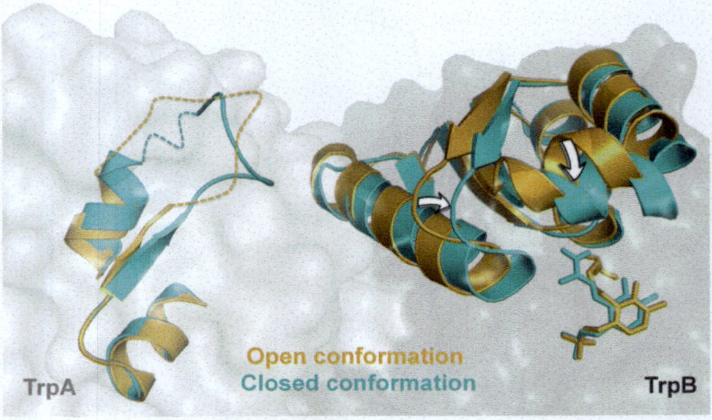

Fig. 1.2 Structural transitions in the conformational switch from open to closed states in *Salmonella typhimurium* TrpS. The open conformation is in orange (PDB ID: 1KFK), and the closed conformation is in *blue* (PDB ID: 2J9X). In TrpA (*light gray*), many residues in loop αL6 are disordered in the open conformation, but most become well ordered upon the switch to the closed conformation. In TrpB (*dark gray*), the *arrows* indicate the direction of the motion of the COMM domain from the open to the closed state. The PLP is represented in sticks

allosteric regulation has hindered the use of isolated TrpB, whose activity in isolation is seriously diminished compared to the full TrpS complex. Consequently, we sought to engineer a stand-alone TrpB for efficient catalysis in the absence of TrpA. We hypothesized that such a simplified biocatalyst could serve as an efficacious starting point for further engineering to expand reactivity.

1.2 Activation of TrpB from *Pyrococcus furiosus* by Directed Evolution

The first task in engineering a stand-alone TrpB catalyst was to identify a suitable parent for directed evolution. Most studies of TrpS have been done with the homolog from *Salmonella typhimurium* (*St*TrpS), a mesophilic organism. We decided to use TrpS from *Pyrococcus furiosus*, a thermophilic archaeon that survives at 100 °C. The ability of this enzyme to function at high temperature (75 °C) enables solubilization of highly hydrophobic substrates such as indole without addition of cosolvent. Another significant advantage of engineering a protein from a thermophilic organism is the possibility of accumulating more mutations that boost activity but may be destabilizing [26].

We wanted to evolve TrpB to catalyze its native reaction, the condensation of indole and Ser to give Trp, more efficiently as a stand-alone enzyme. For this we developed a high-throughput assay that measures the change in absorbance at 290 nm as indole is converted to Trp [27]. The use of a thermostable protein is highly advantageous for screening at this wavelength, because the background absorption from *E. coli* proteins can be reduced through heat treatment at 75 °C, which yields moderately pure *Pf*TrpB enzymes.

We constructed a random mutagenesis library of *Pf*TrpB by error-prone PCR and measured the activity of 528 clones. From this library, we identified 20 clones (3.8% of all variants) with at least a 35% increase in V_{max} relative to the wild-type enzyme [28]. The most active of these contained a single Thr→Ser mutation that increased the catalytic efficiency of *Pf*TrpB on indole by 20-fold, which is even greater than the change induced by TrpA binding. Twelve activating mutations were recombined, and screening yielded a new variant, *Pf*TrpB^{4D11}, that retained the T292S mutation and incorporated four more (E17G, I68V, F274S, T321A). This enzyme, which has a further 2.6-fold increase in catalytic efficiency, was used as the parent for a final round of random mutagenesis, from which we identified *Pf*TrpB^{0B2} harboring the single additional P12L mutation that increased the catalytic efficiency to 3.3×10^5 M^{-1} s^{-1} with indole, 83-fold higher than *Pf*TrpB and threefold higher than the allosterically activated *Pf*TrpS complex [28].

We next investigated whether the increased activity of this stand-alone TrpB was achieved through the same mechanism as the binding of TrpA. Several lines of evidence suggested this was the case. We performed our screening under saturating conditions of each substrate and nevertheless observed a coupled decrease in the K_M for each substrate. *Pf*TrpA binding not only causes a ~threefold increase in k_{cat} but also a decrease in K_M values for Ser and indole, by twofold and fourfold, respectively, suggesting a similar mechanism of activation at work during both effector binding and mutational activation.

To further understand how *Pf*TrpA regulates *Pf*TrpB, we used X-ray crystallography and UV-vis spectroscopy to probe conformational changes and the steady-state distribution of intermediates in the active site. Importantly, our data showed that the structural and spectroscopic properties of *Pf*TrpB and *Pf*TrpS are very similar to those reported in the distantly-related but well-studied enzyme from *Salmonella typhimurium* [25]. The UV-vis spectrum of the internal aldimine (i.e., when the PLP is bound to the catalytic lysine) has a λ_{max} of ~412 nm (Fig. 1.3a). Ser binding to *Pf*TrpB is associated with the large-scale conformational rearrangement of the COMM domain into a partially closed state and accumulation of an external aldimine intermediate, which shifts λ_{max} to 428 nm (Fig. 1.3a). When *Pf*TrpA is bound, the Ser-bound spectrum shifts to a characteristic λ_{max} at 350 nm, consistent with stabilization of the amino acrylate intermediate. Structural studies with *St*TrpS show that this species is stabilized by a fully closed state of the COMM domain (Fig. 1.2), which also has an increased affinity for indole [12]. As can be seen in Fig. 1.3b, the spectrum after addition of Ser to *Pf*TrpB^{0B2} corresponds to amino acrylate intermediate, further supporting the hypothesis that mutations increase activity through the same mechanism as allosteric effector binding, i.e., stabilization of the closed conformational state.

Each of these experiments probed changes in *Pf*TrpB during its native catalytic cycle with Ser and indole. We found that *Pf*TrpA binding also increases the relative rate of product formation upward of 200-fold with different indole analogs. Therefore, it was of interest to know whether mutations were also activating for ncAA synthesis. We found that our stand-alone *Pf*TrpB^{0B2} catalyst was also broadly activated for eight different indole analogs in C-C and C-N bond-forming reactions

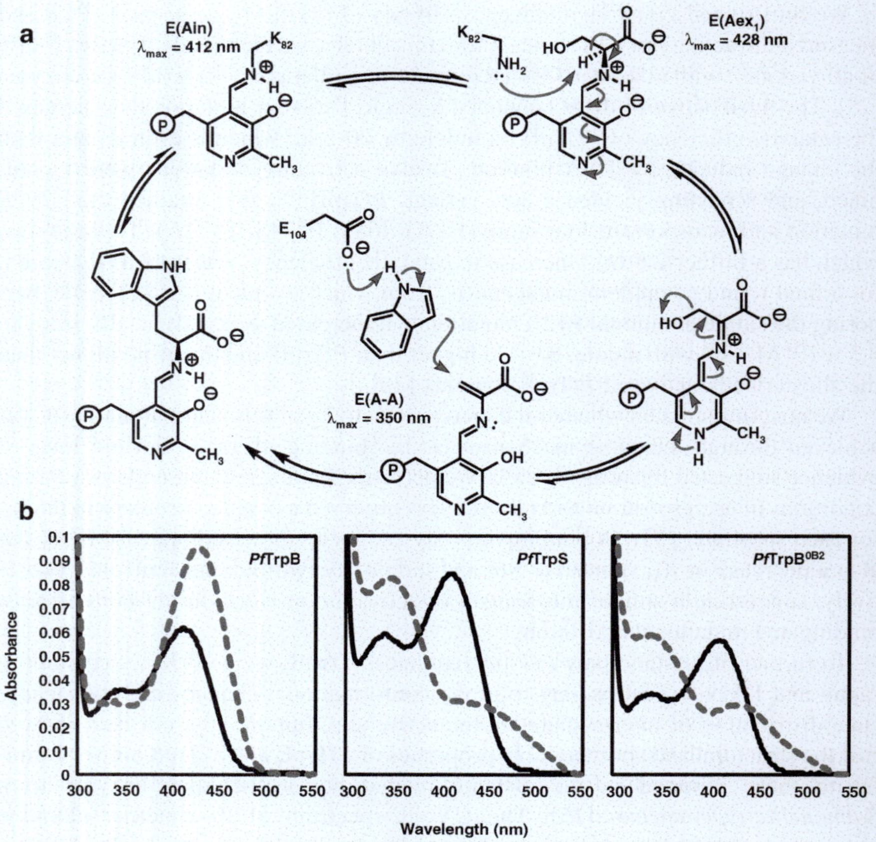

Fig. 1.3 Spectroscopic signatures of the TrpB catalytic cycle. (**a**) Different intermediates in the catalytic cycle of TrpB are shown with their corresponding λ_{max}. Protonation states are assigned according to Caulkins et al. [29] from study of *St*TrpS. Spectroscopic data are broadly similar for the two enzymes, supporting this assignment. (**b**) UV-vis spectra of *Pf*TrpB in isolation, in the native complex, and in the stand-alone engineered protein *Pf*TrpB^{0B2}. Spectra recorded with 20 μM of protein (*black lines*) and after addition of 20 mM Ser (*gray dashes*). The shifts in λ_{max} indicate that *Pf*TrpB accumulates E(Aex$_1$) at steady state, whereas *Pf*TrpS and *Pf*TrpB^{0B2} accumulate the E(A-A) intermediate

[28]. This trend was similar to the rate enhancement induced by *Pf*TrpA binding, with some differences emerging between the two enzyme systems. *Pf*TrpB^{0B2} was moderately faster at C-N bond-forming reactions with indole and indazole, whereas *Pf*TrpS was moderately faster with 5-bromoindole. Hence, mutations that increased activity with indole were broadly activating with other substrates. Again, *Pf*TrpB^{0B2} resembles the catalytic features described for *Pf*TrpS.

Given the considerable interest in understanding allosteric phenomena [30–34], a compelling question arose: are activating mutations confined to a distinct allosteric pathway or interface? The first round of evolution identified 27

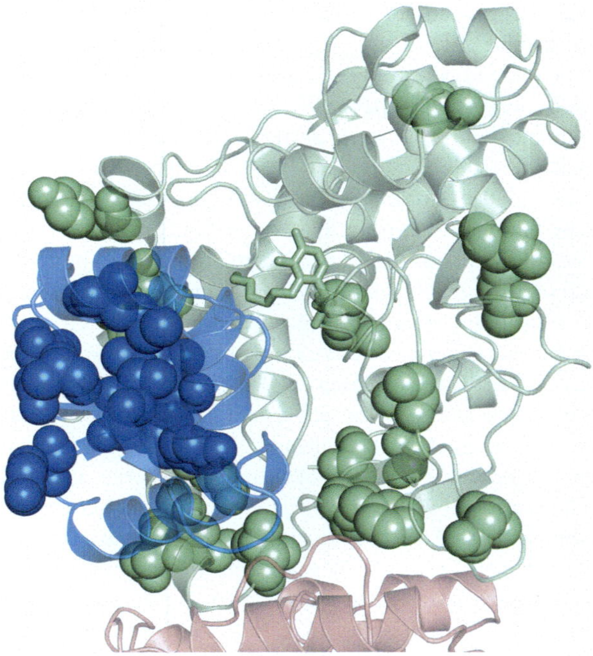

Fig. 1.4 Sites where mutation of *Pf*TrpB is associated with increased activity. In the structure of *Pf*TrpS (PDB ID: 5E0K), TrpA is shown in *pink*, and TrpB is shown in *green* and *blue*. PLP is depicted in sticks. Residues where mutations were found that gave validated increases in V_{max} during high-throughput screening are shown as spheres. There is an abundance of sites in the COMM domain (colored in *blue*) and at the subunit interface that give rise to increases in activity

mutations distributed across 20 improved clones. Although the effect of each mutation has not been measured individually, the data provide a broad picture of *Pf*TrpB activation (Fig. 1.4). It was described that some residues undergo switch-like behavior upon effector binding in *St*TrpS, but just one of the 27 possible activating mutations (N166D) was found along the homologous route for *Pf*TrpB [12, 35, 36]. Taking a more generous view of what an allosteric "pathway" might look like, 17 of the 27 activating mutations (63%) were found at either the protein-protein interface or within the COMM domain. These regions comprise 31% of the total *Pf*TrpB sequence, indicating modest enrichment within the areas previously thought to control allosteric signaling in TrpB. While the mutations have a positive effect on *Pf*TrpB catalysis, it is not clear how or why they influence the rate of the reaction. Indeed, almost 40% of activating mutations are outside of any region that one might *a priori* think has an impact on allosteric signaling or catalysis.

From the study of *Pf*TrpB activation, it is clear that mutations can reproduce complex conformational changes induced by effector binding. Similar effects were previously shown in a handful of other examples from the literature. Shi and Kay identified mutations for the activation of the bacterial HslV protease [37]. The proteolytic activity of HslV is increased ~200-fold upon binding of its partner protein HslU, which is essential for its role in cellular protein degradation [38]. A sensitive NMR analysis was used to monitor chemical shift perturbations in Ile, Leu, Met, Val, and Thr residues in HslV upon HslU binding. From these data, a pair of helices that undergo conformational changes at the HslU binding site was identified. A small panel of conservative mutations at positions within these helices was constructed, and, strikingly, six of the mutations increased catalytic efficiency, the largest by ~20-fold. NMR analysis showed that increases in activity were correlated with substantial chemical shift perturbations similar to the effects of native effector binding.

Another example is the activation of LovD, an acyl transferase that transfers an α-methylbutyrate group that is covalently tethered to an acyl carrier protein, LovF [39, 40]. With substrate surrogates that are not bound to LovF, the acyltransferase activity of LovD is greatly reduced, indicating that the carrier protein also serves as an allosteric activator. Tang and collaborators employed nine rounds of directed evolution to increase activity on a nonnatural substrate in the absence of the protein effector as well as increase the thermal stability and tolerance to organic solvent [40]. They assessed the aggregate effect of these mutations on LovD conformational dynamics with molecular dynamics (MD) simulations. Their results suggested that the activity of LovD is altered upon LovF binding through the stabilization of a closed and catalytically active conformation. Interestingly, simulations suggested that engineered LovD enzymes experience comparable conformation changes in the absence of their effector. Testing this hypothesis experimentally would have been exceptionally difficult without further modification of LovD, as there is no chromophore (like PLP for TrpB) that enables one to probe the steady state of the catalytic cycle directly.

1.3 Activation of TrpB from Other Species by Transfer of Mutations

With our stand-alone *Pf*TrpB in hand, we wished to test whether the mutational activation of TrpB could be generalized to TrpBs from other organisms. All known TrpBs are subject to allosteric regulation by TrpA [12, 41, 42], and it is plausible that the allosteric mechanisms are conserved across TrpS homologs. Furthermore, our interest in expanding the substrate scope of TrpB might be helped by assessing other TrpB homologs, since enzyme homologs often exhibit different substrate scopes [43, 44]. For example, native *St*TrpS is a poor catalyst for N-alkylation, whereas *Pf*TrpS is moderately proficient. However, we did not wish to repeat the effort required to activate *Pf*TrpB (screening ~3100 clones) for the other homologs. Instead we tested whether activating mutations in *Pf*TrpB^{0B2} have the same effects when transferred to TrpBs from other species.

Successful transfer of beneficial mutations among homologous proteins has been reported numerous times, although not for allosteric properties, as far as we know. For instance, multiple sequence alignments of homologs allow the identification of consensus residues that tend to be thermostabilizing within the protein family [45]. Also, mutations that alter nicotinamide cofactor specificity can be transferred to homologous enzymes [46]; in this case, structural and sequence analyses of an entire protein family provided specialized "recipes" to change specificity from NADPH to NADH. However, engineering allostery may be significantly more complex than transferring mutations that enhance thermostability, where the mutational effects are largely additive, or specificity, where the effects are usually more localized. Allosteric regulation occurs through a dynamic mechanism that transfers information between the allosteric binding site and the enzyme active site and involves many, if not all, residues in the protein. Often, experimental evidence establishes that a residue that participates in transmitting this information is not conserved across different homologs. This holds even when the allosteric mechanisms are superficially similar, and evolution frequently causes homologous proteins to develop different allosteric mechanisms [47].

To test the transferability of the allostery-mimicking mutations, we selected diverse TrpB homologs with differing sequence identities and well separated in the phylogenetic tree [48]. The closest homolog to *Pf*TrpB tested was from *Archaeoglobus fulgidus*, *Af*TrpB (72% sequence identity), which is a thermophilic archaeon. We also selected the TrpB from *Thermotoga maritima* (*Tm*TrpB, 64% sequence identity), a thermophile that belongs to the bacterial domain of life. Lastly, we chose the TrpB from *Escherichia coli* (*Ec*TrpB, 57% sequence identity), a mesophile. The sequence alignment of the homologs showed some differences at the positions of the activating mutations in *Pf*TrpB^{0B2}. Importantly, two mutations in *Pf*TrpB^{0B2} were already present as the wild-type residues in two of the homologs, A321 in *Tm*TrpB (mutation T321A in *Pf*TrpB^{0B2}) and S297 in *Ec*TrpB (mutation T292S in *Pf*TrpB^{0B2}).

Making the 0B2 mutations in the homologs led to varied levels of catalytic efficiency. *Af*TrpB^{0B2} was indeed activated, with a ~20-fold increase in catalytic efficiency with respect to wild type. Similar to *P. furiosus* enzymes, the UV-vis spectrum after addition of Ser to *Af*TrpB showed a λ_{max} at 428 nm, reflecting accumulation of the external aldimine intermediate. After addition of Ser to the 0B2 mutant, the spectrum was similar to *Af*TrpS, with a λ_{max} at 350 nm [49] corresponding to the amino acrylate intermediate. Hence, these mutations, which mimic the binding of the allosteric protein partner in *Pf*TrpB, appear to have the same effect in *Af*TrpB. However, this was not the case for the other two mutant homologs, where the catalytic efficiencies of *Tm*TrpB^{0B2} and *Ec*TrpB^{0B2} decreased by more than 40% relative to the wild-type enzymes.

To determine whether a subset of the mutations could activate the other two homologs, we made and screened recombination libraries of the 0B2 mutations for *Tm*TrpB and *Ec*TrpB. With this strategy, we found three activating mutations (P19G, I69V, and T292S) for *Tm*TrpB. T292S was the most activating single mutation, conferring a sixfold increase in catalytic efficiency. All the variants containing this mutation showed a UV-vis spectrum after addition of Ser similar to the native *Tm*TrpS complex [49]. The most activated *Tm*TrpB variant harbored the three mutations and exhibited a tenfold increase in catalytic efficiency.

Activation of the last, *E. coli* homolog was more challenging: the recombination library of the 0B2 mutations in *Ec*TrpB produced no increased activity. The T292S mutation had a prominent effect in the other TrpB homologs, but Ser is the native residue in the equivalent position of *Ec*TrpB. We made a saturation mutagenesis library at this site but again did not find any activated variants. In a final attempt to activate this homolog, we returned to the initial random mutagenesis performed on *Pf*TrpB, where 27 activating mutations were found [28]. We chose to test the mutations of the double mutant *Pf*TrpB$^{M144T\ N166T}$ because these residues are highly conserved across all known TrpBs and are located in the COMM domain. These mutations activated *Ec*TrpB, giving more than a twofold increase in catalytic efficiency. We tested the effects of these two mutations in the other homologs and found that all were activated, with a two- to fivefold increase in catalytic efficiency. The UV-vis spectra after addition of Ser to *Pf*TrpB$^{M144T\ N166D}$ or *Ec*TrpB$^{M149T\ N171D}$ did not have a λ_{max} at 350 nm and instead showed an accumulation of the external aldimine species. However, after addition of Ser to the corresponding *Af*TrpB and *Tm*TrpB mutants, the spectra showed a λ_{max} at 350 nm, characteristic of the amino acrylate species. These results suggest that this set of mutations also mimics the allosteric activation exerted by TrpA binding, although in some of the homologs, this activation does not completely reach TrpS-like behavior [49]. Similarly, in the initial evolution of *Pf*TrpB, the activating T292S mutation alone was not sufficient to shift the spectrum to the amino acrylate species, which required four more mutations.

Once we accumulated this panel of stand-alone TrpB enzymes, we sought to compare their substrate profiles with substituted indoles. We were particularly interested in accessing Trp derivatives with substituents in the 5 position because this site is often substituted in biologically relevant Trp-based compounds. For instance, the halogenase PyrH chlorinates tryptophan in the 5 position during the biosynthesis of the antifungal antibiotic pyrroindomycin B [8]. In the biosynthesis of the neurotransmitter serotonin and the hormone melatonin, a tryptophan hydroxylase mono-oxygenates tryptophan in the 5 position [1]. Halogenation or borylation in this position would generate analogs that could serve as intermediates for other biologically active compounds through cross-coupling reactions [50–52]. Different TrpS homologs have shown a substantial decrease in activity with 5-substituted indoles bearing a substituent bulkier than fluorine [17, 18].

We tested our panel of activated stand-alone biocatalysts with 5-bromoindole. Remarkably, one of the new activated TrpB, *Tm*TrpB$^{M145T\ N167D}$, was ~sixfold faster than *Pf*TrpB^{0B2}, the first enzyme we engineered (Table 1.1). We tested eight more 5-substituted indoles with the two enzymes and saw that *Tm*TrpB$^{M145T\ N167D}$ was always faster than *Pf*TrpB^{0B2}, from 1.4- to 7.5-fold (Table 1.1). As shown in Table 1.1, the yields for 5-chlorotryptophan and 5-bromotryptophan are 93% and 88%, respectively. Previous use of *St*TrpS for these two reactions reported yields of 61% for 5-chlorotryptophan and only 33% for 5-bromotryptophan [17]. Moreover, the chemical syntheses of these compounds described to date involve multiple steps, generate racemic products, and have final yields below 50% [6, 53–55]. Hence, our panel of stand-alone TrpBs is a useful set of biocatalysts for the synthesis of ncAAs and should be a fertile starting point for the further development of biocatalysts to make more synthetically challenging ncAAs.

Table 1.1 Yields and total turnovers of the reactions catalyzed by *Tm*TrpB$^{M145T\ N167D}$ to obtain 5-substituted tryptophan derivatives. The rates relative to *Pf*TrpB^{0B2} are also included

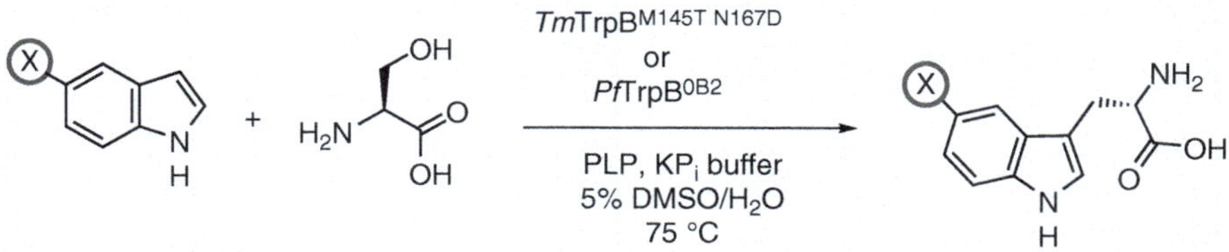

		Reaction catalyzed by *Tm*TrpB$^{M145T\ N167D}$	
X	Rate relative to *Pf*TrpB^{0B2}	Yield (%)	Total turnovers
Cl	3.0	93	9300
Br	5.6	88	4400
NO_2	7.5	25	1250
$B(OH)_2$	1.8	38	1900
CHO	1.9	32	1600
CN	4.5	49	2450
OH	1.4	93	9300
CH_3	1.4	91	9100
OCH_3	1.5	76	7600

1.4 Directed Evolution of *Pf*TrpB for the Synthesis of β-Branched Amino Acids

β-Branched amino acids are particularly desirable because the additional substituent at C_β alters the conformational properties of peptides and small molecules that bear it. For example, the clinically important antibiotic daptomycin features a β-methylglutamate residue, and the activity of the drug is greatly diminished without this modification [56, 57]. Unfortunately, chemical synthesis of β-branched amino acids is particularly challenging, owing to the need for both enantio- and diastereoselective transformations.

One strategy to access β-branched ncAAs would be to substitute Ser in the reaction with Thr. While the nucleophile scope of TrpB has been well explored and our panel of stand-alone variants can catalyze an array of substitutions, synthesis of Trp analogs by replacing Ser had not been reported. We screened our stand-alone *Pf*TrpB catalysts for activity with a variety of Ser analogs and found that the natural amino acid L-threonine (Thr) can replace Ser, yielding (2*S*,3*S*)-β-methyltryptophan (β-MeTrp) in a single step (Fig. 1.5). Previous attempts to perform reactions with Thr required the use of the strong nucleophile benzyl mercaptan, which proceeded with greatly diminished activity compared to Ser, and the stereochemistry was unknown [20].

β-MeTrp is an intermediate in the natural biosynthetic pathways to maremycin and streptonigrin [7, 9]. Studies have shown that this intermediate is produced in

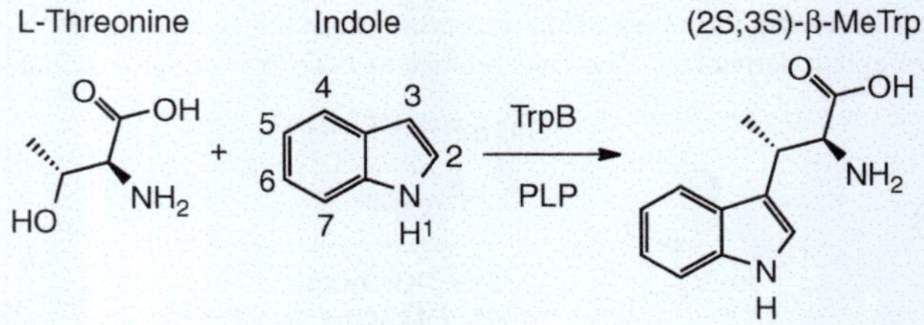

Fig. 1.5 β-Substitution of Thr with indole yields β-MeTrp. This reaction is catalyzed at trace levels by *Pf*TrpB, but the level increases with directed evolution. Activity was also observed with a variety of indole analogs

four steps from Trp, using three different enzymes with *S*-adenosylmethionine as the methyl source. Chemical synthetic routes to this ncAA have relatively good selectivity but require multiple steps, making it challenging to apply for the production of diverse β-MeTrp analogs. Therefore, we sought to evolve our platform for the synthesis of these challenging β-branched amino acids using Thr as a nonnative substrate.

We first explored the *Pf*TrpB lineage of stand-alone enzymes for activity with Thr. Wild-type *Pf*TrpB yielded just 66 turnovers in 24 h. The single mutant *Pf*TrpB^{T292S} showed ~sixfold enhanced activity, and reactions with variant *Pf*TrpB^{4D11} (containing T292S, E17G, I68V, F274S, and T321A mutations) yielded 660 turnovers. Interestingly, *Pf*TrpB^{0B2} gave slightly slower rates than *Pf*TrpB^{4D11}, despite having higher activity with Ser [28]. Therefore, we selected *Pf*TrpB^{4D11} for directed evolution to increase activity with Thr.

We screened 352 clones from a random mutagenesis library of *Pf*TrpB^{4D11} and identified six missense mutations in five clones with increased V_{max}. The most active variant, *Pf*TrpB^{4G1}, has a single additional mutation, F95L. Recombination of all of the mutations and screening resulted in variant *Pf*TrpB^{2B9}, which has mutations I16V, F95L, and V384A and significantly enhanced activity with Thr. Interestingly, the I16V mutation is adjacent to E17G present in the parent, and the F95L mutation is adjacent to the COMM domain of TrpB. Combined, these mutations provide at least a 6000-fold boost in productivity compared to the wild-type *Pf*TrpB enzyme.

This engineered enzyme has several positive features as a biocatalyst. As we screened for V_{max}, some of the increases in activity in cell lysates were due to boosts in the expression of soluble enzyme. Hence, the protein can now be prepared at ~350 mg/L of culture, facilitating larger-scale reactions. The enzyme also retains high activity at elevated temperatures, allowing for high substrate loading. However, reactions with a single equivalent of each substrate give only 44% conversion to products, which corresponds to 2220 total turnovers (TTN). UV-vis spectroscopy showed that an abortive deamination reaction is occurring when Thr is incubated with *Pf*TrpB^{2B9}. With the addition of more equivalents of Thr (up to ten), we observed >99% conversion based on indole and up to 8200 TTN. Additionally, this reaction proceeds with >99% enantiomeric and diastereomeric excess, highlighting the exquisite selectivity of the enzyme.

Using these reaction conditions, we explored the nucleophile scope of this new β-substitution reaction. Indoles with methylation at the 2 and 6 positions (numbering of indole is shown in Fig. 1.5) were well tolerated by the enzyme, yielding 6400 and 1100 turnovers, respectively. We also observed product formation with 4- and 5-fluoroindole, with ~3.4-fold lower TTN for the 4-fluoroindole, which is more electron deficient at C-3 than the 5-fluoroindole analog. However, we did not observe product formation using 5-chloro-, 5-bromo-, or 6-hydroxyindoles, demonstrating a reduced substrate scope in the new reaction.

We observed that 7-azaindole reacted with 220 TTN and also found a secondary product corresponding to N-alkylation. This regioselectivity is identical to that of indazole, which we found reacts exclusively in an N-alkylation reaction that proceeds with 500 TTN. Lastly, we observed a productive reaction with thiophenol in 1300 TTN, demonstrating that the enzyme can catalyze stereoselective C-C, C-N, and C-S bond-forming reactions.

The development of this enzymatic platform for the synthesis of β-branched ncAAs was greatly facilitated by previous efforts to engineer a stand-alone enzyme. At all stages, only a single enzyme was necessary to catalyze the reaction, and this simplified system supported expression of the catalyst at high levels, ~350 mg of *Pf*TrpB^{2B9} per L culture, which will facilitate production of these important synthons [58].

1.5 Summary and Outlook

We have engineered the heterodimeric enzyme TrpS into a single enzyme platform for the synthesis of ncAAs from simple starting materials. This was accomplished by identifying mutations in TrpB that recapitulate the effects that are induced by its allosteric binding partner, TrpA. We identified dozens of mutations that are activating in *Pf*TrpB and showed that some of them could be transferred to homologs to produce an array of catalysts with varied properties. These new stand-alone TrpB enzymes can be evolved readily for altered function, as we demonstrated for the β-substitution of Thr. All of this was accomplished with mutations that were identified by screening random mutant libraries, and the mutations are distributed throughout the protein. Hence, each of the catalysts has a unique active site that can be engineered to increase activity with a particular nucleophile or electrophile. We believe this approach will continue to yield superior catalysts for the biocatalytic production of desirable ncAAs.

Acknowledgments We gratefully thank Sabine Brinkmann-Chen for a critical reading of this chapter. We also thank David K. Romney for his helpful discussions during the elaboration of the chapter. Javier Murciano-Calles acknowledges financial support from the Alfonso Martín Escudero Foundation. This work was funded through the Jacobs Institute for Molecular Engineering for Medicine and Ruth Kirschstein NIH Postdoctoral Fellowship F32GM110851 (to Andrew R. Buller).

References

1. Zhang J, Wu C, Sheng J, Feng X (2016) Molecular basis of 5-hydroxytryptophan synthesis in *Saccharomyces cerevisiae*. Mol Biosyst 12(5):1432–1435
2. Ikeda M et al (1965) Studies on the biosynthesis of nicotinamide adenine dinucleotide: II. A role of picolinic carboxylase in the biosynthesis of nicotinamide adenine dinucleotide from tryptophan in mammals. J Biol Chem 240(3):1395–1401
3. Stepanova AN et al (2008) TAA1-Mediated auxin biosynthesis is essential for hormone cross-talk and plant development. Cell 133(1):177–191
4. Tao Y et al (2008) Rapid synthesis of auxin via a new tryptophan-dependent pathway is required for shade avoidance in plants. Cell 133(1):164–176
5. Barry SM et al (2012) Cytochrome P450-catalyzed L-tryptophan nitration in thaxtomin phytotoxin biosynthesis. Nat Chem Biol 8(10):814–816
6. Kieffer ME, Repka LM, Reisman SE (2012) Enantioselective synthesis of tryptophan derivatives by a tandem Friedel-Crafts conjugate addition/asymmetric protonation reaction. J Am Chem Soc 134(11):5131–5137
7. Kong D et al (2016) Identification of (2S,3S)-β-methyltryptophan as the real biosynthetic intermediate of antitumor agent streptonigrin. Sci Rep 6:20273
8. Zehner S et al (2005) A regioselective tryptophan 5-halogenase is involved in pyrroindomycin biosynthesis in *Streptomyces rugosporus* LL-42D005. Chem Biol 12(4):445–452
9. Zou Y et al (2013) Stereospecific biosynthesis of β-methyltryptophan from L-tryptophan features a stereochemical switch. Angew Chem Int Ed 52(49):12951–12955
10. Lang K, Chin JW (2014) Cellular incorporation of unnatural amino acids and bioorthogonal labeling of proteins. Chem Rev 114(9):4764–4806
11. Patel R (2013) Biocatalytic synthesis of chiral alcohols and amino acids for development of pharmaceuticals. Biomolecules 3(4):741
12. Dunn MF (2012) Allosteric regulation of substrate channeling and catalysis in the tryptophan synthase bienzyme complex. Arch Biochem Biophys 519(2):154–166
13. Corr MJ, Smith DRM, Goss RJM (2016) One-pot access to L-5,6-dihalotryptophans and L-alknyltryptophans using tryptophan synthase. *Tetrahedron* 72(45):7306–7310
14. Ferrari D, Niks D, Yang L-H, Miles EW, Dunn MF (2003) Allosteric communication in the tryptophan synthase bienzyme complex: roles of the β-subunit aspartate 305–arginine 141 salt bridge. Biochemistry 42(25):7807–7818
15. Goss RJM, Newill PLA (2006) A convenient enzymatic synthesis of L-halotryptophans. Chem Commun 47:4924–4925
16. Perni S, Hackett L, Goss RJ, Simmons MJ, Overton TW (2013) Optimisation of engineered *Escherichia coli* biofilms for enzymatic biosynthesis of L-halotryptophans. AMB Express 3(1):1–10
17. Smith DRM et al (2014) The first one-pot synthesis of L-7-iodotryptophan from 7-iodoindole and serine, and an improved synthesis of other L-7-halotryptophans. Org Lett 16(10):2622–2625
18. Tsoligkas AN et al (2011) Engineering biofilms for biocatalysis. Chembiochem 12(9):1391–1395
19. Winn M, Roy AD, Grüschow S, Parameswaran RS, Goss RJM (2008) A convenient one-step synthesis of L-aminotryptophans and improved synthesis of 5-fluorotryptophan. Bioorg Med Chem Lett 18(16):4508–4510
20. Esaki N, Tanaka H, Miles EW, Soda K (1983) Enzymatic synthesis of S-substituted L-cysteines with tryptophan synthase of *Escherichia coli*. Agric Biol Chem 47(12):2861–2864
21. Ferrari D, Yang LH, Miles EW, Dunn MF (2001) β-D305A Mutant of tryptophan synthase shows strongly perturbed allosteric regulation and substrate specificity. Biochemistry 40(25):7421–7432
22. Niks D et al (2013) Allostery and substrate channeling in the tryptophan synthase bienzyme complex: evidence for two subunit conformations and four quaternary states. Biochemistry 52(37):6396–6411

23. Barends TRM et al (2008) Structure and mechanistic implications of a tryptophan synthase quinonoid intermediate. Chembiochem 9(7):1024–1028
24. Lai J et al (2011) X-ray and NMR Crystallography in an enzyme active site: the indoline quinonoid intermediate in tryptophan synthase. J Am Chem Soc 133(1):4–7
25. Ngo H et al (2007) Allosteric regulation of substrate channeling in tryptophan synthase: modulation of the L-serine reaction in stage I of the β-reaction by α-site ligands. Biochemistry 46(26):7740–7753
26. Bloom JD, Labthavikul ST, Otey CR, Arnold FH (2006) Protein stability promotes evolvability. Proc Natl Acad Sci U S A 103(15):5869–5874
27. Lane AN, Kirschner K (1983) The catalytic mechanism of tryptophan synthase from *Escherichia coli*. Eur J Biochem 129(3):571–582
28. Buller AR et al (2015) Directed evolution of the tryptophan synthase β-subunit for stand-alone function recapitulates allosteric activation. Proc Natl Acad Sci 112(47):14599–14604
29. Caulkins BG et al (2015) Catalytic roles of beta Lys87 in tryptophan synthase: N-15 solid state NMR studies. Biochim Biophys Acta Protein Proteomics 1854(9):1194–1199
30. Sol A, Tsai C-J, Ma B, Nussinov R (2009) The origin of allosteric functional modulation: multiple pre-existing pathways. Structure 17(8):1042–1050
31. Gerek ZN, Ozkan SB (2011) Change in allosteric network affects binding affinities of PDZ domains: analysis through perturbation response scanning. PLoS Comput Biol 7(10):e1002154
32. McLeish Tom CB, Cann Martin J, Rodgers Thomas L (2015) Dynamic transmission of protein allostery without structural change: spatial pathways or global modes? Biophys J 109(6):1240–1250
33. Suel GM, Lockless SW, Wall MA, Ranganathan R (2003) Evolutionarily conserved networks of residues mediate allosteric communication in proteins. Nat Struct Mol Biol 10(1):59–69
34. Woods KN, Pfeffer J (2016) Using THz spectroscopy, evolutionary network analysis methods, and MD simulation to map the evolution of allosteric communication pathways in c-type lysozymes. Mol Biol Evol 33(1):40–61
35. Raboni S, Bettati S, Mozzarelli A (2005) Identification of the geometric requirements for allosteric communication between the α- and β-Subunits of tryptophan synthase. J Biol Chem 280(14):13450–13456
36. Weyand M, Schlichting I, Herde P, Marabotti A, Mozzarelli A (2002) Crystal structure of the βSer178 → Pro mutant of tryptophan synthase: a "knock-out" allosteric enzyme. J Biol Chem 277(12):10653–10660
37. Shi L, Kay LE (2014) Tracing an allosteric pathway regulating the activity of the HslV protease. Proc Natl Acad Sci 111(6):2140–2145
38. Yoo SJ et al (1996) Purification and characterization of the heat shock proteins HslV and HslU that form a new ATP-dependent protease in *Escherichia coli*. J Biol Chem 271(24):14035–14040
39. Gao X et al (2009) Directed evolution and structural characterization of a simvastatin synthase. Chem Biol 16(10):1064–1074
40. Jiménez-Osés G et al (2014) The role of distant mutations and allosteric regulation on LovD active site dynamics. Nat Chem Biol 10(6):431–436
41. Hettwer S, Sterner R (2002) A novel tryptophan synthase β-subunit from the hyperthermophile *Thermotoga maritima*: quaternary structure, steady-state kinetics, and putative physiological role. J Biol Chem 277(10):8194–8201
42. Hiyama T, Sato T, Imanaka T, Atomi H (2014) The tryptophan synthase β-subunit paralogs TrpB1 and TrpB2 in *Thermococcus kodakarensis* are both involved in tryptophan biosynthesis and indole salvage. FEBS J 281(14):3113–3125
43. Dunn MR, Otto C, Fenton KE, Chaput JC (2016) Improving polymerase activity with unnatural substrates by sampling mutations in homologous protein architectures. ACS Chem Biol 11(5):1210–1219
44. Khanal A, Yu McLoughlin S, Kershner JP, Copley SD (2015) Differential effects of a mutation on the normal and promiscuous activities of orthologs: implications for natural and directed evolution. Mol Biol Evol 32(1):100–108

45. Lehmann M, Wyss M (2001) Engineering proteins for thermostability: the use of sequence alignments versus rational design and directed evolution. Curr Opin Biotechnol 12(4):371–375
46. Brinkmann-Chen S et al (2013) General approach to reversing ketol-acid reductoisomerase cofactor dependence from NADPH to NADH. Proc Natl Acad Sci U S A 110(27):10946–10951
47. Kuriyan J, Eisenberg D (2007) The origin of protein interactions and allostery in colocalization. Nature 450(7172):983–990
48. Merkl R (2007) Modelling the evolution of the archaeal tryptophan synthase. BMC Evol Biol 7(1):1–20
49. Murciano-Calles J, Romney DK, Brinkmann-Chen S, Buller AR, Arnold FH (2016) A panel of TrpB biocatalysts derived from tryptophan synthase through the transfer of mutations that mimic allosteric activation. Angew Chem Int Ed 55(38):11577–11581
50. Durak LJ, Payne JT, Lewis JC (2016) Late-stage diversification of biologically active molecules via chemoenzymatic C–H functionalization. ACS Catal 6(3):1451–1454
51. Pathak TP, Miller SJ (2013) Chemical tailoring of teicoplanin with site-selective reactions. J Am Chem Soc 135(22):8415–8422
52. Roy AD, Grüschow S, Cairns N, Goss RJM (2010) Gene expression enabling synthetic diversification of natural products: chemogenetic generation of pacidamycin analogs. J Am Chem Soc 132(35):12243–12245
53. Blaser G, Sanderson JM, Batsanov AS, Howard JAK (2008) The facile synthesis of a series of tryptophan derivatives. Tetrahedron Lett 49(17):2795–2798
54. Konda-Yamada Y et al (2002) Convenient synthesis of 7′ and 6′-bromo-D-tryptophan and their derivatives by enzymatic optical resolution using D-aminoacylase. Tetrahedron 58(39):7851–7861
55. Ma C, Liu X, Yu S, Zhao S, Cook JM (1999) Concise synthesis of optically active ring-A substituted tryptophans. Tetrahedron Lett 40(4):657–660
56. Fowler VG et al (2006) Daptomycin versus standard therapy for bacteremia and endocarditis caused by *Staphylococcus aureus*. N Engl J Med 355(7):653–665
57. Nguyen KT et al (2006) Combinatorial biosynthesis of novel antibiotics related to daptomycin. Proc Natl Acad Sci U S A 103(46):17462–17467
58. Herger M, van Roye P, Romney DK, Brinkmann-Chen S, Buller AR, Arnold FH (2016) Synthesis of β-branched tryptophan analogs with an engineered variant of tryptophan synthase. J Am Chem Soc 138(27):8388–8391

Engineering Therapeutic Enzymes

2

Stefan Lutz, Elsie Williams, and Pravin Muthu

Abstract

Biologics constitute a rapidly growing category of pharmaceutical drug products. With over 100 clinically approved therapeutics and many more in development, protein-based compounds represent an important subset among these biomacromolecular drug candidates, ranging from antibodies, anticoagulants, growth factors, hormones, interferons, and interleukins to enzymes. While recombinant gene technology has traditionally played a key role in development and production of these therapeutics, protein engineering offers an additional dimension for tailoring the biochemical and biophysical properties of proteins to the specific needs of clinical applications. Given the tremendous potential of protein engineering methods to alter and improve the function of biocatalysts, our review focuses on recent examples highlighting the advances and challenges in applying these techniques toward the engineering of therapeutic enzymes. More specifically, our review will focus on three categories of therapeutic enzymes: pharmaceutical enzymes where the protein itself constitutes the therapeutic agent, prodrug-activating enzymes where the protein indirectly triggers a clinical effect, and diagnostic enzymes where a protein's superior selectivity and specificity offer advantages over traditional analytical methods.

2.1 Introduction

Proteins are involved in the vast majority of biological processes inside and outside the cellular environment. They have continuously been selected and optimized by Darwinian evolution to facilitate specific chemical reactions over millions of years,

S. Lutz (✉) • E. Williams • P. Muthu
Department of Chemistry, Emory University, 1515 Dickey Drive, Atlanta, GA 30084, USA
e-mail: sal2@emory.edu

M. Alcalde (ed.), *Directed Enzyme Evolution: Advances and Applications*,
DOI 10.1007/978-3-319-50413-1_2

enabling them to perform structural, catalytic, and regulatory functions with impressive efficiency, specificity, and selectivity. Recognizing the tremendous potential and value of these biomacromolecules, efforts by scientists and engineers to harness their exquisite performance to manufacture medicinal, industrial, and consumer products date back over 100 years. While the scope of early studies was often limited by low availability of particular proteins, the development of recombinant DNA technology in the early 1970s represented a paradigm shift, allowing not only for the deliberate manipulation of genes that encode for particular proteins but also for the reproducible and scalable heterologous expression of targeted proteins. These capabilities marked a new era for fundamental and applied studies of native proteins, and they also set the stage for a more creative, synthetic approach toward studying and tailoring protein structure and function a decade later. As discussed here and in other parts of this book, the generation of protein variants via site-directed and random mutagenesis, as well as in vitro recombination, again revolutionized the field, providing an effective strategy toward probing the contributions of individual amino acids or protein domains to overall function. At the same time, these methods could be applied for altering the functional properties of proteins, in principle enabling the tailoring of proteins toward almost any desirable target function.

Given their key role in the chemistry of life, proteins are also often intimately involved in cellular malfunction that can lead to disease, as well as processes targeted by drugs and toxins. In the spirit of "fighting fire with fire," these same proteins offer opportunities for the development of new, potentially highly effective treatments and diagnostics to overcome, compensate for, or simply monitor perturbations in a biological system. In fact, the exploitation of native and engineered proteins for such therapeutic applications has been highly successful [1, 2]. In 2011, almost 100 proteins were approved for clinical use in the United States and European Union with sales exceeding US$100bln. When sorted by molecular mechanisms, a clear majority of these protein-based therapeutics fall into the category of noncovalent binders, consisting of genuine unmodified proteins and monoclonal antibodies. Meanwhile, the category of proteins affecting covalent bonds listed 21 enzymes [2]. This latter category represents a particularly exciting group of therapeutics as enzymes, not limited to stoichiometric interactions, can due to their catalytic activity dramatically amplify their impact on host cells. Consequently, biological efficacy is enhanced even at low doses of the therapeutics, while costs and undesirable side effects are minimized. However, therapeutic enzymes also present unusual challenges as they are exogenous proteins, introduced into a host cell to catalyze a reaction of clinical benefit. Such function comes with unique design criteria distinct from general drug and protein design. On one hand, these enzymes can be tuned to levels of specificity and selectivity typically not achievable with traditional small-molecule drugs. On the other hand, they must minimize interference with native cellular functions, remain functional under physiological conditions (e.g., blood plasma), exhibit suitable pharmacokinetic profiles, and elicit no or minimal immunogenic response from patient. These additional constraints significantly increase the complexity and challenge for engineering therapeutic enzymes. It is for these reasons that the last two decades have seen serious efforts to identify novel biocatalysts for therapeutic applications that far exceed the abovementioned list of enzymes with clinical approval.

In this article, we are reviewing developments of therapeutic enzymes with a particular focus on studies involving engineering of the actual protein. Additional opportunities for functional improvements of proteins including posttranslational modifications such as PEGylation or fusion to other proteins are not discussed. While some of the enzymes discussed have been approved or are currently in clinical testing, others remain under development in the laboratory or have been discontinued. We believe that they all represent interesting examples of the challenges and opportunities for tailoring enzymes for therapeutic applications. In addition, many studies provide valuable insights into the protein's structure-function relationships. For our discussion, we have organized these enzymes into three subcategories: pharmaceutical enzymes, prodrug-activating enzymes, and diagnostic enzymes.

2.1.1 Pharmaceutical Enzymes

The subcategory of pharmaceutical enzymes represents proteins that themselves constitute the therapeutic agent. Representatives include proteases, esterases, and nucleases, as well as amino acid-processing enzymes. The clinical efficacy of these enzymes typically benefits from enhanced catalytic activity and pharmacokinetics, as well as reduced immunogenicity. While improvements of activity and stability are relatively straightforward and are routinely addressed by directed evolution methods [3, 4], immunogenicity remains the most common reason for drug failure. Historically, the majority of preclinical research on therapeutic enzymes were done with nonhuman proteins and often resulted in termination of clinical trials due to an observed immune response [5]. In some cases, simply altering the protein origin to the human homolog was sufficient. For example, initial studies on DNaseI as a therapeutic agent to treat cystic fibrosis were done with the bovine enzyme [6]. Pulmozyme® (Genentech) is its human homolog and displayed the intended therapeutic effect without observed immunogenicity [7]. For many other enzymes, the reduction of immune response is not straightforward and remains a significant challenge. While various approaches to identify potential epitopes have been explored, no general solution to protein immunogenicity exists to date.

2.1.2 Enzymes for Suicide Gene Therapy

A more recent application of therapeutic enzymes is toward the development of improved chemotherapies through the introduction of foreign genes. For decades, chemotherapy has been a component of standard of treatment for cancer patients [8]. However, the high systemic toxicity of the therapeutic agents remains a dose-limiting constraint. Gene-directed enzyme-prodrug therapy (GDEPT) involves the intratumoral delivery and expression of so-called suicide genes, encoding for enzymes which, in combination with a prodrug, elicit a cytotoxic effect [9]. First introduced by Mootlen et al. [10], GDEPT places strict constraints on the therapeutic enzyme, requiring high activity and specificity for the unnatural prodrug but no significant interference with native cellular functions. For these reasons, wild-type enzymes were generally found to be poor GDEPT candidates [11]. Protein engineering of native enzymes has been proposed for GDEPT and has been advanced in four specific cases: 2′-deoxyribonucleoside kinases and purine nucleoside phosphorylase that operate in conjunction with nucleoside analog prodrugs, cytosine deaminase that utilizes prodrugs originally developed as fungicides, and nitroreductases that activate custom prodrugs. Beyond these four extensively studied enzyme-prodrug systems, more than 20 other GDEPT systems deploying natural enzymes are described in the literature [12, 13].

2.1.3 Diagnostic Enzymes

The inherent selectivity and specificity of enzymes can offer significant advantages over traditional analytical methods for the detection of analytes and biomarkers. The functional properties of such enzyme-based biosensors not only allow for direct identification of targeted molecules in complex biological samples; the sensor-target interactions also impacts the catalytic performance of the enzyme, amplifying the effect of a single, bound analyte by manyfold and dramatically boosting sensor sensitivity. Besides the need for increased stability which is typically addressed by enzyme immobilization, engineering of diagnostic enzymes has largely focused on changing and optimizing the specificity for particular analytes, as exemplified by variants of cholinesterases for the detection of organophosphate and carbamates found in pesticides and nerve agents.

2.2 Proteases

Proteases play key roles in regulating fundamental biological processes, approximately 2% of the human genome encodes for protease enzymes, and they are the molecular targets for many established drug classes [14]. Perhaps not surprising then is the variety of medical conditions for which researchers are pursuing engineered proteases as therapeutic enzymes. These include leukemia [15], vitreomacular adhesion [16], and, as will be expanded upon below, cardiovascular disorders, multiple neurological disorders, and celiac disease.

2.2.1 Proteases as Anticoagulant Enzymes for Treatment of Cardiovascular Disorders

2.2.1.1 Urokinase Plasminogen Activator and Tissue Plasminogen Activator

The use of proteases to treat cardiovascular disorders is the most established therapeutic use of this enzyme class. Urokinase plasminogen activator (u-PA) was the first enzyme to be approved for therapeutic use by the FDA in 1978 [17] not only ushering in proteases as a class of drugs for thrombolytics but demonstrating the commercial viability of enzymes as drugs. u-PA is able to break down blood clots as its proteolytic activity cleaves the zymogen plasminogen into the active serine protease plasmin. Active plasmin then cleaves fibrin, a fibrous protein that is a crucial competent of blood clots, eventually dissolving the clot. While u-PA was a significant advance for proteolytic-based therapy, it is a tissue plasminogen activator (t-PA), first trailed in humans in 1984 [18] and approved for myocardial infarction and eventually for stroke, that is now more widely used. t-PA also converts plasminogen to active plasmin, but it preferentially cleaves plasminogen that is bound to fibrin, allowing t-PA to be systemically delivered but still effect local fibrinolysis at indicated areas [19]. However, t-PA is rapidly inactivated by endogenous inhibitors, most importantly plasminogen activator inhibitor-1 (PAI-1), as part of a natural feedback regulation to prevent against rampant fibrinolysis [20], and recombinant infused t-PA is reported to have a biological half-life of only 6 min [21].

Improving both the biological half-life and specificity toward fibrin bound plasminogen were clear aims for which to engineer second-generation t-PA proteases. In 2000, the FDA approved tenecteplase (TNKase®) developed by Genentech. This improved variant contains three modifications from the wild-type t-PA sequence [22]. The first modification was a replacement of the residues at positions 296–299 with Ala, identified via a strategy in which charged residues close together in the primary sequence of t-PA were systematically substituted for Ala [23]. This variant shows sharply decreased PAI-1 inhibition and increased specificity and activity for plasmin bound to fibrin. Additional mutations T103N [24] and N117Q modified N-linked glycosylation sites and reduced plasma clearance [22]. These sequence modifications resulted in a modest improvement of half-life, from 6 to 18 min in clinical trials [21]. Unfortunately, not all attempts to improve t-PA bioavailability have been successful. The variant lanoteplase showed both an increased half-life and a decreased PA1-I binding [25]; however, when tested in large-scale clinical trial, this appears to have triggered an increase in intracranial bleeding that halted development of the protease [25], highlighting the unfortunate reality that even when an engineering campaign successfully meets an identified target, the resulting therapeutic enzyme is not assured of in vivo clinical success.

2.2.1.2 Engineered Thrombin

Thrombin is a protease that can trigger both procoagulation and anticoagulation processes. In an unbound state, thrombin proteolytically converts fibrinogen into fibrin allowing the formation of blood clots. Fibrin glue, a combination of fibrinogen and native thrombin enzyme, has a long-standing use as a surgical aid to wound healing [26]. However, upon binding to the endothelial cell receptor thrombomodulin, thrombin "switches" to an anticoagulant activity as it can now activate the zymogen protein C, converting it into activated protein C (APC) that continues the protease cascade and triggers downstream anticoagulation pathways. Multiple engineering efforts have focused on separating out these anti- and procoagulant activities to develop thrombin as an anticoagulant protease for potential therapeutic use [27–29]. One variant of interest is a double mutant W215A/E217A that has demonstrated efficacy as an anticoagulant in both preclinical primate [30] and rodent models [31] of ischemic stroke and thrombosis. This rationally designed variant combined two single mutations (W215A [32] and E217A [27]) that had previously been characterized as interfering with proteolysis of fibrinogen but not protein C. The double mutant shows a 20,000-fold lower activity for fibrin yet still activates protein C comparably to wild-type enzyme.

2.2.2 Procoagulant Protease Engineering

Factor VII (FVII) is a protease in the procoagulant protease cascade required to form blood clots. Recombinant FVIIa, marketed as NovoSeven® by Novo Nordisk, was approved by the FDA for the treatment of hemophilia in 1999. While the wild-type formulation has been demonstrated as a safe and effective treatment for procoagulation, there is a potential market for variants with increased and sustained procoagulant activity. Persson et al. consolidated rationally identified substitutions from previous works, resulting in a triple mutant (V158D/E296V/M298Q) of FVIIa [33]. The activity of the wild-type FVIIa is stimulated by association with tissue factor (TF), which is locally present at sites of vascular injury. The triple mutant has amino acid substitutions that force the protease to adopt a conformation similar to the TF-induced structure, increasing the TF-independent activity of the enzyme. Unfortunately, in phase III clinical trials, several patients developed anti-drug antibodies eliciting a neutralizing response, and development of this variant has been halted [34]. Harvey et al. also used protein engineering to improve FVIIa for a separate property [35]. The wild-type FVIIa has a relatively low affinity for the membranes of platelets, reducing its in vivo efficacy. The group was guided by previous literature, suggesting that membrane affinity of FVIIa was mediated through the the γ-carboxyglutamic acid (Gla) domain [36]. Site-saturation mutagenesis of this 40 amino acid residue domain was carried out and a light-scattering assay used to evaluate proteins for increase binding to phospholipid vesicles. The resulting lead variant contained four substitutions (P10Q/K32E/D33F/A34E) and progressed to phase III clinical trials for the treatment of hemophilia. This study was also halted, as patients again developed neutralizing antibodies toward the engineered protease [37].

The clinical outcome of the modified FVIIa variants highlights a difficulty of engineering enzymes as therapeutic agents. Both studies were halted as a precautionary measure due to an unforeseen immune response, resulting from genetic modification of a human wild-type sequence. The development of non-immunogenic enzyme variants with alternate properties remains a consistent challenge for the engineering of therapeutic enzymes.

2.2.3 Alzheimer's Disease

A role for therapeutic proteases in the treatment of multiple neurological disorders has been investigated. One such disorder is Alzheimer's disease. While the molecular basis for Alzheimer's is not fully understood, the disease is characterized by buildup of β-amyloid (Aβ) peptide-based plaques in the brain. Many therapeutic strategies in advanced clinical trials seek to target these Aβ plaques through immunotherapy or inhibition of the secretases that produce them from amyloid precursor protein [38]. An alternative strategy is the use of proteases to break down Aβ plaques – either through pharmacological upregulation of endogenous protease activity [39] or gene therapy-based approaches designed to increase expression of Aβ plaque degrading proteases [40, 41]. A large concern, and a general hurdle for many protease-based therapies, is off-target toxicity arising from the often broad substrate specificity of proteases. As such, increasing specificity is a principle aim of many protease engineering campaigns, such as those focused on increasing the specificity of neprilysin [42, 43], a human zinc metalloprotease able to degrade Aβ plaques. Site-directed mutagenesis of active site and solvent accessible residues was carried out, and beneficial single mutations were identified by screening for increased activity with Aβ peptide sequences and decreased activity on eight known native peptide substrates. These were then screened combinatorially resulting in a lead variant with two mutations, G399V/G174K, that showed a 20-fold increase in k_{cat}/K_m relative to wild-type neprilysin and a 2.6- to 3200-fold decrease in k_{cat}/K_m on a panel of native peptide substrates [44]. In separate work, a member of the trypsin-like serine protease superfamily, human kallikrein 7 (hK7), has also demonstrated an in vitro ability to cleave peptides at a sequence in the core of the $A\beta$ peptide, and Guerrero et al. reported on efforts to increase this catalytic activity and specificity toward $A\beta$ by engineering the protease to prefer phenylalanine upstream of the peptide cleavage site instead of the native tyrosine preference [45]. In a yeast-based expression system, a randomized 10^7-sized mutant library was screened using a protease-activated fluorescent reporter. The resulting lead hK7 variant showed a 10–30-fold increase in selectivity toward the $A\beta$ peptide sequence versus native peptide substrates. However, this was due primarily to the loss of activity toward tyrosine-containing native peptide sequences rather than increased $A\beta$ peptide activity. The variant demonstrated reduced toxicity toward both the yeast expression system and neuronal cells co-cultured with purified protein.

2.2.4 Botulinum Neurotoxins for Neurological and Non-neurological Indications

In 1989, botulinum neurotoxin serotype A (BoNT/A) from *Clostridium botulinum* was approved for use by the FDA for the treatment of strabismus (cross eyes) and blepharospams (eyelid spasms), although it is now perhaps most commonly known for its cosmetic use in attenuating frown lines. BoNT/A is currently also used in a wide number of approved and off-label indications including dystonias, migraines, overactive bladder, excessive sweating or drooling, and muscle conditions associated with diseases such as cerebral palsy, Parkinson's disease, and multiple sclerosis [46, 47]. The molecular basis of the neuro-inhibitory effect of BoNTs is due to a targeted protease catalytic activity. The mature form of BoNTs contains three domains – a light chain zinc endopeptidase domain attached via a disulfide bond to a heavy chain-binding domain and translocation domain. These allow the BoNTs to bind to peripheral cholinergic nerve terminals and eventually enter the nerve cell cytosol where the zinc endopeptidase then cleaves specific N-ethylmaleimide-sensitive fusion protein attachment protein receptor (SNARE) proteins. As these SNARE proteins are essential for docking and fusion of synaptic vesicles, this impairment of SNARE function inhibits acetylcholine release and consequently neurotransmitter signal transduction. While no recombinant BoNTs are currently approved as therapeutics, there are ongoing efforts to engineer BoNTs. In-depth coverage of this body of work is beyond the scope of this review but has recently been comprehensively reviewed by Masuyer et al. [48]. BoNT modifications have been made with a wide range of aims including to improve safety of current treatments [49], to increase efficacy [50, 51], or to modify cell targeting so as to expand the therapeutic use of BoNTs to other neurons or to entirely non-neurological conditions. This engineering has involved both amino acid substitutions in the catalytic protease domain and exploitation of the modular nature of the multi-domain BoNTs to "switch out" different translocations and/or binding domains both from other BoNT serotypes [51, 52] or to replace them with completely unrelated cell-binding proteins [53, 54], including antibodies [55] (Table 2.1).

Examples of BoNT domain engineering include the work of Wang et al. in which a BoNT/A variant was generated with an inactive protease domain and an artificially attached active endopeptidase domain from a serotype E BoNT [52]. This hybrid variant was designed to combine the long-acting delivery of BoNT/A proteins with the protease action of BoNT/E, which by cleaving at a different site than BoNT/A in its SNARE target (SNAP25) gives greater potential for pain alleviation than the activity of native BoNT/A. In this example the hybrid BoNT was still targeted toward a neuronal cell; however, it has been demonstrated that BoNTs can be retargeted toward non-neuronal cell types when attached to an appropriate protein partner. These potential therapeutic proteins are designed on the basis that SNARE proteins are essential not just for release of neurotransmitter vesicles but also in vesicle-based secretion pathways of non-neuronal cell types. That BoNTs protease activity could be retargeted was first demonstrated by an in vitro chemical attachment of the endopeptidase domain of BoNT/A to a wheat germ agglutinin protein

Table 2.1 Summary of engineered proteases

Therapeutic use	Enzyme	Modification	Properties and examples of clinical use	Therapeutic product	Reference
Anticoagulant	Urokinase	(None)	Fibrinolysis for treatment of pulmonary embolism	Abbokinase (approved 1983)	
	t-PA	(None)	Fibrinolysis for treatment of pulmonary embolism, acute ischemic stroke and myocardial infarction	Alteplase/Activase® (approved 1987)	
	t-PA	T103N/N117Q/K296A/H297A/ R298A/R299A	Reduced in vivo inhibition and increased half-life	Tenecteplase/TNKase (approved 2000)	Keyt (1994) [22]
	t-PA	N117G, deletion of fibronectin and epidermal growth factor domain	Increased half-life	Lanoteplase (discontinued)	Larsen (1991) [56]
	Thrombin	W215A/E217A	Decreased procoagulant fibrinogen activation but maintains protein C activation, triggering anticoagulant protease cascade		Cantwell (2000) [30]
Procoagulant	Thrombin	(None)	Activates fibrinogen, promoting coagulation at damaged tissues for wound treatment	Recothrom® (approved 2008)	
	FVIIa	(None)	Induced by tissue factor, a membrane signal protein, to promote coagulation for hemorrhage treatment	NovoSeven® (approved 1999)	
	FVIIa	V158D/E296V/M298Q	Mutations force FVIIa into tissue factor activated conformation	NN1731 (discontinued)	Persson (2001) [33]
	FVIIa	P10Q/K32E/D33F/A34E	Improved affinity for membranes and tissue factor activation	BAY 86–6150 (discontinued)	Mahlangu (2016) [37]
Alzheimer's disease	Neprilysin	G399V/G174K	Increased specificity for proteolysis of β-amyloid peptide sequences		Webster (2014) [43]
	Human Kallikrein 7		Increased specificity for proteolysis of β-amyloid peptide sequences		Guerrero (2016) [45]

(continued)

Table 2.1 (continued)

Therapeutic use	Enzyme	Modification	Properties and examples of clinical use	Therapeutic product	Reference
Neuromuscular disorders	BoNT/A	(None)	Cleaves SNAP-25 complex, inhibiting neurotransmission for treatment of muscle dystonias and multiple other medical conditions	Botox® (approved 1989)	
	BoNT/A	L256E/V258P and A308L	Altered specificity, instead cleaving SNAP-23 complex to potentially inhibit pathogenic secretion in non-neuronal cells		Sikorra (2016) [57]
	BoNT/E	K224D	Altered specificity, instead cleaving SNAP-23 complex to potentially inhibit pathogenic secretion in non-neuronal cells		Chen (2009) [58]
	BoNT/A and E hybrid	Fusion of BoNT/E protease domain and BoNT/A translocation and binding domains	Longer lasting pain attenuation		Wang (2011) [52]
	BoNT/D-GRRH hybrid	Fusion of BoNT/D protease and translocation domain to GRRH binding peptide	Inhibition of pituitary growth hormone hypersecretion in animal model of acromegaly		Somm (2012) [59]
Celiac disease	Endoproteases EP-B2 and SC-PEP	(none)	Combination of two unmodified glutamine and proline-specific proteases to break down dietary gliadin	ALV003 (phase II)	Tye-Din (2010) [60]
	SC-PEP	I581V/F459Y/M511L/I406V	Improved activity at low pH and resistance to pepsin degradation		Ehren (2008) [61]
	Kumamolisin	V119D/S262K/N291D/D293T/G319S/D358G/D368H	Improved specificity for (PQ) dipeptide motif		Gordon (2012) [62]

that has a generic ability to be internalized in vitro by multiple neuronal cell types and by pancreatic B cells. Once internalized, the BoNT/A-containing protein attenuated insulin secretion [53]. Following this, more specific cell targeting for therapeutic benefit has been investigated. For example, in pursuit of a therapy to decrease the pituitary growth hormone hypersecretion seen in acromegaly, an adult-onset endocrine disease, the endopeptidase and translocation domain of BoNT serotype D was attached to a peptide that binds to growth hormone-releasing hormone receptors (GRRH). This led to in vitro internalization of the hybrid protein into GRRH-containing cells and cleavage of the target intracellular SNARE. Rats injected with the hybrid protein showed markedly attenuated growth hormone secretion and synthesis [59]. The potential expansion of BoNT-based therapies has also been pursued by amino acid substitution of residues within the endopeptidase domain to change the specificity toward the targeted SNARE from the neuronal SNAP25 cleaved by BoNT/A to that of the non-neuronal SNAP23 [57, 58]. In the first engineering effort of this type, Chen and Barberi used rational engineering to identify one amino acid substitution, K224D, that could be made to the endopeptidase domain of BoNT/E to enable it to cleave not just SNAP25 but also SNAP23.

2.2.5 Digestive Proteases

In a separate context, proteases have also been investigated for the treatment of celiac disease. Celiac disease is a chronic illness in which dietary gluten triggers an autoimmune response and subsequent inflammation and damage of the intestine in genetically susceptible individuals. The current treatment is elimination of gluten from the patient's diet entirely, as no curative therapy exists [63]. More specifically this inflammatory immune response is triggered by peptides that result from the incomplete digestion of gliadin – gliadin being a significant protein component of gluten. Due to its high proline and glutamine content, gliadin is unusually resistant to breakdown by human digestive proteases [64]. A potential therapeutic treatment is the digestion of these immunogenic peptides by exogenously delivered proteases. Currently the enzyme mix ALV003, a combination of EP-B2 (a protease expressed in germinating barley seeds) and SC-PEP (a proline-specific protease from the bacteria *Sphingomonas capsulata*), is in phase II clinical trials as an oral enzyme therapy [60]. A therapeutically useful protease would need to be delivered orally, be stable and active in the low pH stomach environment, be resistant to degradation from endogenous proteases, and show sufficient activity and specificity toward the disease-triggering proline- and glutamine-rich oligopeptides, such that these oligopeptides are degraded even when present with other dietary proteins that could compete as substrates. In addition to examination of native proteases – such as with ALV003 – there have been efforts to use enzyme engineering to improve upon some of these features.

Ehren et al. focused on SC-PEP, the proline-specific protease being trialed as part of ALV003 [61]. Their main aims were to improve SC-PEP activity in the physiologically relevant, acidic environment (pH 4.5) and to increase resistance to pepsin degradation. They used structural data, published literature, and sequence analysis of 100 PEP homologs to identify 30 target amino acid substitutions that were then combined to generate a small enzyme library of 47 SC-PEP variants – each carrying three to five individual amino acid substitutions. These variants were tested for protease activity at pH 4.5 and in the presence of pepsin. The contribution of each individual amino acid substitution to these factors was scored and this information used to select the mutations that were then recombined to build a second library of 48 variants. Of these, 48% were more resistant to pepsin degradation than wild-type SC-PEP, and 54% showed a greater than 10% increase in degradation of the model peptide at pH 4.5. This demonstrates the increase in overall library quality that can be gained from empirical testing of even a small number of mutations: in the first round of library testing, only 13% of library members demonstrated increased pepsin resistance and 14.8% showed improved activity.

In a separate approach, Gorden et al. engineered the protease kumamolisin from the acidophilic bacterium *Alicyclobacillus sendaiensis* [62]. Database searches identified this enzyme as a promising initial candidate for engineering as it is highly active at pH 2–4 and a temperature of 37 °C, cleaves peptides after the dipeptide recognition sequences Pro/Arg or Pro/Lys, and exhibits slight activity for the motif common in gliadin, Pro/Gln. The RosettaDesign software was used to model the tetrapeptide Pro/Gln/Leu/Pro into the binding pocket of the crystal structure of kumamolisin. This leads to the design of a library of 261 variants carrying various substitutions selected from examining 75 residues lining the enzyme active site. While 20% of the library showed a decreased ability to cleave the model substrate, 30% retained wild-type activity and 50% showed an activity increase. The most active variant (V119D/S262K/N291D/D293T/G319S/D358G/D368H) showed 100-fold greater proteolytic activity for a model Pro-Gln-rich peptide as compared to the native enzyme and roughly 800-fold increased specificity as it no longer cleaved a model peptide substrate containing the native Pro/Arg motif. In an extension of this work, the same group again used Rosetta software to further enhance the activity of this improved kumamolisin variant against two gliadin peptides known to be highly immunogenic – one rich in Pro/Gln/Gln, the other in Pro/Gln/Leu [65].

2.2.6 Protease Engineering Strategies

In regard to protease engineering in general, a recent review highlights a range of powerful high-throughput strategies that have been used to engineer protease activity [66]. These include phage-based systems [67] and FACS with protease-activated fluorescent reporters [44, 45, 68] or antibiotic resistance proteins [69]. While not directly focused on engineering of proteases for therapeutic applications, it should be noted that these strategies developed from tuning substrate specificities or increasing catalytic efficiency of model proteases such as the *E. coli* OmpT [70] or the tobacco protease TEV [69, 71, 72] may in the future be applied to the challenges of engineering proteases as therapeutics. Equally interesting is the work of researchers using enzyme engineering tools to elucidate the mutational paths through which proteases can evolve resistance to protease inhibitor drugs [73].

2.3 Ribonucleases

Ribonucleases (RNases) have been investigated as anticancer therapeutics for over 60 years [74]. Differences in surface structures between normal and malignant cells are thought to be responsible for the selective intake of exogenous RNases and consequent interference with RNA metabolism. Specifically, the enzymes' cytotoxic properties are conferred by rapid RNA degradation in targeted cells, stalling cell cycle progression and leading to apoptosis. To date, the RNase therapeutic to advance furthest in clinical trials is ranpirnase (Onconase), an unmodified enzyme from the RNase A superfamily isolated from the *Rana pipiens* frog. Onconase showed promising results when used to treat malignant mesothelioma in phase II trials but subsequently failed in advanced tests [75, 76]. Nevertheless, it is currently still under clinical investigation as a possible antiviral for the treatment of human papillomavirus.

To increase the therapeutic effects with respect to current and future therapeutic applications of RNases, protein engineering has been used to generate variants with decreased binding affinities toward cytosolic ribonuclease inhibitor protein (RI) [77]. RI is a 50-kDa horseshoe-shaped protein that binds extremely tightly to many RNase A family members and greatly diminishes RNase cytotoxicity [78]. Engineered RNases that maintain catalytic activity but exhibit decreased binding to human RI include mutants of the human enzymes RNase 1 and RNase 5 (angiogenin), as well as bovine pancreatic RNase A [79–84]. Employing rational design strategies, the examination of crystal structures of RNase-RI complexes allowed researchers to identify key residues at the protein-protein interface. Amino acid substitutions at these positions reduce the binding affinity of RI and consequently minimize RNase inhibition. More recently, these engineered human RNases have also been investigated in combination with cancer-targeting antibodies as non-immunogenic, cytotoxic warheads [84, 85].

2.4 Amino Acid-Degrading Enzymes

A number of enzymes have been investigated as potential anticancer agents due to their ability to metabolize amino acids. The mechanistic basis underlying the therapeutic effect is a depletion of the endogenous amino acid pool. The resulting auxotrophy in one or more specific amino acids forces the tumor to rely on exogenous sources to supplement the limiting amino acids or else halt protein synthesis, which triggers apoptosis in the malignant cells (for recent reviews see [86, 87]). Of particular interest for this resource-depletion strategy are enzymes for the degradation of L-arginine (Arg) and L-asparagine (Asn).

2.4.1 Asparagine-Degrading Enzymes

Asparaginase II from *E. coli* (EcAII) is currently a key component in the treatment of childhood acute lymphoblastic leukemia (ALL). Marked under the trade name Elspar [88], EcAII hydrolyzes Asn to Asp and ammonia, which results in depletion of serum levels of Asn and triggers lymphoblast apoptosis. Consequently, overall survival rates of childhood acute lymphoblastic leukemia are relatively high at close to 90% [89], yet immunogenicity of the bacterial enzyme can cause undesirable side effects [90] and renders some patients ineligible or unable to continue treatment. In attempts to alleviate these effects, Cantor et al. pursued an engineering strategy aimed at generating an EcAII mutant with decreased immune response while retaining catalytic activity [91]. Starting with in silico analysis of EcAII to predict three putative T-cell epitopes, four-residue fragments within each epitope were then subjected to saturation mutagenesis to disrupt interactions with major histocompatibility complex (MHC-II). The resulting mutant libraries were overexpressed in *E. coli* and enriched for enzymes retaining Asn hydrolase activity using a neutral drift strategy. More specifically, the selection of *E. coli* expressing active EcAII variants was based on the host cell's ability to hydrolyze Asn to Asp, which was required for concomitant expression of GFP. After targeting each predicted epitope, the lead EcAII mutant showed eight amino acid changes. Although catalytic performance for the mutant was slightly impaired as reflected by the threefold drop in specific activity due to an increase in the apparent Michaelis-Menten constant, this variant did display decreased immunogenicity. Tests in a transgenic mouse model that expressed human leukocyte antigen-II molecules indicated a tenfold reduction in anti-EcAII IgG levels as compared to wild-type enzyme. Separately, increased therapeutic benefit has also been pursued via mutagenesis studies on asparaginase homologs from other prokaryotes [92]. These studies included investigating the contribution of secondary glutaminase activity in *Helicobacter pylori* asparaginase mutants to cytotoxicity, as well as on increasing the catalytic activity and cytotoxicity of the thermostable *Pyrococcus furiosus* asparaginase [93]. Finally, efforts have been reported to generate a thermostable variant of the *Erwinia chrysanthemi* asparaginase [94]. The wild-type enzyme is currently used as a second-line therapeutic treatment for ALL [95].

2.4.2 Arginine-Degrading Enzymes

Arginine depletion has also been investigated, in particular for the treatment of hepatocellular carcinomas and melanomas lacking the enzyme argininosuccinate synthetase 1 [87]. Of interest are a number of prokaryotic arginine deiminases (ADIs) that hydrolyze Arg to citrulline and ammonium. Most clinically advanced is a mycobacterium ADI in PEGylated form [96–99]. To date however the majority of engineering efforts have focused on the more recently characterized highly cytotoxic PpADI from *Pseudomonas plecoglossicida*, thoroughly reviewed by Han et al. [100, 101]. Targets for protein engineering have included improving catalytic performance and thermostability, as well as shifting the enzyme's pH optimum from its native pH 6 closer to physiological conditions to prevent dramatic activity losses observed for wild-type PpADI [102]. These engineering efforts have been aided by a newly developed screening assay that can assess PpADI activities at physiologically relevant low levels of Arg (~100 μM) [103]. Briefly, PpADI mutants were expressed in an *E. coli* strain which controls GFP expression via a promoter that can be repressed by Arg. Host cells carrying improved ADI variants hence could easily be identified as they showed enhanced GFP expression and could be enriched by FACS. This high-throughput assay allowed for the screening of about eight million variants over three iterative rounds of directed evolution to arrive at a mutant with an almost three orders of magnitude higher $k_{cat}/S_{0.5}$ compared to wild-type enzyme. Furthermore, these engineered ADIs possessed significant catalytic activity at physiological arginine levels (100 μM), whereas the native enzyme showed no detectable activity at this substrate concentration.

Addressing concerns over immunogenicity of prokaryotic ADIs in therapeutic applications, human enzymes that can metabolize Arg have also been investigated. While mammals do not have direct ADI homologs, the human enzyme arginase I (hArgI) that hydrolyzes Arg to urea and ornithine was identified as a possible substitute [104]. However, preliminary studies of the native enzyme revealed K_m values for arginine in the millimolar range. Furthermore, the pH optimum of hArgI near pH 9.5 is outside the physiological range [105]. A less conventional but effective strategy to overcome these limitations has focused on the enzyme's cofactor rather than amino acid mutagenesis [106]. Native hArgI requires a Mn^{2+} cofactor to generate a metal-activated water for attack on the guanidinium carbon of Arg. Cofactor replacement with Co^{2+} resulted in a p*Ka* shift by one pH unit and generated an enzyme variant with a tenfold increase in k_{cat}/K_m at pH 7.4. Separately, serum half-life was markedly improved by PEGylation of the enzymes [107].

2.5 Deoxyribonucleoside Kinases

Beyond applications of engineered enzymes as therapeutic agents themselves, their potential role as prodrug-activating catalysts has flourished over the last decade. One of the examples are 2′-deoxyribonucleoside kinases (dNKs), enzymes that are part of the nucleoside salvage pathway and play a critical role in therapeutic applications of nucleoside analog (NA) prodrugs [108]. The salvage pathway enables mammalian cells to recycle scavenged 2′-deoxynucleosides from the environment via transmembrane uptake and three consecutive, dNK-catalyzed phosphoryl-transfer steps [109]. The process complements de novo nucleotide biosynthesis to supply triphosphate anabolites for DNA replication and other cellular functions. In addition, the salvage pathway is responsible for the intracellular activation of NAs, synthetic mimics (Fig. 2.1) of the natural DNA building blocks that upon phosphorylation to their corresponding triphosphates turn into competitive substrates for low-fidelity DNA polymerases and reverse transcriptase found in cancer cells and viruses, respectively [110].

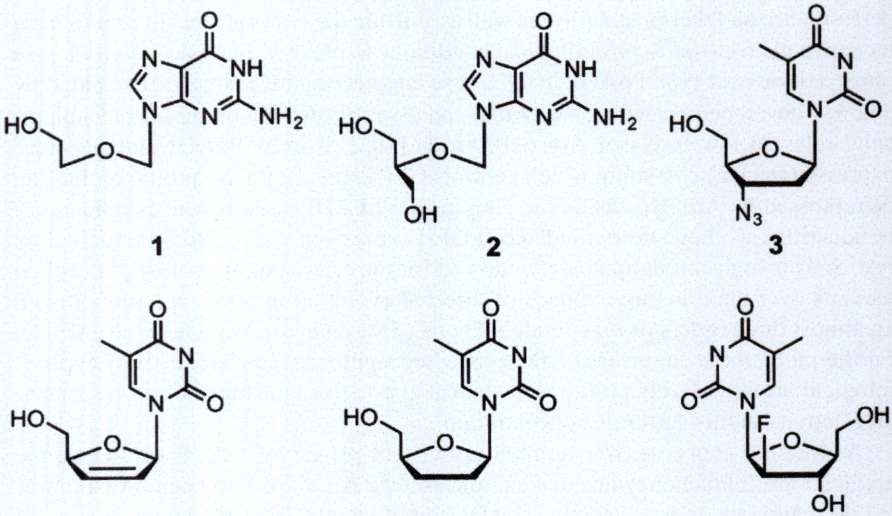

Fig. 2.1 Nucleoside analogs as prodrugs for antiviral and cancer therapy: aciclovir (**1**), ganciclovir (**2**), AZT (**3**), d4T (**4**), ddT (**5**), and L-FMAU (**6**)

Problems with NA activation arise due to the high substrate specificity of the host's endogenous kinases in general and the initial phosphorylation by dNKs in particular. Inefficient phosphorylation of NAs not only reduces the potency of existing prodrugs but can result in the accumulation of cytotoxic reaction intermediates. It is also responsible for the failure of a large number of potential NA prodrug candidates in vivo [111]. One solution to overcome the shortcomings of the phosphorylation cascade has been the coadministration of prodrugs with exogenous, broad-specificity kinases via suicide gene therapy [10, 112]. While biochemical and preclinical experiments have demonstrated the effectiveness of the strategy in principle, the studies also uncovered limitations arising from the dNKs' significantly lower activity for NAs compared to their performance with the native substrates. In addition, the broad substrate specificity of exogenous kinases raised concerns as it interferes with the tightly regulated 2′-deoxyribonucleoside metabolism [113]. Studies have hence focused on identifying engineered NA kinases for the selective and efficient activation of these prodrugs.

2.5.1 Thymidine Kinase from Herpes Simplex Virus

Herpes simplex virus type-1 thymidine kinase (HSVtk) was the first candidate for NA prodrug activation and remains one of the most commonly used enzymes in advanced (preclinical) studies [114–116]. Besides its native role in viral replication, HSVtk is a natural candidate for application in conjunction with NA prodrugs due to its thousandfold higher activity and elevated substrate promiscuity compared to mammalian dNKs [117]. In fact, the phosphorylative activation of aciclovir (ACV, **1**) and ganciclovir (GCV, **2**) (Fig. 2.1) by wild-type HSVtk made these NA prodrugs early therapeutics for herpes infections [118]. Nevertheless, the enzyme's effectiveness in antiviral treatment is limited by its two to three orders of magnitude preference for native metabolites over NAs. Engineering HSVtk for greater prodrug affinity could significantly enhance the clinical value of this enzyme.

In early studies, Loeb and coworkers conducted a series of directed evolution experiments to increase the NA specificity of HSVtk. Targeting the regions flanking two putative nucleoside binding sites (positions 162–164 and 171–173) by random mutagenesis, HSVtk variants with increased sensitivity for **1** and **2** were identified via genetic selection of million-member libraries using the thymidine kinase-deficient *E. coli* strain KY895 [119–121]. These initial mutation studies improved dNK activity and shifted substrate specificities upon amino acid substitution at Phe161 alone, as well as in combination with changes in neighboring positions (Table 2.2) [121]. In vitro assays with two selected candidates, variants 30 and 75, show a dramatic rise in the activity ratio for NA over thymidine, and these functional gains could be confirmed in mammalian cell culture experiments [139]. A follow-up study by Black et al. to reevaluate the five amino acid replacements in variants 30 and 75 identified a further optimized variant termed SR39 [124]. In vitro and in vivo, variant SR39 exhibited the greatest sensitivity to **2** among engineered HSVtks, increasing cytotoxicity by ~300-fold over wild-type HSVtk. Variant SR39 by itself, as well as in combination with other enzymes such as guanylate kinase [140], has since demonstrated its effectiveness in various cancer models, making it a benchmark for suicide gene therapy [141].

Guided by structural information and emerging NA drug resistance, several other groups have explored amino acid replacements in other positions of HSVtk to find variants with improved activity and specificity for NAs. Drake and coworkers explored the impact of amino acid replacements at Gln125. Conservative amino acid substitutions (Q125N) at that position showed moderate specificity changes in favor of **2** over thymidine [123]. Meanwhile, Balzarini and coworkers revisited the highly conserved positions 167 and 168 [125, 142]. The introduction of bulky amino acid side chains including A167Y/F and A168H not only created an effective steric block for native thymidine but resulted in unexpectedly preservation and even enhancement of activity for **2**. Beyond these mutagenesis studies focused on the active site of HSVtk, Christians et al. performed a DNA shuffling experiment to enhance the activity of the viral kinase for 3′-azidothymidine (AZT, **3**) [122]. The study used homologous thymidine kinases from HSV type-1 and type-2, as well as their chimeras as parents for in vitro recombination and screening for increase **3** sensitivity in *E. coli* KY895. Four iterative rounds of laboratory evolution identified lead kinase variant (Cycle4 TK), a patchwork of multiple sequence fragments from the two parents and a few additional amino acid changes originating from random mutations. Functionally, Cycle4 TK was found to be a generalist with almost identical kinetic parameters for **3** and thymidine, representing a roughly tenfold improvement and a 25-fold decline in catalytic performance for **3** and thymidine, respectively.

Table 2.2 Overview of engineered 2′-deoxynucleoside kinases as prodrug-activating catalysts in antiviral and cancer therapy and reporter for noninvasive PET imaging techniques

Enzyme	Product/reference	Modification	Observed property
Thymidine kinase (herpes simplex virus type 1)	Wild type	(None)	Highly active and promiscuous 2′-deoxyribonucleoside kinase
	Black (1993) [120]	P155A/F161V; F161I/C	Improved activity for natural pyrimidine nucleosides, ACV, GCV, and AZT
	Black (1996) [121]	L159I/I160L/F161A/A168Y/L169F and I160L/F161L/A168V/L169M	Variants 30 and 75: lower thymidine activity but enhanced activity for ACV and GCV
	Christians (1999) [122]	Chimeric enzyme	Cycle4 TK: improved activity for AZT
	Hinds (2000) [123]	Q125N	Lower thymidine activity but enhanced activity for GCV
	Black (2001) [124]	L159I/I160F/F161L/A168F/L169M	Variants SR39: superior activity for GCV
	Balzarini (2006) [125]	A167F and A168H	Eliminates thymidine activity but enhanced activity for GCV
2′-Deoxyribonucleoside kinase (*Drosophila melanogaster*)	Munch-Petersen (1998) [126]	(None)	Highly active, broad-specificity kinase
	Knecht (2000) [127]	N45D/N64D	Variant MuD: increased specificity for AZT and ddC
	Knecht (2002) [128]	V84A/M88R/A110D	Increased specificity for purine nucleosides
	Knecht (2007) [129]	N45D/N64D/N210D/ L239P	Variants B5 and B10: increased specificity for purine analogs
	Gerth (2007) [130]	Chimeric enzyme	Improved activity for d4T
	Solaroli (2007) [131]	M88R	Increased cytotoxicity of purine NAs
	Liu (2009) [132]	T85M/E172V/Y179F/H193Y	Variant R4.V3: increased catalytic performance for ddT
	Liu (2010) [133]	L66F/Y70M/E172I/Y175W	Variant RosD7: increased catalytic performance for ddT

(continued)

Table 2.2 (continued)

Thymidine kinase type-2 (*Homo sapiens*)	Wild type	(None)	High specificity for thymine and uracil 2′-deoxynucleosides
	Campbell (2012) [134]	N93D/L109F	Improved activity for L-FMAU
2′-Deoxycytidine kinase (*Homo sapiens*)	Wild type	(None)	Moderate activity for 2′-deoxycytidine and purine nucleosides
	Sabini (2007) [135]	A100V/R104M/D133A	Broader substrate specificity including thymidine
	Iyidogan (2008) [136]	D47E/R104Q/D133G/N163I/F242L	Variant epTK6: highly promiscuous 2′-deoxyribonucleoside kinase
	Hazra (2008) [137]	R104M/D133A	Improved activity for L-nucleosides
	Muthu (2014) [138]	R104M/V130T/D133A/L191A	Variant B6-II: improved activity for L-FMAU

2.5.2 2′-Deoxyribonucleoside Kinase from *Drosophila melanogaster*

The 2′-deoxyribonucleoside kinase from *Drosophila melanogaster* (DmdNK) is expressed in fruit fly embryos and possesses the highest catalytic activity and broadest substrate specificity among known mammalian dNKs [126]. Although primarily a pyrimidine kinase, DmdNK also shows significant activity for purine nucleosides. The wild-type enzyme tolerates a variety of 2′-deoxyribose derivatives, making it an appealing candidate for therapeutic applications in suicide gene therapy [143]. A number of protein engineering studies have further tweaked its functional performance toward efficient and specific activation of NAs.

In searching for DmdNK variants with increased sensitivity toward multiple NAs, Knecht et al. followed earlier HSVtk engineering approaches by deploying random mutagenesis and in vivo selection in *E. coli* KY895 [127]. The lead candidate emerging from these experiments was DmdNK variant MuD with two amino acid changes (N45D/N64D). Expression of MuD in the bacterial host raised its sensitivity to **3** and 2′,3′-dideoxycytidine (ddC) by >300-fold and >tenfold, respectively. Kinetic analysis suggested that these functional gains could largely be attributed to the enzyme's loss of function for native 2′-deoxynucleosides while leaving kinetic parameters for NAs mostly unchanged. A rationale for the observed changes was not immediately obvious; both Asn are surface residues and distal to the active site. A subsequent crystallographic study suggested that N64D destabilizes the neighboring LID region, a conformationally flexible segment that contributes to substrate binding via hydrogen-bonding interactions with the substrates 3′-OH group [144]. In separate experiments, the same group focused on raising enzyme activity for purine substrates by employed rational design based on structural comparisons of HSVtk and DmdNK [128]. The work identified a fruit fly enzyme variant with changes in three active site residues (V84A/M88R/A110D) that showed slightly improved kinetic performance for purine nucleosides. This enzyme and related DmdNK variants were later tested for activation of purine NAs, demonstrating >100-fold enhanced cytotoxicity of purine arabinofuranosides in vitro and in cell cultures [131]. In follow-up work, DNA shuffling experiments led to several new variants with improved activity for purine NAs cladribine and fludarabine **9** (Fig.2.2) [129]. Although causing declines in specific activity, the amino acid changes resulted in variants with favorable catalytic performance with NAs relative to the natural substrates. The benefit of such functional changes was confirmed in cancer models for two lead variants: B5 and B10 (Table 2.2). These two variants exhibited up to thousandfold greater sensitivity for the two purine NAs. However, the studies also revealed significant variability in kinase performance in different cell lines, highlighting the limitations of laboratory evolution experiments that rely on an *E. coli*-based surrogate system over screening kinase variants in a specific cancer cell line. Beyond the potential functional role of the previously reported amino acid replacements at positions 64 and 84, the basis for the observed contributions of positions 210 and 239 is not obvious as they are located in the conformationally poorly defined C-terminal region. Nevertheless,

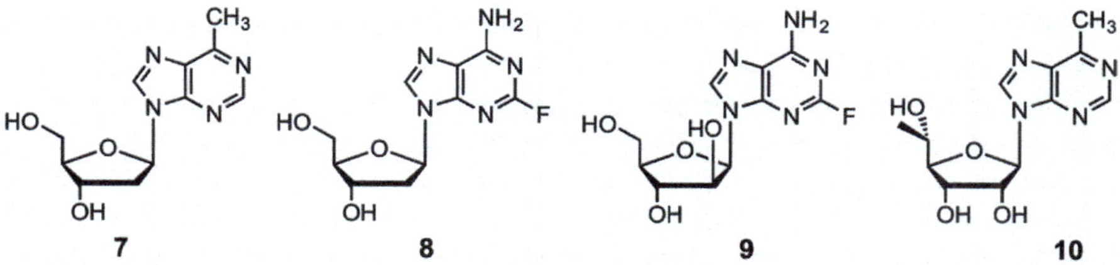

Fig. 2.2 Purine nucleoside analogs as prodrugs for PNP-based anticancer therapy: MePR (**7**), FdA (**8**), fludarabine (**9**), and Me(*talo*)PR (**10**)

the region's importance to enzyme function was independently confirmed by a study using nonhomologous recombination to generate chimera of DmdNK and human thymidine kinase 2 [130]. Multiple hybrid enzymes with C-terminal sequences from the fruit fly enzyme were found that exhibited significantly improved activity for the anti-HIV prodrug 2′,3′-didehydro-3′-deoxythymidine (d4T, **4**).

More recently, two novel methods were applied to dNK engineering by our group. To overcome the functional bias and limitations related to the use of the auxotrophic *E. coli* KY895, we initially developed a FACS-based approach to screen directed evolution libraries of dNKs for 2′,3′-dideoxythymidine (ddT, **5**) activity [132]. DmdNK libraries were created by random mutagenesis and DNA shuffling, followed by expression of the corresponding kinase variants in *E. coli*. Upon incubation with a fluorescent analog of ddT, host cells expressing a functional kinase accumulated the fluorophor and could be isolated by FACS. Four iterative rounds of directed evolution yielded top-performing variant R4.V3 (Table 2.2) that showed a 20-fold preference for **5** over thymidine. Secondly, we experimented with in silico redesign of DmdNK using RosettaDesign software to accelerate the discovery process and eliminate the need for high-throughput screening [133]. Active site modeling with ddT suggested several alternative solutions for tuning the enzyme's substrate specificity. Experimental evaluation of these computational designs led to variant RosD7 which showed roughly eight-fold preference for ddT over native substrate and required testing of only about two dozen DmdNK variants.

While DmdNK, HSVtk, and their variants are strong candidates for potential applications in suicide gene therapy, their nonhuman origin remains a significant limitation due to concerns over immunogenic response. Engineering efforts have therefore increasingly focused on tailoring the properties of human dNKs.

2.5.3 *2′-Deoxycytidine Kinase from Homo sapiens*

Among the four native dNKs in human, 2′-deoxycytidine kinase (dCK) is the most promiscuous enzyme and catalyzes the phosphorylation of 2′-deoxycytidine and both purine homologs, as well as several NAs [110]. Even though dCK has relatively poor catalytic activity compared to DmdNK and HSVtk, its broad substrate specificity and inherently reduced immunogenicity make it an attractive target for protein engineering.

Initial efforts to improve the catalytic efficiency of dCK were based on crystallographic data and previous results from DmdNK engineering studies. Sabini et al. focused on three residues, Ala100, Arg104, and Asp133, that correspond to positions 84, 88, and 110 in DmdNK [135]. These residues form an elaborate hydrogen-bonding network responsible for the effective discrimination of thymidine as a substrate. A simple amino acid swap in these positions (A100V/R104M/D133A) broadened the substrate specificity of dCK. Besides improved activity for 2′-deoxycytidine, the dCK variant showed unprecedented activity for thymidine and several additional NAs. In a separate study, our group and others implemented a combination of rational design and random mutagenesis to more systematically explore the impact of amino acid replacements in these three positions on the substrate specificity of dCK [136, 137]. Besides confirming the key role of positions 104 and 133 in respect to substrate specificity, these experiments identified a number of alternate, functionally more advantageous amino acid substitutions. Of particular interest was variant epTK6 (Table 2.2) which showed superior catalytic performance with a diverse collection of NAs.

2.5.4 From Suicide Gene Therapy to PET Reporter Systems

The development of tailored dNKs for prodrug activation is not the only strategy to overcome the often rate-limiting initial phosphorylation step for NAs. Significant efforts have also been devoted to the development of synthetic alternatives. While direct cellular uptake of phosphorylated nucleosides is not feasible, esterification of NA monophosphates can temporarily mask the electrostatic charges of the phosphate group and facilitate diffusion across the cell membrane [145]. Upon cellular entry, ester hydrolysis removes the masking groups and liberates the corresponding monophosphate anabolite. As this phosphate ester strategy has proven increasingly effective for the delivery of various NA prodrugs [146], efforts to engineer dNKs have gradually shifted to other applications, including the chemo-enzymatic synthesis of nucleoside triphosphate and the development of effective reporter systems for positron emission tomography (PET) imaging [147].

PET is a versatile method for noninvasive visualization of biological processes in living subjects and is based on two-component systems; an isotopically labeled small-molecule reporter (typically ^{18}F) and a reporter gene expressed within target cells that will interact and trap the radionuclide reporter whose accumulation can be detected via PET. Engineered dNKs, applied in combination with ^{18}F-labeled NAs, offer a powerful strategy to monitor, for example, whole-body tissue distribution of viral vectors in gene and oncolytic therapies, as well as adoptive cell-based therapies [148, 149]. The concept was first implemented with wild-type HSVtk or variant SR39 as reporter genes in combination with [18F]-labeled probes such as

2′-fluoro-2′-deoxy-1-β-L-arabinofuranosyl-5-methyluracil (L-FMAU, **6**) and found widespread application in cell culture studies, as well as in translational work with animals and humans [150–152]. While phosphorylation of **6** by highly specific, endogenous dNKs occurs only at very low levels, the herpes enzymes activate it albeit with moderate efficiency. Over the last decade, the development of novel PET reporter kinases has hence focused on enzymes with high activity and specificity for such PET probes. In addition, engineering efforts have shifted to human dNKs to minimize risks associated with immune responses in clinical applications.

Initial experiments with human PET kinase/reporter systems largely relied on previously reported dCK variants [153]. As discussed above, these variants with amino acid changes in positions 104 and 133 not only exhibit activity for thymidine and 2′-deoxyuridine but were found to also effectively phosphorylate NAs with L-ribose sugars, a key feature of many potential PET probes [136, 137]. Side-by-side comparison of these dCK variants and HSVtk SR39 in cell culture and animal model studies suggested an on par performance of some human kinase variants [154]. While these findings were encouraging, kinetic data also indicated that in spite of functional improvements for **6**, these enzymes still favored native metabolites. This limits their clinical value as substrate competition for the active site reduces imaging sensitivity and requires higher doses of radionuclide probes. The next generation of high-contrast PET reporter systems demands bioorthogonal dNKs, enzymes that effectively activate L-isomeric NA probes while discriminating against native 2′-deoxynucleosides.

Attempting to meet those demands, we employed a semi-rational engineering strategy to create an L-selective dCK variant [138]. Leveraging preexisting kinetic data, we used computational design to create a small, focused library of 16 dCK variants and evaluated them with D- and L-nucleosides. Lead variant B6-II showed a two- and tenfold preference for **6** over D-2′-deoxycytidine and D-thymidine, respectively. These early successes have led to a second round of dCK engineering, using design-of-experiment methodology to systematically evaluate roughly 200 dCK sequences with up to 11 substitutions. In this ongoing study, kinase variants with up to tenfold improved activity for **6**, and almost 100-fold preference for L-nucleosides has been identified (Muthu and Lutz unpublished). In separate experiments, Lavie and coworkers have explored engineered variants of human mitochondrial thymidine kinase type 2 (TK2) as an alternative to human dCK in PET imaging. TK2 is a native pyrimidine nucleoside kinase with moderate catalytic activity for NAs including L-nucleosides [155], yet technical challenges with heterologous protein expression have so far prohibited the enzyme's structural analysis and greatly limited protein engineering efforts. Building on results from PET studies with wild-type TK2 [156], Campbell et al. used a TK2 homology model to identify two key functional residues in the enzyme activity site [134]. Amino acid substitutions in positions Asn93 and Leu109 (N93D/L109F) not only lowered the TK2 variant's turnover of native D-nucleosides but also raised its activity for **6** in vitro, as well as in cell culture and mice studies.

2.6 Cytosine Deaminases

Cytosine deaminases (CDAs) are another family of nucleoside-metabolizing enzymes that have undergone tailoring by protein engineering for potential therapeutic applications. CDAs are part of the pyrimidine salvage pathway, catalyzing the hydrolytic deamination of cytosine to uracil and ammonia. More important from a medicinal standpoint, they are only found in prokaryotes and fungi. The native enzymes accept 5-fluorocytosine (5FC) as a substrate, converting the nucleobase analog into 5-fluorouracil, which is a very potent inhibitor of thymidine synthase [157]. Inhibition of thymidine synthase in turn disrupts de novo production of thymidine monophosphate and consequently stalls nucleic acid metabolism. The small size and high cell permeability of 5FC makes this prodrug and its CDA-based activation a highly promising anticancer treatment.

Protein engineering of CDAs has mostly focused on increased catalytic activity and enzyme stability. In early work, Mahan et al. used structural information to identify a lid region in *E. coli* codA (residues 310–320) which undergoes a conformational rearrangement as part of the enzyme's catalytic cycle [158]. Targeting the region by alanine-scanning mutagenesis, variant D314A showed a 20-fold decrease for native cytosine activity and a simultaneous twofold increase for 5-fluorocytosine activity. In a follow-up study, the same authors used the increased 5FC sensitivity of *E. coli* host cells expressing improved enzyme variants as a negative selection assay to evaluate a randomized whole-gene library [159]. The study found one variant with greatly enhanced activity for 5FC, carrying two amino acid replacements (Q102R/D314G). Further analysis revealed that only the substitution at position 314 contributed to the observed functional changes. This result prompted the authors to prepare a site-saturation mutagenesis library for a comprehensive evaluation of that position, yielding D314S as an additional option for CDA variants with enhanced 5FC activation. These latter engineering studies were complemented with crystallographic analysis of selected enzyme variants to better understand the functional improvements. The observed structural changes suggest small but important alterations in the active site organization, as well as possible changes to the dynamics of the lid region. More recently, CDA variants with improved selectivity for 5FC were reported by Fuchita et al. [160]. The authors employed targeted mutagenesis in two regions of codA (residues 149–159 and 310–320) and evaluated the million-member library through negative selection. In the top-performing CDA variant, three amino acid substitutions (V152A/F316C/D317G) caused an approximately 20-fold shift in substrate preference from cytosine to 5FC. Kinetic studies of the variant revealed a combination of changes to K_M and k_{cat} values. Meanwhile, crystallographic analysis confirmed that none of the three residues was in direct contact with the substrate but seemed to affect enzyme activity through subtle conformational changes of neighboring residues.

In addition to the studies of bacterial deaminases, yeast CDA makes an attractive prodrug-activating enzyme due to its overall superior catalytic performance.

However, its potential in therapeutic applications is compromised by the native enzymes' limited stability. To this end, Korkegian et al. took a computational route to improve the stability of CDA from *Saccharomyces cerevisiae* [161]. Using RosettaDesign to predict amino acid substitutions for stabilizing the protein structure, their experiments identified two neighboring substitutions (A23L and I140L) and a third, distant change (V108I) which together improved the packing of the protein's hydrophobic core. Improved core packing in the redesigned enzyme raised its temperature of unfolding (T_m) by 10 °C and translated into a 30-fold increased half-life at 50 °C without compromising catalytic performance. In a separate study by the same team, CDA variants with improved activity for 5FC were sought via randomized mutagenesis of 11 amino acid residues [162]. These positions were selected based on their location in highly conserved regions of the protein. After diversification by primer-based codon scrambling, the resulting protein library was tested for improved variants in a two-stage genetic complementation assay, using a codA-deficient *E. coli* strain. Positive selection of library members via growth on cytosine minimal media was followed by negative selection for 5FC sensitive variants. The experiments identified three distinct variants with single amino acid substitutions at either position 92, position 93, or position 98, yet only D92E conferred increased 5FC sensitivity in subsequent mammalian cell culture. The kinetics of purified yeast CDA (D92E) showed little change in the Michaelis-Menten parameters, yet biophysical studies indicated an increased T_m by 4 °C which could be rationalized by the residues' position at the homodimer interface. Interestingly, the detected gains in activity in vivo are likely due to elevated protein stability. However, the limits of such a strategy for improving enzyme activity could be seen when the D92E substitution was combined with the computationally designed CDA variant. While the quadruple mutant showed additive behavior in regard to protein stability, the activity of the variant declined.

In summary, recent protein engineering studies have demonstrated the potential for functional improvements in CDAs. Concerns over the immunogenicity of ectopically expressed CDAs in patients remain to be addressed and have hampered clinical trials. Additionally, 5FC by itself is a potent bactericide, and oral administration of the prodrug could affect a patient's intestinal flora, potentially resulting in unintended side effects. Nevertheless, clinical studies of dNK-CDA conjugates have been reported [163, 164] (Table 2.3).

2.7 Purine Nucleoside Phosphorylases

In purine metabolism, purine nucleoside phosphorylase (PNP) catalyzes the reversible phosphorolysis of inosine and guanosine into the corresponding nucleobase and ribose-1-phosphate [172]. PNPs can be divided into two categories based on their structures: trimeric and hexameric. Trimeric PNPs are found in both prokaryotes and eukaryotes and exhibit high specificity for 6-oxopurine nucleobases. Hexameric PNPs are exclusive to prokaryotes and accept a broader range of nucleobase substrates including 6-amino and 6-oxopurines. Given these differences in substrate

Table 2.3 Non kinase-based GDEPT systems modified by protein engineering

Enzyme	Product/ reference	Modification	Observed property
Cytosine deaminase (*E. coli*)	Wild type	(None)	Robust enzyme with catalytic activity for cytosine >> 5-fluorocytosine (5FC)
	Mahan (2004) [158]	D314A/G/S	Shift in catalytic activity favoring 5FC over cytosine due to altered lid dynamics
	Fuchita (2009) [160]	V152A/F316C/D317G	Shift in catalytic activity favoring 5FC over cytosine via distal amino acid changes
Cytosine deaminase (*S. cerevisiae*)	Wild type	(None)	Fragile enzyme with high catalytic activity for 5-fluorocytosine (5FC)
	Korkegian (2005) [161]	A23L/V108L/I140L	Increased Tm by 10 °C due to improved hydrophobic core packing; unchanged catalytic performance. Increased Tm by 10 °C due to improved hydrophobic core packing; unchanged catalytic performance
	Stolworthy (2008) [162]	D92E	Increased Tm by 4 °C due to stabilization of homodimer interface; unchanged catalytic performance
Purine nucleoside phosphorylase (*E. coli*)	Wild type	(None)	Moderate catalytic performance on a broad range of substrates
	Bennett (2003) [165]	M64V	>100-fold improved catalytic activity for nucleoside analog due to reshaping of active site binding pocket
Purine nucleoside phosphorylase (human)	Wild type	(None)	Activity for 6-oxopurine substrates
	Stockler (1997) [166]	N243D/E201Q	Altered substrate specificity for 6-aminopurine substrates
Purine nucleoside phosphorylases (*S. sofataricus*)	Wild type	(None)	Six- and 12-fold improved catalytic activity for fludarabine compared to *E. coli* PNP. Six- and 12-fold improved catalytic activity for fludarabine compared to *E. coli* PNP

Table 2.3 (continued)

Enzyme	Product/ reference	Modification	Observed property
Nitroreductases (*E. coli*)	Wild type (NfsB)	(None)	Molecular target for several bactericides
	NfsB: Grove (2003) [167]	F125K	Improved activity for CB1954
	NfsB: Guise (2007) [168]	T41Q/N71S/F124T	Improved activity for CB1954
	YieF: Barack (2006) [169]	T160N/Q175L/Y230A/ Y239N	Improved activity for CB1954
	NfsB: Jaberipour (2010) [170]	T41L/F70A	Improved activity for CB1954
	FRaseI: Swe (2012) [171]	A120V/F124G	Improved activity for CB1954

specificity, initial efforts to utilize PNP in therapeutic applications focused on the hexameric class. Conceptionally, PNPs are delivered into a tumor via suicide gene therapy where their broader substrate specificity is exploited to catalyze the cleavage of nontoxic nucleoside prodrugs into cytotoxic purine analogs [173]. The hexameric PNP from *E. coli* (DeoD) was found to be sufficiently promiscuous to activate a number of nucleoside prodrugs including 9-β-D-[2′-deoxyribofuranosyl]-6-methylpurine (MePR, **7**) and 2-fluoro-2′-deoxy-adenosine (FdA, **8**), as well as the ribose analog 9-β-D-arabino-furanosyl-2-fluoroadenine (FaraA, **9**). No phosphorolysis of these prodrugs was detected with trimeric human PNP.

Despite these initial successes, PNP-based prodrug activation faces a major challenge due to the prodrug toxicity to a patient's intestinal microbiome. In early studies, these undesirable side effects on the gut flora were targeted by protein engineering of DeoD. Two strategies to enhance the enzyme's therapeutic efficacy were pursued: improving catalytic activity to allow for lower prodrug doses and altering substrate specificity to activate prodrugs not converted by any wild-type PNP. Specifically, molecular modeling of **7** and derivatives (such as Me (*talo*)PR; **10**) in DeoD revealed the importance of sugar puckering in productive substrate binding and identified a steric clash between Met64 and the C6′-methyl group of **10**. While this structural incompatibility makes **10** a poor substrate for native PNPs, the substitution of Met64 with several smaller hydrophobic amino acids led to variant Met64Val possessing an enlarged active site binding pocket [165]. Detailed kinetic analysis of the Met64Val variant showed that catalytic performance with **10** had improved by over two orders of magnitude, while activity for **7** was largely unchanged. These in vitro gains also translated into greater potency in vivo. Treatment of mice bearing D54 tumors with **10** in the presence of this PNP variant raised the cytotoxicity of the prodrug by at least tenfold without exhibiting any significant toxicity in non-PNP-transduced cells.

Separately, fludarabine (**9**) has demonstrated potential as a prodrug in PNP-based suicide gene therapy. To date, studies have largely focused on nucleoside analog activation by native DeoD, demonstrating in vitro efficacy in cell cultures and mouse models [174, 175]. However, the modest activity of *E. coli* PNP clearly limits the prodrug potency, requiring administration in high doses that often cause serious side effects. A recent report of two PNPs from the hyperthermophilic archaean *Sulfolobus solfataricus* offers an interesting new starting point for future engineering efforts to improve phosphorolysis of **9** [176]. Compared to DeoD, the two enzymes possess six- and 12-fold higher catalytic efficiency, respectively. Sequence-structure comparison may allow for the identification of critical functional residues and interactions that can guide the redesign of the *E. coli* enzyme by rational methods and directed evolution.

Finally, the high substrate specificity of trimeric enzymes in general, and the human PNP in particular, can be altered by protein engineering. Relying on crystallographic data, Glu201 and Asn243 were identified as the two key residues responsible for the strict preference for 6-oxopurine nucleoside substrates [166]. A change in the hydrogen bond donor/acceptor pattern through site-directed mutagenesis at these two positions (E201Q and N243D) resulted in the predicted reversal of substrate preference. The high catalytic activity of the unnatural human 6-aminopurine nucleoside phosphorylase and its reduced immunogenicity compared to the bacterial PNPs offer an attractive starting point for further engineering and optimization.

2.8 Nitroreductases

A prominent non-nucleoside-based GDEPT system is the use of bacterial nitroreductase enzymes to activate DNA-damaging nitroaromatic prodrugs [177]. Bacterial nitroreductases are a class of flavin-dependent oxidoreductases with broad substrate specificities that have made them of interest for a number of biotechnological applications including bioremediation [178] and transgenic cell labeling and ablation [179]. Most GDEPT work has focused on the nitroreductase NfsB from *E. coli* and its activation of the prodrug 5-(aziridin-1-yl)-2,4-dinitrobenzamide (CB1954), which when reduced is converted to a highly cytotoxic DNA cross-linker [180]. Phase I and II trials of CB1954 in conjunction with a replication-defective adenovirus expressing *E. coli* NfsB showed that treatment was safe, and a decline in levels of prostate-specific antigen in some patients suggested some effect; however, overall it was deemed that greater therapeutic efficacy would be desirable [181].

In an initial engineering effort, Grove et al. examined a previously solved crystal structure of NfsB [182] and used this to target six residues around the active site for saturation mutagenesis, with the resulting single mutants screened for improved CB1954 activation via monitoring CB1954-dependent *E. coli* growth inhibition [167]. Fourteen of these beneficial single mutants were then combined to generate a small 53-member library of double mutants [170]. The most effective mutants, T41L/N71S and T41L/F70A, were reported to be 14–17-fold more potent than wild-type

NfsB_Ec at sensitizing SKOV3 human cancer cells to CB1954. While these efforts focused on fairly low-throughput *E. coli* growth inhibition-based screens, a substantially larger library (~1,000,000 variants) made from simultaneous randomization of multiple active site residues was screened using a system in which nitroreductase variants were expressed in *E. coli* from chromosomally inserted bacteriophages. When active nitroreductase variants reduced CB1954, the subsequent DNA damage triggered entry into the phage lytic cycle, bursting the *E. coli* and allowing collection of the bacteriophage carrying improved nitroreductases [168]. Other groups have focused on engineering nitroreductases other than NfsB, such as the NfsB homolog FRaseI from *Vibrio fischeri* [171] or an unrelated *E. coli* nitroreductase YieF. While YieF was not improved via a CB1954-focused campaign, a variant Y6 (T160N/Q175L/Y230A/Y239N) that had been evolved by successive rounds of random mutagenesis for improved chromate reduction also serendipitously showed an increased ability to sensitize HeLa cells to CB1954 [169].

2.9 Cholinesterases: Bridging Protein Therapeutics and Biosensors

Cholinesterases (ChE) are a broad family of enzymes that selectively hydrolyze cholinesters into choline and the corresponding carboxylic acid. In humans, there are two types of ChEs: acetylcholinesterase (AChE) and butyrylcholinesterase (BuChE) [183]. AChE is the primary ChE in the body and catalyzes the breakdown of the neurotransmitter acetylcholine. Found at neuromuscular junctions and synaptic clefts, the membrane-associated enzyme is responsible for hydrolysis of the signaling molecule which terminates synaptic transmission. In contrast, BuChE is more ubiquitously found throughout the body and possesses broad specificity for choline-based esters. Its natural biological function is largely unknown. Because of their essential functions, inhibition of ChE by drugs or poisons including organophosphates and carbamates (Fig. 2.3) has been the subject of many studies including protein engineering. These engineering studies initially focused on probing the human cholinesterases to explain and predict the consequences of exposure to inhibitors [184]. More recently, efforts have shifted toward alternate applications of ChEs as therapeutic agents, both as potential drug candidates and as next-generation biosensors.

2.9.1 Butyrylcholinesterase

The absence of an identifiable critical physiological function largely resulted in scientific neglect of BuChE for over 50 years [185]. However, the enzyme's fate changed with the discovery of its effectiveness as a stoichiometric scavenger for nerve agents. In addition, animals injected with exogenous BuChE at concentrations exceeding endogenous levels by a thousandfold showed no adverse effects.

These findings have driven the development of prophylactic treatments for individuals at risk of exposure to toxins including organophosphates and carbamates [186–189].

Beyond these therapeutic applications of native enzyme, BuChE was observed to hydrolyze cocaine, albeit with low efficiency [190]. A number of protein engineering studies have subsequently focused on generating BuChE variants with enhanced substrate specificity for possible treatment of general cocaine addiction and acute toxicity. These efforts have generally been limited to rational design studies rather than directed evolution as protein production of human BuChE requires mammalian expression systems for proper glycosylation and tetramer assembly, significantly complicating the experimental evaluation of large libraries of enzyme variants by high-throughput methods.

Focusing on the natural amino acid variability in the active site of BuChEs from different mammals, Xie et al. investigated the functional impact of specific amino acid substitutions in about a dozen positions of the human enzyme [191]. All but one variant showed lower hydrolytic activity for cocaine. The substitution of Ala328 for Tyr raised the variant's activity by fourfold over wild-type enzyme. The moderate functional gain could be rationalized by computational modeling, showing conformational changes that eliminate steric constraints in the active site, permitting for more effective cocaine binding. Separately, MD simulations and in silico design studies focused on amino acid positions Ala328 and Tyr332 for enhanced cocaine hydrolysis by BuChE [192–194]. These works were motivated by kinetic data for native BuChE showing a thousandfold faster conversion of unnatural (+)-cocaine over its natural enantiomer [195]. Molecular modeling implicated the two amino acids in reducing steric hindrance and forming critical cation-π interactions for productive binding of the unnatural enantiomer but not for natural (−)-cocaine. Simple inversion of the amino acid side chains (A328Y/Y332A or A328Y/Y332G) resulted in 10–20-fold higher catalytic performance of the BuChE variant with (−)-cocaine. Independently, amino acid variations at positions 328 and 332 in combination with two additional unspecified substitutions were reported as AME_{359} [196]. This BuChE variant showed fourfold higher activity than previous double mutants, and while details of the work were not disclosed at the time, Pan et al. later listed its amino acid replacements as F227A/S287G/A328W/Y332M [197]. That latest study deployed molecular dynamics simulations to capture the effects of amino acid changes on the enzyme's ability to stabilize the transition state of BuChE-catalyzed cocaine hydrolysis. These modeling efforts led to a BuChE variant with four amino acid changes, A199S, S287G, A328W, and Y332G. Experimental evaluation of this variant yielded a cocaine hydrolase with almost 500-fold improved catalytic performance over wild-type enzyme. Given the functional improvements summarized above, administration of BuChE variants in patients effectively reduces the half-life of cocaine from 45 to 90 min to just a few seconds [198]. Tests with these engineered enzymes for acute cocaine toxicity have shown promise for clinical efficacy in animal models [199, 200] and phase I clinical trials in humans [198].

2.9.2 Acetylcholinesterase

AChE serves as a key component in neurotransmission. As such, it is a primary molecular target of organophosphates and carbamates and represents the basis for the biological activity of many pesticides and nerve agents. Similarly to BuChE, the administration of exogenous AChE has been demonstrated to be an effective stoichiometric scavenger to counter acute toxicity yet is not practical as a potential therapeutic due to the enzyme's critical biological function and potential serious side effects associated with high doses of AChE. Instead, AChE has found utility as a biosensor to detect pesticides, nerve agents, and related compounds [201]. These AChE-based biosensors offer a highly tunable, selective, and robust solution for the identification of single or multiple chemical agents directly from field samples; an unmet need in environmental safety as the detection of organophosphates and carbamates is traditionally based on GC-MS methods [202]. While delivering accurate results, these analytical techniques require time-consuming sample preparation, specialized and expensive equipment, and highly trained personnel. An enzyme-based sensory system, capitalizing on the biocatalyst's high substrate specificity, selectivity, and potential for signal amplification, offers an attractive alternative for detection of low levels of analytes. This concept was first demonstrated by two groups in the early 1980s, using immobilized AChE from the electrical eel (*Electrophorus electricus*) to monitor the presence of organophosphate and carbamate pesticides in water samples [203, 204]. Eel AChE was chosen for its high catalytic activity combined with commercial availability of the enzyme. The presence of pesticides at picomolar concentration (parts per billion) could be detected via the reduction in ChE activity.

Building on the idea of an AChE-based biosensor, Villatte et al. screened a small collection of AChE homologs from different species for their sensitivity to organophosphate and carbamate insecticides (Fig. 2.3) [205]. Among these wild-type enzymes, the homolog from fruit fly (*Drosophila melanogaster*) showed the highest sensitivity to inhibitors, exceeding the responsiveness of eel AChE by up to eight-fold for 14 out of the 19 tested insecticides. Further active site analysis of the fruit

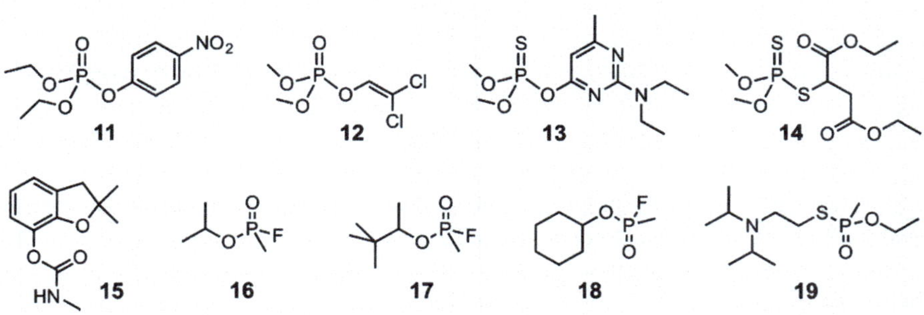

Fig. 2.3 Chemical structures of organophosphates and carbamates used as insecticides and nerve agents: paraoxon (**11**), dichlorvos (**12**), pirimiphos-methyl (**13**), malathion (**14**), carbofuran (**15**), sarin (**16**), soman (**17**), cyclohexylsarin (**18**), and VX (**19**)

fly enzyme by computational modeling indicated a critical π-stacking interaction between the enzyme and inhibitors. The model suggested tighter inhibitor binding upon substitution of Phe for Tyr408, one of the hydrophobic residues in the substrate binding pocket. Experimental evaluation of this amino acid replacement showed moderate improvements in inhibitor sensitivity, raising the AChE variant's responsiveness to 12-fold over eel AChE. These functional gains led to more comprehensive engineering efforts of the fruit fly enzyme via a series of site-directed mutagenesis experiments [206]. Seeking to identify variants with enhanced inhibitor sensitivity, the study targeted sites known to control insecticide toxicity, key positions in the active site identified by inhibitor docking studies, and functionally important residues as determined by alanine scanning. Each of these strategies yielded multiple variants with increased sensitivity over fruit fly AChE, yet gains were typically <tenfold. Possible synergistic effects from combinations of individual beneficial changes were also explored to further improve biosensor performance. Bundling of favorable amino acid substitutions in two positions (E69Y/Y71D) led to a 300-fold increased sensitivity for dichlorvos (**12**) compared to the parental enzyme. The two residues are located in a loop region at the active site entrance thought to control access to the catalytic center of the enzyme. Amino acid changes are likely to cause changes to the loop's conformational flexibility and consequently impact its biological function. Besides increased inhibitor binding, improvements of overall AChE stability were investigated in a separate study [207]. The fruit fly enzyme contains eight Cys residues; seven of them form intra- or intermolecular disulfide bonds. Replacement of the eighth, unpaired Cys, which is located at position 290 in the core structure of the enzyme, with Val resulted in small to moderate (<threefold) improved lifetime at elevated temperatures and in the presence of chemical denaturants. The functional gains could be rationalized through reduced disulfide exchange and increased hydrophobic core packing.

More recently, a separate, monomeric AChE from parasitic nematode *Nippostrongylus brasiliensis* (ACheE B) has emerged as a promising protein engineering template for the generation of biosensors with high sensitivity and broad specificity [208]. ACheE B can be heterologously expressed in *Pichia pastoris*, a significant advantage for engineering efforts compared to the previously discussed AChEs which depend on insect or mammalian expression systems for suitable protein yields [209]. Guided by previous engineering studies, amino acid changes at ten positions in or near the active site were explored by single- and double-site variations. The study targeted primarily residues with large hydrophobic side chains, replacing them with smaller counterparts to widen the active site cleft and enlarge the substrate binding pocket. The resulting AChE B variants were then screened against a collection of 14 potential inhibitors and identified several candidates with enhanced sensitivity for organophosphonates. Functional gains typically ranged in the three- to tenfold range but showed up to 100-fold improvements for pirimiphos-methyl (**13**), raising the sensitivity of the nematode enzyme to levels comparable with the fruit fly and eel AChE standards in the field. Interestingly, the impact of individual amino acid substitutions on the inhibition patterns of these three enzymes showed some notable differences in spite of their high sequence homology. These differences highlight the challenge of accurately predicting the impact of amino acid substitutions. Overall, ACheE B is an attractive alternative platform for biosensor applications, given its ability to be expressed in yeast, as well as its on par functional performance with and superior stability over current AChE standards.

While none of the engineering efforts have generated a true AChE generalist with broad and uniform inhibitor sensitivity, the assembly of sensor arrays with multiple enzymes was reported to cover a broader spectrum and complex mixtures of analytes. To accurately detect and quantify paraoxon (**11**) and carbofuran (**15**) in binary mixtures, Bachmann and Schmid initially built multielectrode thick-film sensors that carried four different wild-type AChEs (eel, bovine, rat, and fruit fly) [210]. Analyzing the responses from individual enzymes with the help of machine learning algorithms, they were able to reliably measure nanomolar concentrations of each compound. The same team subsequently replaced the native enzymes with four engineered variants of fruit fly AChE [211]. These variants were carefully selected based on their divergent sensitivity for a variety of organophosphate and carbamate pesticides. The resulting biosensor showed roughly fourfold lower detection limits than its counterpart based on wild-type enzymes and underscores the potential benefits of lab-customized AChEs with tailored inhibitor sensitivity.

2.9.3 Other Esterases

Cholinesterase represents an important target for organophosphate-based pesticide and nerve agents (Fig. 2.3) and can be deployed as stoichiometric scavengers to counter the toxic efforts of these inhibitors. However, they fail to be effective (catalytic) tools for destroying these compounds. Nature offers an alternative for catalytic detoxification of organophosphates though. Environmental exposure of microbes by decades of pesticide use has resulted in the emergence of enzymes that break down these molecules. Among the most extensively studied enzymes in this category are phosphotriesterases (PTEs), β/α-barrel proteins isolated from *Flavobacterium* sp. and *Pseudomonas diminuta* [213–215]. PTEs have no known natural substrates but very effectively hydrolyze a broad range of pesticides including paraoxon (**11**). They also show low to moderate activity for neurotoxic warfare agents such as soman (**17**) [216].

Over the last 20 years, a number of PTE engineering studies have attempted to improve the catalytic degradation of organophosphates (Table 2.4). By and large, these studies have focused on the small and large substrate binding pockets in the active site, consisting of residues Ile106, Trp131, Phe132, Phe306, Ser308, and Phe309, as well as His254, His257, Leu303, and Met317, respectively (Fig. 2.4). In one of the first papers, Raushel and coworkers employed site-directed mutagenesis to prepare 11 PTE variants with polar residues in the small binding pocket [217]. The modification accelerated cleavage of phosphorus-fluorine bonds by tenfold. The gain was rationalized by favorable interactions of the fluoride with the proton donors and more productive substrate orientation in the active site. Continuing their exploration of the active site by rational design, they performed a systematic

Table 2.4 Protein engineering of organophosphate hydrolases

Enzyme	Product/reference	Modification	Observed property
Phosphotriesterase (*P. diminuta* and Flavobac. sp)	Wild type	(None)	Effective hydrolase for paraoxon and related insecticides
	Watkins (1997) [217]	F132Y/H and F132H/F306H	Improved activity for diisopropyl fluorophosphate
	Chen-Goodspeed (2001) [218]	I106G/A; F132G/A; H254F/Y; H257Y/W; L271F/Y; S308G/A; M317F/Y/W	Multiple PTE variants with enhanced activity and relaxed or reversed stereoselectivity
	Cho (2002) [219]	A14T/A80V/K185R/H257Y/I274N	Variant 22A11: improved activity for methyl parathion
	Griffiths (2003) [220]	I106T/F132L	Variant h5: twofold improvement for paraoxon hydrolysis
	Hill (2003) [221]	H254G/H257W/L303T	~1000-fold increased activity for soman hydrolysis
	Cho (2004) [222]	A14T/L17P/A80V/V116I/ K185R/A203T/I274N/P342S	Variant B3561: improved activity for broad range of substrates including chlorpyrifos
	Tsai (2010) [223]	H254Q/H257F; H257Y/L303T; H254G/H257W/L303T	~10–500-fold improved activity for various G and V-type nerve agents
	Bigley (2013) [224]	I106C/F132V/H254Q/H257Y/A270V/L272M/I274N/S308L	Variant L7ep-3a: ~100-fold improved activity for VX nerve agent
	Cherny (2013) [225]	K77A/A80V/F132E/T173N/G208D/H254G/I274N	Variant G5-C23: 100 to 1000-fold improved catalytic performance for V-type nerve agents
Phosphotriesterase (*Agrobacterium radiobacter*)	Wild type	(None)	Effective hydrolase for dimethyl organophosphates, phosmet, and fenthion
	Jackson (2009) [226]	W131H/F132A	Increased activity for chlorfenvinphos hydrolysis
	Naqvi (2014) [227]	S308L/Y309A	~5000-fold increased activity for degradation of malathion

Table 2.4 (continued)

Enzyme	Product/reference	Modification	Observed property
Paraoxonase 1 (New Zealand rabbit liver)	Aharoni (2004) [228]	I126Y/M130L/ K137S/L142V/ A301G/A320V/ M341L/V343I	Variant G3C9: highly expressed in *E. coli* with wild-type-like catalytic performance. Variant G3C9: highly expressed in *E. coli* with wild-type-like catalytic performance
		G3C9 + L69V/ E218D or G19R/ S193P/N287D/ V346A	Variant G3C9.10 and G3C9.49: 40–50-fold higher catalytic efficiency for paraoxon hydrolysis
	Amitai (2006) [229]	G3C9 + L69V, H115W or V346A	Increased activity for various toxic organophosphates by ~100–300-fold
	Gupta (2011) [230]	G3C9 + L69G/ S111T/H115W/ H134R/F222S/T332S	Variant 4E9: highly efficient hydrolase for degradation of G-agents

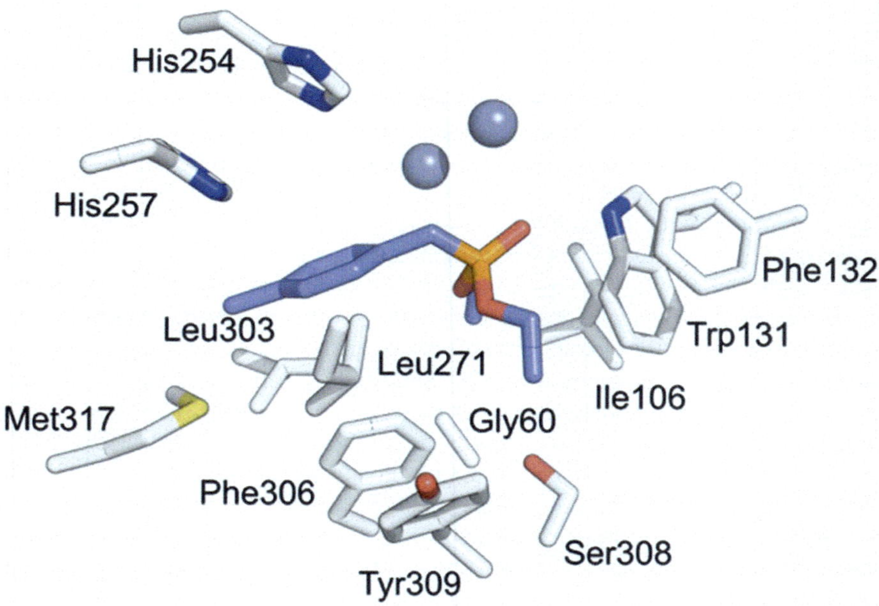

Fig. 2.4 Key amino acid residues in substrate binding pockets of PTE with bound substrate analog (PDB access#: 1dpm [212]). The diethoxy-p-nitrophenyl-phosphonate substrate analog is colored in blue, while the position of the two zinc atoms is shown as gray spheres

evaluation of the functional impact of altering amino acids in both pockets [218]. Their study demonstrated that significant control over reactivity and stereoselectivity of PTE variants could be gained by shrinking or enlarging these two pockets. Rather than being limited to a few specific amino acid substitutions, Hill et al. targeted His254, His257, and Leu303 by multisite-saturation mutagenesis to explore the enzyme's hydrolytic activity with a chromogenic analog of **17** [221]. Substitutions in these positions led to a triple variant with an almost thousandfold enhanced catalytic efficiency. While a rationale for the functional gains was not immediately obvious from x-ray crystallography, MD simulations suggested an overall optimization of substrate orientation in the active site [223]. Merging the list of target sites and multisite-saturation mutagenesis, the same group re-evaluated their PTE libraries against a larger collection of chromogenic nerve agent analogs [231]. As in the previous study by Hill et al., candidates with substitutions in the large binding pocket emerged as the best-performing variants, showing one or two orders of magnitude improved hydrolytic activity for analogs of **16**, **17**, and **18**.

To expand the search for favorable amino acid changes beyond a few selected sites chosen by rational design, the development of surface display, in vitro compartmentalization, and genetic complementation methods for high-throughput screening of PTEs set the stage for large-scale, in vitro directed evolution experiments. Chen and coworkers generated large combinatorial libraries of PTE variants by DNA shuffling and deployed an *E. coli* cell surface display system to find an improved hydrolase for the paraoxon analog methyl parathion [219]. Two rounds of in vitro evolution yielded 22A11, a variant with five amino acid replacements that showed a 25-fold higher specific activity for the test substrate than its parental PTE. The location of the substitutions suggests that their impact on enzyme function results from a combination of direct and indirect effects; H257Y alters the size and shape of the substrate binding pocket, while A80V, L185R, and I274N are surface-exposed residues that might affect protein folding and stability. Using wild-type PTE and 22A11 as templates, the authors subsequently used the same laboratory evolution scheme to search for a hydrolase of chlorpyrifos, another major pesticide [222]. The study identified variant B3561, a descendent of 22A11 with a total of eight amino acid changes including four substitutions originally found in the parental sequence. The catalytic efficiency of B3561 for chlorpyrifos was increased by >700-fold over wild-type PTE (twofold over 22A11) and was on par with the hydrolysis rate of paraoxon by wild PTE. Separately, Griffiths and Tawfik adopted their in vitro compartmentalization method for PTE library screening, using microbead-based surface immobilization in combination with FACS [220, 232]. The method was deployed to analyze four multisite-saturation mutagenesis libraries, targeting key positions in the small binding pocket of PTE for variants with improved paraoxon hydrolysis. The kinetic analysis of lead variant h5 with substitutions in positions 106 and 132 resulted in a 70-fold increase in catalytic rate but compromised substrate binding, enhancing the overall catalytic performance of the variant by a moderate twofold over wild-type PTE. Finally, Raushel and coworkers exploited an in vivo selection scheme for library analysis, using an *E. coli* strain whose growth was controlled by an artificial phosphonate assimilation pathway

[233]. The selection of large PTE libraries with several phosphonate substrates modeled after G-type nerve agents (**16**–**18**) yielded numerous enzyme variants with improved catalytic performance, yet none of these hits surpassed the previously reported variant with H257Y/L303T [231].

The moderate successes of large-scale directed evolution experiments led to a trend reversal in PTE engineering. More recent studies have instead been relying on smaller, more focused libraries that can be screened by medium-throughput, microtiterplate-based colorimetric assays. Besides changes in engineering strategy, target substrates have also shifted to include the degradation of V-type nerve agents including VX (**19**). In contrast to **11** and G-type nerve toxins, organophosphates classified as V-agents possess a thiolo leaving group, which makes them poor substrates for existing hydrolases. Starting with PTE variant (H254Q/H257F), Raushel and coworkers employed iterative rounds of multisite-saturation and random mutagenesis to identify variant L7ep-3a with six additional amino acid changes and a 100-fold enhanced catalytic efficiency for **19** [224]. The x-ray structure analysis revealed substantial remodeling of the active site for more effective substrate interactions [234]. Meanwhile, Cherny et al. merged traditional laboratory-based directed evolution with in silico remodeling of PTE to search for a novel hydrolase with broad specificity for V-type nerve agents [225]. Key positions in the enzyme active site were identified based on the results from previous engineering studies and supplemented with modeling data obtained from RosettaDesign software. These amino acid variations were sampled iteratively over five rounds and led to the identification of several lead candidates including PET variant G5-C23 whose catalytic performance and stereoselectivity for **19** matched the previously reported variant L7ep-3a. In addition, G5-C23 also shows broad substrate specificity for two other V-type nerve agents (RVX and CVX) and G-type neurotoxins (**16**–**18**).

Besides extensive engineering studies on PTE from *Pseudomonas diminuta*, similar efforts have been undertaken to tailor functional homologs from *Agrobacterium radiobacter* and mammalian sources. The native hydrolase from *Agrobacterium* (OpdA) showed superior catalytic activity over PTE for several organophosphates, making it a potential candidate for medical treatment of pesticide poisoning [235]. To further expand the enzyme's substrate specificity, Jackson et al. employed computational docking to locate steric constraints that interfered with productive substrate binding in the active site [226, 227]. Initially, modifications in the small binding pocket (W131H/F132A) all but eliminated the enzyme's stereoselectivity for the *E*- and *Z*-isomers of chlorfenvinphos and raised the catalytic efficiency by roughly ten- and 500-fold, respectively. Subsequently, unfavorable sterics deemed responsible for the low-level hydrolysis of **14** by OpdA was eliminated by multisite-saturation mutagenesis (CASTing). The leading OpdA variant S308L/Y309A possessed an expanded binding pocket which allowed for a favorable substrate binding geometry and resulted in ~5000-fold rate enhancement. In mammals, members of the serum paraoxonase (PON) family primarily catalyze the hydrolysis of esters and lactones yet were found to exhibit residual activity for organophosphates. To improve the latter activity, Aharoni et al. first used DNA shuffling of four native PON1s from human, rat, mouse, and rabbit to select for highly expressed recombinants [228]. The emerging lead candidate G3C9 was largely derived from rabbit but carried eight amino acid substitutions that appear to facilitate more effective protein folding and increase stability due to reorganization of hydrophobic regions of the enzyme. The hydrolytic activity of G3C9 for a variety of organophosphates was then further increased by random mutagenesis [228, 229]. To further advance the functional performance of these PON1 against selected G-type nerve agents, the authors used six rounds of targeted and random mutagenesis in combination with plate-based and FACS-based screening to analyze their library with a fluorogenic coumarin analog of **18** [230]. Variant 4E9 emerged as one of the top-performing hydrolases, containing six amino acid changes located primarily in the enzyme active site. Beyond its high efficiency in cyclohexylsarin degradation, 4E9 showed similarly high catalytic activity for related G-type nerve agents. Extending their functional studies to animal models, the authors also demonstrated significantly increased survival rates for mice prophylactically treated with this enzyme.

Conclusions

The utilization of protein-based biologics in general and engineered enzymes in particular offers tremendous opportunities for the development of highly effective and specific therapies. Whether the enzyme itself is the drug, is responsible for the activation of prodrugs, or serves as detection device for disease markers, the functional versatility of proteins and their evolvability makes these biological macromolecules a powerful and almost limitless resource for clinical applications. Over the last two decades, strategies to engineer therapeutic enzymes have largely mirrored the advances in the field of protein engineering. While early efforts focused on few, site-specific changes, the introduction of high-throughput screening techniques led to the expansion of library size and more comprehensive sampling of proteins by directed evolution. However, the first rule for directed evolution that "you get what you select for" has proven particularly true for therapeutic enzymes. Screening large libraries of enzyme variants in bacterial surrogate systems rather than their eventual (mammalian cell) host is technically much simpler and affordable but has, in more than once instance, resulted in underperforming lead candidates when subsequently tested in clinical settings. It is with that background that newly emerging protein engineering strategies based on computational design, machine learning algorithms, and greater mechanistic and structural insights offer exciting future opportunities for the development of the next-generation protein-based biologics. These methods provide critical guidance for protein redesign, enabling the generation of small, focused libraries whose functional diversity can be assessed in complex assays that more accurately approximate the cellular environment and maximize the functional gains. Building on the past successes and failures of engineered enzymes in therapeutic applications reviewed here, the future holds much promise for tailored enzymes to detect and tackle disease and environmental threats to human health.

References

1. Leader B, Baca QJ, Golan DE (2008) Protein therapeutics: a summary and pharmacological classification. Nat Rev Drug Discov 7(1):21–39
2. Dimitrov DS (2012) Therapeutic proteins. Methods Mol Biol 899:1–26
3. Kurtzman AL, Govindarajan S, Vahle K, Jones JT, Heinrichs V, Patten PA (2001) Advances in directed protein evolution by recursive genetic recombination: applications to therapeutic proteins. Curr Opin Biotechnol 12(4):361–370
4. Vasserot AP, Dickinson CD, Tang Y, Huse WD, Manchester KS, Watkins JD (2003) Optimization of protein therapeutics by directed evolution. Drug Discov Today 8(3):118–126
5. McCafferty J, Glover DR (2000) Engineering therapeutic proteins. Curr Opin Struct Biol 10(4):417–420
6. Shak S, Capon DJ, Hellmiss R, Marsters SA, Baker CL (1990) Recombinant human DNase I reduces the viscosity of cystic fibrosis sputum. Proc Natl Acad Sci U S A 87(23):9188–9192
7. Quan JM, Tiddens HA, Sy JP, McKenzie SG, Montgomery MD, Robinson PJ, Wohl ME, Konstan MW, Pulmozyme Early Intervention Trial Study G (2001) A two-year randomized, placebo-controlled trial of dornase alfa in young patients with cystic fibrosis with mild lung function abnormalities. J Pediatr 139(6):813–820
8. Corrie PG (2008) Cytotoxic chemotherapy: clinical aspects. Medicine 36(1):24–28
9. Greco O, Dachs GU (2001) Gene directed enzyme/prodrug therapy of cancer: historical appraisal and future prospectives. J Cell Physiol 187(1):22–36
10. Moolten FL (1986) Tumor chemosensitivity conferred by inserted herpes thymidine kinase genes: paradigm for a prospective cancer control strategy. Cancer Res 46(10):5276–5281
11. Encell LP, Landis DM, Loeb LA (1999) Improving enzymes for cancer gene therapy. Nat Biotechnol 17(2):143–147
12. Duarte S, Carle G, Faneca H, De Lima MCP, Pierrefite-Carle V (2012) Suicide gene therapy in cancer: where do we stand now? Cancer Lett 324(2):160–170
13. Zarogoulidis, P, Darwiche, K, Sakkas, A, Yarmus, L, Huang, H, Li, Q, Freitag, L, Zarogoulidis, K, Malecki, M (2013) Suicide gene therapy for cancer – current strategies. J Genet Syndr Gene Ther 4 pii:16849
14. Hamada Y, Kiso Y (2016) New directions for protease inhibitors directed drug discovery. Biopolymers 106(4):563–579
15. Kurokawa M, Ito T, Yang CS, Zhao C, Macintyre AN, Rizzieri DA, Rathmell JC, Deininger MW, Reya T, Kornbluth S (2013) Engineering a BCR-ABL-activated caspase for the selective elimination of leukemic cells. Proc Natl Acad Sci U S A 110(6):2300–2305
16. Noppen B, Fonteyn L, Aerts F, De Vriese A, De Maeyer M, Le Floch F, Barbeaux P, Zwaal R, Vanhove M (2014) Autolytic degradation of ocriplasmin: a complex mechanism unraveled by mutational analysis. Protein Eng Des Sel 27(7):215–223
17. Craik CS, Page MJ, Madison EL (2011) Proteases as therapeutics. Biochem J 435(1):1–16
18. Collen D, Lijnen HR (2009) The tissue-type plasminogen activator story. Arterioscler Thromb Vasc Biol 29(8):1151–1155
19. Hoylaerts M, Rijken DC, Lijnen HR, Collen D (1982) Kinetics of the activation of plasminogen by human tissue plasminogen activator. Role of fibrin. J Biol Chem 257(6):2912–2919
20. Andreasen PA, Egelund R, Petersen HH (2000) The plasminogen activation system in tumor growth, invasion, and metastasis. Cell Mol Life Sci 57(1):25–40
21. Semba CP, Sugimoto K, Razavi MK, Society of C, Interventional R (2001) Alteplase and tenecteplase: applications in the peripheral circulation. Tech Vasc Interv Radiol 4(2):99–106
22. Keyt BA, Paoni NF, Refino CJ, Berleau L, Nguyen H, Chow A, Lai J, Pena L, Pater C, Ogez J et al (1994) A faster-acting and more potent form of tissue plasminogen activator. Proc Natl Acad Sci U S A 91(9):3670–3674
23. Bennett WF, Paoni NF, Keyt BA, Botstein D, Jones AJ, Presta L, Wurm FM, Zoller MJ (1991) High resolution analysis of functional determinants on human tissue-type plasminogen activator. J Biol Chem 266(8):5191–5201

24. Refino CJ, Paoni NF, Keyt BA, Pater CS, Badillo JM, Wurm FM, Ogez J, Bennett WF (1993) A variant of t-PA (T103N, KHRR 296-299 AAAA) that, by bolus, has increased potency and decreased systemic activation of plasminogen. Thromb Haemost 70(2):313–319
25. Llevadot J, Giugliano RP (2000) Pharmacology and clinical trial results of lanoteplase in acute myocardial infarction. Exp Opin Inv Drugs 9(11):2689–2694
26. Lundblad RL, Bradshaw RA, Gabriel D, Ortel TL, Lawson J, Mann KG (2004) A review of the therapeutic uses of thrombin. Thromb Haemost 91(5):851–860
27. Gibbs CS, Coutre SE, Tsiang M, Li WX, Jain AK, Dunn KE, Law VS, Mao CT, Matsumura SY, Mejza SJ et al (1995) Conversion of thrombin into an anticoagulant by protein engineering. Nature 378(6555):413–416
28. Tsiang M, Paborsky LR, Li WX, Jain AK, Mao CT, Dunn KE, Lee DW, Matsumura SY, Matteucci MD, Coutre SE, Leung LL, Gibbs CS (1996) Protein engineering thrombin for optimal specificity and potency of anticoagulant activity in vivo. Biochemistry 35(51):16449–16457
29. Marino F, Pelc LA, Vogt A, Gandhi PS, Di Cera E (2010) Engineering thrombin for selective specificity toward protein C and PAR1. J Biol Chem 285(25):19145–19152
30. Gruber A, Cantwell AM, Di Cera E, Hanson SR (2002) The thrombin mutant W215A/E217A shows safe and potent anticoagulant and antithrombotic effects in vivo. J Biol Chem 277(31):27581–27584
31. Berny-Lang MA, Hurst S, Tucker EI, Pelc LA, Wang RK, Hurn PD, Di Cera E, McCarty OJ, Gruber A (2011) Thrombin mutant W215A/E217A treatment improves neurological outcome and reduces cerebral infarct size in a mouse model of ischemic stroke. Stroke 42(6):1736–1741
32. Arosio D, Ayala YM, Di Cera E (2000) Mutation of W215 compromises thrombin cleavage of fibrinogen, but not of PAR-1 or protein C. Biochemistry 39(27):8095–8101
33. Persson E, Kjalke M, Olsen OH (2001) Rational design of coagulation factor VIIa variants with substantially increased intrinsic activity. Proc Natl Acad Sci U S A 98(24):13583–13588
34. Lentz SR, Ehrenforth S, Karim FA, Matsushita T, Weldingh KN, Windyga J, Mahlangu JN, adept™2 investigators (2014) Recombinant factor VIIa analog in the management of hemophilia with inhibitors: results from a multicenter, randomized, controlled trial of vatreptacog alfa. J Thromb Haemost 12 (8):1244–1253
35. Harvey SB, Stone MD, Martinez MB, Nelsestuen GL (2003) Mutagenesis of the gamma-carboxyglutamic acid domain of human factor VII to generate maximum enhancement of the membrane contact site. J Biol Chem 278(10):8363–8369
36. Neuenschwander PF, Morrissey JH (1994) Roles of the membrane-interactive regions of factor VIIa and tissue factor. The factor VIIa Gla domain is dispensable for binding to tissue factor but important for activation of factor X. J Biol Chem 269(11):8007–8013
37. Mahlangu J, Paz P, Hardtke M, Aswad F, Schroeder J (2016) TRUST trial: BAY 86-6150 use in haemophilia with inhibitors and assessment for immunogenicity. Haemophilia. doi:10.1111/hae.12994
38. Qian X, Hamad B, Dias-Lalcaca G (2015) The Alzheimer disease market. Nat Rev Drug Discov 14(10):675–676
39. Jacobsen JS, Comery TA, Martone RL, Elokdah H, Crandall DL, Oganesian A, Aschmies S, Kirksey Y, Gonzales C, Xu J, Zhou H, Atchison K, Wagner E, Zaleska MM, Das I, Arias RL, Bard J, Riddell D, Gardell SJ, Abou-Gharbia M, Robichaud A, Magolda R, Vlasuk GP, Bjornsson T, Reinhart PH, Pangalos MN (2008) Enhanced clearance of Abeta in brain by sustaining the plasmin proteolysis cascade. Proc Natl Acad Sci U S A 105(25):8754–8759
40. Leissring MA, Farris W, Chang AY, Walsh DM, Wu X, Sun X, Frosch MP, Selkoe DJ (2003) Enhanced proteolysis of beta-amyloid in APP transgenic mice prevents plaque formation, secondary pathology, and premature death. Neuron 40(6):1087–1093
41. Spencer B, Marr RA, Rockenstein E, Crews L, Adame A, Potkar R, Patrick C, Gage FH, Verma IM, Masliah E (2008) Long-term neprilysin gene transfer is associated with reduced levels of intracellular Abeta and behavioral improvement in APP transgenic mice. BMC Neurosci 9:109

42. Sexton T, Hitchcook LJ, Rodgers DW, Bradley LH, Hersh LB (2012) Active site mutations change the cleavage specificity of neprilysin. PLoS One 7(2):e32343
43. Webster CI, Burrell M, Olsson LL, Fowler SB, Digby S, Sandercock A, Snijder A, Tebbe J, Haupts U, Grudzinska J, Jermutus L, Andersson C (2014) Engineering neprilysin activity and specificity to create a novel therapeutic for Alzheimer's disease. PLoS One 9(8): e104001
44. Eckman EA, Adams SK, Troendle FJ, Stodola BA, Kahn MA, Fauq AH, Xiao HD, Bernstein KE, Eckman CB (2006) Regulation of steady-state beta-amyloid levels in the brain by neprilysin and endothelin-converting enzyme but not angiotensin-converting enzyme. J Biol Chem 281(41):30471–30478
45. Guerrero JL, O'Malley MA, Daugherty PS (2016) Intracellular FRET-based screen for redesigning the specificity of secreted proteases. ACS Chem Biol 11(4):961–970
46. Dressler D (2012) Clinical applications of botulinum toxin. Curr Opin Microbiol 15(3):325–336
47. Orsini M, Leite MA, Chung TM, Bocca W, de Souza JA, de Souza OG, Moreira RP, Bastos VH, Teixeira S, Oliveira AB, Moraes Bda S, Matta AP, Jacinto LJ (2015) Botulinum neurotoxin type A in neurology: update. Neurol Int 7(2):5886
48. Masuyer G, Chaddock JA, Foster KA, Acharya KR (2014) Engineered botulinum neurotoxins as new therapeutics. Annu Rev Pharmacol Toxicol 54:27–51
49. Vazquez-Cintron E, Tenezaca L, Angeles C, Syngkon A, Liublinska V, Ichtchenko K, Band P (2016) Pre-clinical study of a novel recombinant botulinum neurotoxin derivative engineered for improved safety. Sci Rep 6:30429
50. Rummel A, Mahrhold S, Bigalke H, Binz T (2011) Exchange of the H(CC) domain mediating double receptor recognition improves the pharmacodynamic properties of botulinum neurotoxin. FEBS J 278(23):4506–4515
51. Wang J, Zurawski TH, Bodeker MO, Meng J, Boddul S, Aoki KR, Dolly JO (2012) Longer-acting and highly potent chimaeric inhibitors of excessive exocytosis created with domains from botulinum neurotoxin A and B. Biochem J 444(1):59–67
52. Wang J, Zurawski TH, Meng J, Lawrence G, Olango WM, Finn DP, Wheeler L, Dolly JO (2011) A dileucine in the protease of botulinum toxin A underlies its long-lived neuroparalysis: transfer of longevity to a novel potential therapeutic. J Biol Chem 286(8):6375–6385
53. Chaddock JA, Purkiss JR, Friis LM, Broadbridge JD, Duggan MJ, Fooks SJ, Shone CC, Quinn CP, Foster KA (2000) Inhibition of vesicular secretion in both neuronal and nonneuronal cells by a retargeted endopeptidase derivative of Clostridium botulinum neurotoxin type A. Infect Immun 68(5):2587–2593
54. Fonfria E, Donald S, Cadd VA (2016) Botulinum neurotoxin A and an engineered derivate targeted secretion inhibitor (TSI) A enter cells via different vesicular compartments. J Recept Signal Transduct Res 36(1):79–88
55. Ma H, Meng J, Wang J, Hearty S, Dolly JO, O'Kennedy R (2014) Targeted delivery of a SNARE protease to sensory neurons using a single chain antibody (scFv) against the extracellular domain of P2X(3) inhibits the release of a pain mediator. Biochem J 462(2): 247–256
56. Larsen GR, Timony GA, Horgan PG, Barone KM, Hensen KS, Augus LB, Stoudemire JB (1991) Protein engineering of novel plasminogen activators with increased thrombolytic potency in rabbits relative to activase. J Biol Chem 266(1):8156–8161
57. Sikorra S, Litschko C, Muller C, Thiel N, Galli T, Eichner T, Binz T (2016) Identification and characterization of botulinum neurotoxin a substrate binding pockets and their re-engineering for human SNAP-23. J Mol Biol 428(2 Pt A):372–384
58. Chen S, Barbieri JT (2009) Engineering botulinum neurotoxin to extend therapeutic intervention. Proc Natl Acad Sci U S A 106(23):9180–9184
59. Somm E, Bonnet N, Martinez A, Marks PM, Cadd VA, Elliott M, Toulotte A, Ferrari SL, Rizzoli R, Huppi PS, Harper E, Melmed S, Jones R, Aubert ML (2012) A botulinum toxin-derived targeted secretion inhibitor downregulates the GH/IGF1 axis. J Clin Invest 122(9):3295–3306

60. Tye-Din JA, Anderson RP, Ffrench RA, Brown GJ, Hodsman P, Siegel M, Botwick W, Shreeniwas R (2010) The effects of ALV003 pre-digestion of gluten on immune response and symptoms in celiac disease in vivo. Clin Immunol 134(3):289–295
61. Ehren J, Govindarajan S, Moron B, Minshull J, Khosla C (2008) Protein engineering of improved prolyl endopeptidases for celiac sprue therapy. Protein Eng Des Sel 21(12):699–707
62. Gordon SR, Stanley EJ, Wolf S, Toland A, Wu SJ, Hadidi D, Mills JH, Baker D, Pultz IS, Siegel JB (2012) Computational design of an alpha-gliadin peptidase. J Am Chem Soc 134(50):20513–20520
63. Tack GJ, Verbeek WH, Schreurs MW, Mulder CJ (2010) The spectrum of celiac disease: epidemiology, clinical aspects and treatment. Nat Rev Gastroenterol Hepatol 7(4):204–213
64. Shan L, Molberg O, Parrot I, Hausch F, Filiz F, Gray GM, Sollid LM, Khosla C (2002) Structural basis for gluten intolerance in celiac sprue. Science 297(5590):2275–2279
65. Wolf C, Siegel JB, Tinberg C, Camarca A, Gianfrani C, Paski S, Guan R, Montelione G, Baker D, Pultz IS (2015) Engineering of Kuma030: a gliadin peptidase that rapidly degrades immunogenic gliadin peptides in gastric conditions. J Am Chem Soc 137(40):13106–13113
66. Guerrero JL, Daugherty PS, O'Malley MA (2016) Emerging technologies for protease engineering: new tools to clear out disease. Biotechnol Bioeng. doi:10.1002/bit.26066 [Epub ahead of print]
67. Esvelt KM, Carlson JC, Liu DR (2011) A system for the continuous directed evolution of biomolecules. Nature 472(7344):499–503
68. Hill ME, MacPherson DJ, Wu P, Julien O, Wells JA, Hardy JA (2016) Reprogramming caspase-7 specificity by regio-specific mutations and selection provides alternate solutions for substrate recognition. ACS Chem Biol 11(6):1603–1612
69. Sandersjoo L, Kostallas G, Lofblom J, Samuelson P (2014) A protease substrate profiling method that links site-specific proteolysis with antibiotic resistance. Biotechnol J 9(1):155–162
70. Varadarajan N, Rodriguez S, Hwang BY, Georgiou G, Iverson BL (2008) Highly active and selective endopeptidases with programmed substrate specificities. Nat Chem Biol 4(5):290–294
71. Kostallas G, Samuelson P (2010) Novel fluorescence-assisted whole-cell assay for engineering and characterization of proteases and their substrates. Appl Environ Microbiol 76(22):7500–7508
72. Carrico ZM, Strobel KL, Atreya ME, Clark DS, Francis MB (2016) Simultaneous selection and counter-selection for the directed evolution of proteases in *E. coli* using a cytoplasmic anchoring strategy. Biotechnol Bioeng 113(6):1187–1193
73. Dickinson BC, Packer MS, Badran AH, Liu DR (2014) A system for the continuous directed evolution of proteases rapidly reveals drug-resistance mutations. Nat Commun 5:5352
74. Ledoux L (1955) Action of ribonuclease on two solid tumours in vivo. Nature 176(4470):36–37
75. Vogelzang NJ, Aklilu M, Stadler WM, Dumas MC, Mikulski SM (2001) A phase II trial of weekly intravenous ranpirnase (Onconase), a novel ribonuclease in patients with metastatic kidney cancer. Investig New Drugs 19(3):255–260
76. Mikulski SM, Costanzi JJ, Vogelzang NJ, McCachren S, Taub RN, Chun H, Mittelman A, Panella T, Puccio C, Fine R, Shogen K (2002) Phase II trial of a single weekly intravenous dose of ranpirnase in patients with unresectable malignant mesothelioma. J Clin Oncol 20(1):274–281
77. Rutkoski TJ, Raines RT (2008) Evasion of ribonuclease inhibitor as a determinant of ribonuclease cytotoxicity. Curr Pharm Biotechnol 9(3):185–189
78. Kobe B, Deisenhofer J (1996) Mechanism of ribonuclease inhibition by ribonuclease inhibitor protein based on the crystal structure of its complex with ribonuclease A. J Mol Biol 264(5):1028–1043
79. Leland PA, Schultz LW, Kim BM, Raines RT (1998) Ribonuclease A variants with potent cytotoxic activity. Proc Natl Acad Sci U S A 95(18):10407–10412
80. Rutkoski TJ, Kurten EL, Mitchell JC, Raines RT (2005) Disruption of shape-complementarity markers to create cytotoxic variants of ribonuclease A. J Mol Biol 354(1):41–54

81. Johnson RJ, Chao TY, Lavis LD, Raines RT (2007) Cytotoxic ribonucleases: the dichotomy of Coulombic forces. Biochemistry 46(36):10308–10316
82. Johnson RJ, McCoy JG, Bingman CA, Phillips GN Jr, Raines RT (2007) Inhibition of human pancreatic ribonuclease by the human ribonuclease inhibitor protein. J Mol Biol 368(2):434–449
83. Cremer C, Braun H, Mladenov R, Schenke L, Cong X, Jost E, Brummendorf TH, Fischer R, Carloni P, Barth S, Nachreiner T (2015) Novel angiogenin mutants with increased cytotoxicity enhance the depletion of pro-inflammatory macrophages and leukemia cells ex vivo. Cancer Immunol Immunother 64(12):1575–1586
84. Cong X, Cremer C, Nachreiner T, Barth S, Carloni P (2016) Engineered human angiogenin mutations in the placental ribonuclease inhibitor complex for anticancer therapy: insights from enhanced sampling simulations. Protein Sci. doi:10.1002/pro.2941
85. Riccio G, D'Avino C, Raines RT, De Lorenzo C (2013) A novel fully human antitumor immunoRNase resistant to the RNase inhibitor. Protein Eng Des Sel 26(3):243–248
86. Phillips MM, Sheaff MT, Szlosarek PW (2013) Targeting arginine-dependent cancers with arginine-degrading enzymes: opportunities and challenges. Cancer Res Treat 45(4):251–262
87. Patil MD, Bhaumik J, Babykutty S, Banerjee UC, Fukumura D (2016) Arginine dependence of tumor cells: targeting a chink in cancer's armor. Oncogene. doi:10.1038/onc.2016.37
88. Boissel N, Sender LS (2015) Best practices in adolescent and young adult patients with acute lymphoblastic leukemia: a focus on asparaginase. Young Adult Oncol 4(3):118–128
89. Pulte D, Gondos A, Brenner H (2009) Improvement in survival in younger patients with acute lymphoblastic leukemia from the 1980s to the early 21st century. Blood 113(7):1408–1411
90. Liu C, Kawedia JD, Cheng C, Pei D, Fernandez CA, Cai X, Crews KR, Kaste SC, Panetta JC, Bowman WP, Jeha S, Sandlund JT, Evans WE, Pui CH, Relling MV (2012) Clinical utility and implications of asparaginase antibodies in acute lymphoblastic leukemia. Leukemia 26(11):2303–2309
91. Cantor JR, Yoo TH, Dixit A, Iverson BL, Forsthuber TG, Georgiou G (2011) Therapeutic enzyme deimmunization by combinatorial T-cell epitope removal using neutral drift. Proc Natl Acad Sci U S A 108(4):1272–1277
92. Maggi M, Chiarelli LR, Valentini G, Scotti C (2015) Engineering of *Helicobacter pylori* L-asparaginase: characterization of two functionally distinct groups of mutants. PLoS One 10(2):e0117025
93. Bansal S, Srivastava A, Mukherjee G, Pandey R, Verma AK, Mishra P, Kundu B (2012) Hyperthermophilic asparaginase mutants with enhanced substrate affinity and antineoplastic activity: structural insights on their mechanism of action. FASEB J 26(3):1161–1171
94. Kotzia GA, Labrou NE (2009) Engineering thermal stability of L-asparaginase by in vitro directed evolution. FEBS J 276(6):1750–1761
95. Figueiredo L, Cole PD, Drachtman RA (2016) Asparaginase *Erwinia chrysanthemi* as a component of a multi-agent chemotherapeutic regimen for the treatment of patients with acute lymphoblastic leukemia who have developed hypersensitivity to *E. coli*-derived asparaginase. Expert Rev Hematol 9(3):227–234
96. Feun L, Savaraj N (2006) Pegylated arginine deiminase: a novel anticancer enzyme agent. Expert Opin Investig Drugs 15(7):815–822
97. Glazer ES, Piccirillo M, Albino V, Di Giacomo R, Palaia R, Mastro AA, Beneduce G, Castello G, De Rosa V, Petrillo A, Ascierto PA, Curley SA, Izzo F (2010) Phase II study of pegylated arginine deiminase for nonresectable and metastatic hepatocellular carcinoma. J Clin Oncol 28(13):2220–2226
98. Ott PA, Carvajal RD, Pandit-Taskar N, Jungbluth AA, Hoffman EW, Wu BW, Bomalaski JS, Venhaus R, Pan L, Old LJ, Pavlick AC, Wolchok JD (2013) Phase I/II study of pegylated arginine deiminase (ADI-PEG 20) in patients with advanced melanoma. Investig New Drugs 31(2):425–434
99. Tomlinson BK, Thomson JA, Bomalaski JS, Diaz M, Akande T, Mahaffey N, Li T, Dutia MP, Kelly K, Gong IY, Semrad T, Gandara DR, Pan CX, Lara PN Jr (2015) Phase I trial of arginine deprivation therapy with ADI-PEG 20 plus docetaxel in patients with advanced malignant solid tumors. Clin Cancer Res 21(11):2480–2486

100. Liu Y-M, Sun Z-H, Ni Y, Zheng P, Liu Y-P, Meng F-J (2008) Isolation and identification of an arginine deiminase producing strain *Pseudomonas plecoglossicida* CGMCC2039. World J Microbiol Biotechnol 24(10):2213–2219
101. Han RZ, Xu GC, Dong JJ, Ni Y (2016) Arginine deiminase: recent advances in discovery, crystal structure, and protein engineering for improved properties as an anti-tumor drug. Appl Microbiol Biotechnol 100(11):4747–4760
102. Ni Y, Li Z, Sun Z, Zheng P, Liu Y, Zhu L, Schwaneberg U (2009) Expression of arginine deiminase from *Pseudomonas plecoglossicida* CGMCC2039 in *E. coli* and its anti-tumor activity. Curr Microbiol 58(6):593–598
103. Cheng F, Kardashliev T, Pitzler C, Shehzad A, Lue H, Bernhagen J, Zhu L, Schwaneberg U (2015) A competitive flow cytometry screening system for directed evolution of therapeutic enzyme. ACS Synth Biol 4(7):768–775
104. Ni Y, Schwaneberg U, Sun ZH (2008) Arginine deiminase, a potential anti-tumor drug. Cancer Lett 261(1):1–11
105. Kuhn NJ, Talbot J, Ward S (1991) pH-sensitive control of arginase by Mn(II) ions at submicromolar concentrations. Arch Biochem Biophys 286(1):217–221
106. Stone EM, Glazer ES, Chantranupong L, Cherukuri P, Breece RM, Tierney DL, Curley SA, Iverson BL, Georgiou G (2010) Replacing Mn(2+) with Co(2+) in human arginase i enhances cytotoxicity toward l-arginine auxotrophic cancer cell lines. ACS Chem Biol 5(3):333–342
107. Stone E, Chantranupong L, Gonzalez C, O'Neal J, Rani M, VanDenBerg C, Georgiou G (2012) Strategies for optimizing the serum persistence of engineered human arginase I for cancer therapy. J Control Release 158(1):171–179
108. Jordheim LP, Durantel D, Zoulim F, Dumontet C (2013) Advances in the development of nucleoside and nucleotide analogues for cancer and viral diseases. Nat Rev Drug Discov 12(6):447–464
109. Baldwin SA, Beal PR, Yao SY, King AE, Cass CE, Young JD (2004) The equilibrative nucleoside transporter family, SLC29. Pflugers Arch 447(5):735–743
110. Arner ES, Eriksson S (1995) Mammalian deoxyribonucleoside kinases. Pharmacol Ther 67(2):155–186
111. Shi J, McAtee JJ, Schlueter Wirtz S, Tharnish P, Juodawlkis A, Liotta DC, Schinazi RF (1999) Synthesis and biological evaluation of 2′,3′-didehydro-2′,3′- dideoxy-5-fluorocytidine (D4FC) analogues: discovery of carbocyclic nucleoside triphosphates with potent inhibitory activity against HIV-1 reverse transcriptase. J Med Chem 42(5):859–867
112. Culver KW, Ram Z, Wallbridge S, Ishii H, Oldfield EH, Blaese RM (1992) In vivo gene transfer with retroviral vector-producer cells for treatment of experimental brain tumors. Science 256(5063):1550–1552
113. Song S, Pursell ZF, Copeland WC, Longley MJ, Kunkel TA, Mathews CK (2005) DNA precursor asymmetries in mammalian tissue mitochondria and possible contribution to mutagenesis through reduced replication fidelity. Proc Natl Acad Sci U S A 102(14):4990–4995
114. Trask TW, Trask RP, Aguilar-Cordova E, Shine HD, Wyde PR, Goodman JC, Hamilton WJ, Rojas-Martinez A, Chen SH, Woo SL, Grossman RG (2000) Phase I study of adenoviral delivery of the HSV-tk gene and ganciclovir administration in patients with current malignant brain tumors. Mol Ther 1(2):195–203
115. Ji N, Weng D, Liu C, Gu Z, Chen S, Guo Y, Fan Z, Wang X, Chen J, Zhao Y, Zhou J, Wang J, Ma D, Li N (2016) Adenovirus-mediated delivery of herpes simplex virus thymidine kinase administration improves outcome of recurrent high-grade glioma. Oncotarget 7(4):4369–4378
116. Wheeler LA, Manzanera AG, Bell SD, Cavaliere R, McGregor JM, Grecula JC, Newton HB, Lo SS, Badie B, Portnow J, Teh BS, Trask TW, Baskin DS, New PZ, Aguilar LK, Aguilar-Cordova E, Chiocca EA (2016) Phase II multicenter study of gene-mediated cytotoxic immunotherapy as adjuvant to surgical resection for newly diagnosed malignant glioma. Neuro-Oncology 18(8):1137–1145
117. Wigler M, Silverstein S, Lee LS, Pellicer A, Cheng Y, Axel R (1977) Transfer of purified herpes virus thymidine kinase gene to cultured mouse cells. Cell 11(1):223–232

118. Skorenski M, Sienczyk M (2014) Anti-herpesvirus agents: a patent and literature review (2003 to present). Expert Opin Ther Patents 24(8):925–941
119. Munir KM, French DC, Dube DK, Loeb LA (1992) Permissible amino acid substitutions within the putative nucleoside binding site of Herpes Simplex virus type 1 encoded thymidine kinase established by random sequence mutagenesis [corrected]. J Biol Chem 267(10):6584–6589
120. Black ME, Loeb LA (1993) Identification of important residues within the putative nucleoside binding site of HSV-1 thymidine kinase by random sequence selection: analysis of selected mutants in vitro. Biochemistry 32(43):11618–11626
121. Black ME, Newcomb TG, Wilson HM, Loeb LA (1996) Creation of drug-specific Herpes Simplex virus type 1 thymidine kinase mutants for gene therapy. Proc Natl Acad Sci U S A 93(8):3525–3529
122. Christians FC, Scapozza L, Crameri A, Folkers G, Stemmer WP (1999) Directed evolution of thymidine kinase for AZT phosphorylation using DNA family shuffling. Nat Biotechnol 17(3):259–264
123. Hinds TA, Compadre C, Hurlburt BK, Drake RR (2000) Conservative mutations of glutamine-125 in Herpes Simplex virus type 1 thymidine kinase result in a ganciclovir kinase with minimal deoxypyrimidine kinase activities. Biochemistry 39(14):4105–4111
124. Black ME, Kokoris MS, Sabo P (2001) Herpes Simplex virus-1 thymidine kinase mutants created by semi-random sequence mutagenesis improve prodrug-mediated tumor cell killing. Cancer Res 61(7):3022–3026
125. Balzarini J, Liekens S, Solaroli N, El Omari K, Stammers DK, Karlsson A (2006) Engineering of a single conserved amino acid residue of Herpes Simplex virus type 1 thymidine kinase allows a predominant shift from pyrimidine to purine nucleoside phosphorylation. J Biol Chem 281(28):19273–19279
126. Munch-Petersen B, Piskur J, Sondergaard L (1998) Four deoxynucleoside kinase activities from *Drosophila melanogaster* are contained within a single monomeric enzyme, a new multifunctional deoxynucleoside kinase. J Biol Chem 273(7):3926–3931
127. Knecht W, Munch-Petersen B, Piskur J (2000) Identification of residues involved in the specificity and regulation of the highly efficient multisubstrate deoxyribonucleoside kinase from *Drosophila melanogaster*. J Mol Biol 301(4):827–837
128. Knecht W, Sandrini MP, Johansson K, Eklund H, Munch-Petersen B, Piskur J (2002) A few amino acid substitutions can convert deoxyribonucleoside kinase specificity from pyrimidines to purines. EMBO J 21(7):1873–1880
129. Knecht W, Rozpedowska E, Le Breton C, Willer M, Gojkovic Z, Sandrini MP, Joergensen T, Hasholt L, Munch-Petersen B, Piskur J (2007) *Drosophila* deoxyribonucleoside kinase mutants with enhanced ability to phosphorylate purine analogs. Gene Ther 14(17):1278–1286
130. Gerth ML, Lutz S (2007) Non-homologous recombination of deoxyribonucleoside kinases from human and *Drosophila melanogaster* yields human-like enzymes with novel activities. J Mol Biol 370(4):742–751
131. Solaroli N, Johansson M, Balzarini J, Karlsson A (2007) Enhanced toxicity of purine nucleoside analogs in cells expressing *Drosophila melanogaster* nucleoside kinase mutants. Gene Ther 14(1):86–92
132. Liu L, Li Y, Liotta D, Lutz S (2009) Directed evolution of an orthogonal nucleoside analog kinase via fluorescence-activated cell sorting. Nucleic Acids Res 37(13):4472–4481
133. Liu L, Murphy P, Baker D, Lutz S (2010) Computational design of orthogonal nucleoside kinases. Chem Commun (Camb) 46(46):8803–8805
134. Campbell DO, Yaghoubi SS, Su Y, Lee JT, Auerbach MS, Herschman H, Satyamurthy N, Czernin J, Lavie A, Radu CG (2012) Structure-guided engineering of human thymidine kinase 2 as a positron emission tomography reporter gene for enhanced phosphorylation of non-natural thymidine analog reporter probe. J Biol Chem 287(1):446–454
135. Sabini E, Hazra S, Konrad M, Burley SK, Lavie A (2007) Structural basis for activation of the therapeutic L-nucleoside analogs 3TC and troxacitabine by human deoxycytidine kinase. Nucleic Acids Res 35(1):186–192

136. Iyidogan P, Lutz S (2008) Systematic exploration of active site mutations on human deoxycytidine kinase substrate specificity. Biochemistry 47(16):4711–4720
137. Hazra S, Sabini E, Ort S, Konrad M, Lavie A (2009) Extending thymidine kinase activity to the catalytic repertoire of human deoxycytidine kinase. Biochemistry 48(6):1256–1263
138. Muthu P, Chen HX, Lutz S (2014) Redesigning human 2′-deoxycytidine kinase enantioselectivity for L-nucleoside analogues as reporters in positron emission tomography. ACS Chem Biol 9(10):2326–2333
139. Kokoris MS, Sabo P, Adman ET, Black ME (1999) Enhancement of tumor ablation by a selected HSV-1 thymidine kinase mutant. Gene Ther 6(8):1415–1426
140. Ardiani A, Sanchez-Bonilla M, Black ME (2010) Fusion enzymes containing HSV-1 thymidine kinase mutants and guanylate kinase enhance prodrug sensitivity in vitro and in vivo. Cancer Gene Ther 17(2):86–96
141. Karjoo Z, Chen X, Hatefi A (2016) Progress and problems with the use of suicide genes for targeted cancer therapy. Adv Drug Deliv Rev 99(Pt A):113–128
142. Balzarini J, Liekens S, Esnouf R, De Clercq E (2002) The A167Y mutation converts the Herpes Simplex virus type 1 thymidine kinase into a guanosine analogue kinase. Biochemistry 41(20):6517–6524
143. Slot Christiansen L, Munch-Petersen B, Knecht W (2015) Non-viral deoxyribonucleoside kinases – diversity and practical use. J Genet Genomics 42(5):235–248
144. Welin M, Skovgaard T, Knecht W, Zhu C, Berenstein D, Munch-Petersen B, Piskur J, Eklund H (2005) Structural basis for the changed substrate specificity of *Drosophila melanogaster* deoxyribonucleoside kinase mutant N64D. FEBS J 272(14):3733–3742
145. Hecker SJ, Erion MD (2008) Prodrugs of phosphates and phosphonates. J Med Chem 51(8):2328–2345
146. Thornton P, Kadri H, Micolli A, Mehellou Y (2016) Nucleoside phosphate and phosphonate prodrug clinical candidates. J Med Chem. doi:10.1021/acs.jmedchem.6b00523
147. Eriksson S, Kierdaszuk B, Munch-Petersen B, Oberg B, Johansson NG (1991) Comparison of the substrate specificities of human thymidine kinase 1 and 2 and deoxycytidine kinase toward antiviral and cytostatic nucleoside analogs. Biochem Biophys Res Commun 176(2): 586–592
148. Serganova I, Ponomarev V, Blasberg R (2007) Human reporter genes: potential use in clinical studies. Nucl Med Biol 34(7):791–807
149. Yaghoubi SS, Campbell DO, Radu CG, Czernin J (2012) Positron emission tomography reporter genes and reporter probes: gene and cell therapy applications. Theranostics 2(4):374–391
150. Gambhir SS, Bauer E, Black ME, Liang Q, Kokoris MS, Barrio JR, Iyer M, Namavari M, Phelps ME, Herschman HR (2000) A mutant Herpes Simplex virus type 1 thymidine kinase reporter gene shows improved sensitivity for imaging reporter gene expression with positron emission tomography. Proc Natl Acad Sci U S A 97(6):2785–2790
151. Jacobs A, Tjuvajev JG, Dubrovin M, Akhurst T, Balatoni J, Beattie B, Joshi R, Finn R, Larson SM, Herrlinger U, Pechan PA, Chiocca EA, Breakefield XO, Blasberg RG (2001) Positron emission tomography-based imaging of transgene expression mediated by replication-conditional, oncolytic Herpes Simplex virus type 1 mutant vectors in vivo. Cancer Res 61(7):2983–2995
152. Dempsey MF, Wyper D, Owens J, Pimlott S, Papanastassiou V, Patterson J, Hadley DM, Nicol A, Rampling R, Brown SM (2006) Assessment of 123I-FIAU imaging of herpes simplex viral gene expression in the treatment of glioma. Nucl Med Commun 27(8):611–617
153. Likar Y, Zurita J, Dobrenkov K, Shenker L, Cai S, Neschadim A, Medin JA, Sadelain M, Hricak H, Ponomarev V (2010) A new pyrimidine-specific reporter gene: a mutated human deoxycytidine kinase suitable for PET during treatment with acycloguanosine-based cytotoxic drugs. J Nucl Med 51(9):1395–1403
154. Gil JS, Machado HB, Campbell DO, McCracken M, Radu C, Witte ON, Herschman HR (2013) Application of a rapid, simple, and accurate adenovirus-based method to compare PET reporter gene/PET reporter probe systems. Mol Imaging Biol 15(3):273–281

155. Wang L, Munch-Petersen B, Herrstrom Sjoberg A, Hellman U, Bergman T, Jornvall H, Eriksson S (1999) Human thymidine kinase 2: molecular cloning and characterisation of the enzyme activity with antiviral and cytostatic nucleoside substrates. FEBS Lett 443(2):170–174
156. Ponomarev V, Doubrovin M, Shavrin A, Serganova I, Beresten T, Ageyeva L, Cai C, Balatoni J, Alauddin M, Gelovani J (2007) A human-derived reporter gene for noninvasive imaging in humans: mitochondrial thymidine kinase type 2. J Nucl Med 48(5):819–826
157. Mullen CA, Kilstrup M, Blaese RM (1992) Transfer of the bacterial gene for cytosine deaminase to mammalian cells confers lethal sensitivity to 5-fluorocytosine: a negative selection system. Proc Natl Acad Sci U S A 89(1):33–37
158. Mahan SD, Ireton GC, Stoddard BL, Black ME (2004) Alanine-scanning mutagenesis reveals a cytosine deaminase mutant with altered substrate preference. Biochemistry 43(28):8957–8964
159. Mahan SD, Ireton GC, Knoeber C, Stoddard BL, Black ME (2004) Random mutagenesis and selection of *Escherichia coli* cytosine deaminase for cancer gene therapy. Protein Eng Des Sel 17(8):625–633
160. Fuchita M, Ardiani A, Zhao L, Serve K, Stoddard BL, Black ME (2009) Bacterial cytosine deaminase mutants created by molecular engineering show improved 5-fluorocytosine-mediated cell killing in vitro and in vivo. Cancer Res 69(11):4791–4799
161. Korkegian A, Black ME, Baker D, Stoddard BL (2005) Computational thermostabilization of an enzyme. Science 308(5723):857–860
162. Stolworthy TS, Korkegian AM, Willmon CL, Ardiani A, Cundiff J, Stoddard BL, Black ME (2008) Yeast cytosine deaminase mutants with increased thermostability impart sensitivity to 5-fluorocytosine. J Mol Biol 377(3):854–869
163. Rogulski KR, Kim JH, Kim SH, Freytag SO (1997) Glioma cells transduced with an *Escherichia coli* CD/HSV-1 TK fusion gene exhibit enhanced metabolic suicide and radiosensitivity. Hum Gene Ther 8(1):73–85
164. Chang JW, Lee H, Kim E, Lee Y, Chung SS, Kim JH (2000) Combined antitumor effects of an adenoviral cytosine deaminase/thymidine kinase fusion gene in rat C6 glioma. Neurosurgery 47(4):931–938
165. Bennett EM, Anand R, Allan PW, Hassan AE, Hong JS, Levasseur DN, McPherson DT, Parker WB, Secrist JA 3rd, Sorscher EJ, Townes TM, Waud WR, Ealick SE (2003) Designer gene therapy using an *Escherichia coli* purine nucleoside phosphorylase/prodrug system. Chem Biol 10(12):1173–1181
166. Stoeckler JD, Poirot AF, Smith RM, Parks RE Jr, Ealick SE, Takabayashi K, Erion MD (1997) Purine nucleoside phosphorylase. 3. Reversal of purine base specificity by site-directed mutagenesis. Biochemistry 36(39):11749–11756
167. Grove JI, Lovering AL, Guise C, Race PR, Wrighton CJ, White SA, Hyde EI, Searle PF (2003) Generation of *Escherichia coli* nitroreductase mutants conferring improved cell sensitization to the prodrug CB1954. Cancer Res 63(17):5532–5537
168. Guise CP, Grove JI, Hyde EI, Searle PF (2007) Direct positive selection for improved nitroreductase variants using SOS triggering of bacteriophage lambda lytic cycle. Gene Ther 14(8):690–698
169. Barak Y, Thorne SH, Ackerley DF, Lynch SV, Contag CH, Matin A (2006) New enzyme for reductive cancer chemotherapy, YieF, and its improvement by directed evolution. Mol Cancer Ther 5(1):97–103
170. Jaberipour M, Vass SO, Guise CP, Grove JI, Knox RJ, Hu L, Hyde EI, Searle PF (2010) Testing double mutants of the enzyme nitroreductase for enhanced cell sensitisation to prodrugs: effects of combining beneficial single mutations. Biochem Pharmacol 79(2):102–111
171. Swe PM, Copp JN, Green LK, Guise CP, Mowday AM, Smaill JB, Patterson AV, Ackerley DF (2012) Targeted mutagenesis of the *Vibrio fischeri* flavin reductase FRase I to improve activation of the anticancer prodrug CB1954. Biochem Pharmacol 84(6):775–783

172. Bzowska A, Kulikowska E, Shugar D (2000) Purine nucleoside phosphorylases: properties, functions, and clinical aspects. Pharmacol Ther 88(3):349–425
173. Sorscher EJ, Peng S, Bebok Z, Allan PW, Bennett LL Jr, Parker WB (1994) Tumor cell bystander killing in colonic carcinoma utilizing the *Escherichia coli* DeoD gene to generate toxic purines. Gene Ther 1(4):233–238
174. Martiniello-Wilks R, Wang XY, Voeks DJ, Dane A, Shaw JM, Mortensen E, Both GW, Russell PJ (2004) Purine nucleoside phosphorylase and fludarabine phosphate gene-directed enzyme prodrug therapy suppresses primary tumour growth and pseudo-metastases in a mouse model of prostate cancer. J Gene Med 6(12):1343–1357
175. Lukenbill J, Kalaycio M (2013) Fludarabine: a review of the clear benefits and potential harms. Leuk Res 37(9):986–994
176. Cacciapuoti G, Bagarolo ML, Martino E, Scafuri B, Marabotti A, Porcelli M (2016) Efficient fludarabine-activating PNP from archaea as a guidance for redesign the active site of *E. coli* PNP. J Cell Biochem 117(5):1126–1135
177. Williams EM, Little RF, Mowday AM, Rich MH, Chan-Hyams JV, Copp JN, Smaill JB, Patterson AV, Ackerley DF (2015) Nitroreductase gene-directed enzyme prodrug therapy: insights and advances toward clinical utility. Biochem J 471(2):131–153
178. Rylott EL, Budarina MV, Barker A, Lorenz A, Strand SE, Bruce NC (2011) Engineering plants for the phytoremediation of RDX in the presence of the co-contaminating explosive TNT. New Phytol 192(2):405–413
179. White DT, Mumm JS (2013) The nitroreductase system of inducible targeted ablation facilitates cell-specific regenerative studies in zebrafish. Methods 62(3):232–240
180. Bridgewater J, Springer C, Knox R, Minton N, Michael N, Collins M (1995) Expression of the bacterial nitroreductase enzyme in mammalian cells renders them selectively sensitive to killing by the prodrug CB1954. Eur J Cancer 31(13):2362–2370
181. Patel P, Young JG, Mautner V, Ashdown D, Bonney S, Pineda RG, Collins SI, Searle PF, Hull D, Peers E, Chester J, Wallace DM, Doherty A, Leung H, Young LS, James ND (2009) A phase I/II clinical trial in localized prostate cancer of an adenovirus expressing nitroreductase with CB1954. Mol Ther 17(7):1292–1299
182. Lovering AL, Hyde EI, Searle PF, White SA (2001) The structure of *Escherichia coli* nitroreductase complexed with nicotinic acid: three crystal forms at 1.7 A, 1.8 A and 2.4 A resolution. J Mol Biol 309(1):203–213
183. Vellom DC, Radic Z, Li Y, Pickering NA, Camp S, Taylor P (1993) Amino acid residues controlling acetylcholinesterase and butyrylcholinesterase specificity. Biochemistry 32(1):12–17
184. Schwarz M, Glick D, Loewenstein Y, Soreq H (1995) Engineering of human cholinesterases explains and predicts diverse consequences of administration of various drugs and poisons. Pharmacol Ther 67(2):283–322
185. Lockridge O (2015) Review of human butyrylcholinesterase structure, function, genetic variants, history of use in the clinic, and potential therapeutic uses. Pharmacol Ther 148:34–46
186. Lenz DE, Yeung D, Smith JR, Sweeney RE, Lumley LA, Cerasoli DM (2007) Stoichiometric and catalytic scavengers as protection against nerve agent toxicity: a mini review. Toxicology 233(1–3):31–39
187. Parikh K, Duysen EG, Snow B, Jensen NS, Manne V, Lockridge O, Chilukuri N (2011) Gene-delivered butyrylcholinesterase is prophylactic against the toxicity of chemical warfare nerve agents and organophosphorus compounds. J Pharmacol Exp Ther 337(1):92–101
188. Saxena A, Sun W, Fedorko JM, Koplovitz I, Doctor BP (2011) Prophylaxis with human serum butyrylcholinesterase protects guinea pigs exposed to multiple lethal doses of soman or VX. Biochem Pharmacol 81(1):164–169
189. Terekhov S, Smirnov I, Bobik T, Shamborant O, Zenkova M, Chernolovskaya E, Gladkikh D, Murashev A, Dyachenko I, Palikov V, Palikova Y, Knorre V, Belogurov A Jr, Ponomarenko N, Blackburn GM, Masson P, Gabibov A (2015) A novel expression cassette delivers efficient production of exclusively tetrameric human butyrylcholinesterase with improved pharmacokinetics for protection against organophosphate poisoning. Biochimie 118:51–59

190. Brimijoin S, Gao Y (2012) Cocaine hydrolase gene therapy for cocaine abuse. Future Med Chem 4(2):151–162
191. Xie W, Altamirano CV, Bartels CF, Speirs RJ, Cashman JR, Lockridge O (1999) An improved cocaine hydrolase: the A328Y mutant of human butyrylcholinesterase is 4-fold more efficient. Mol Pharmacol 55(1):83–91
192. Sun H, El Yazal J, Lockridge O, Schopfer LM, Brimijoin S, Pang YP (2001) Predicted Michaelis-Menten complexes of cocaine-butyrylcholinesterase. Engineering effective butyrylcholinesterase mutants for cocaine detoxification. J Biol Chem 276(12): 9330–9336
193. Sun H, Pang YP, Lockridge O, Brimijoin S (2002) Re-engineering butyrylcholinesterase as a cocaine hydrolase. Mol Pharmacol 62(2):220–224
194. Hamza A, Cho H, Tai HH, Zhan CG (2005) Molecular dynamics simulation of cocaine binding with human butyrylcholinesterase and its mutants. J Phys Chem B 109(10): 4776–4782
195. Gatley SJ (1991) Activities of the enantiomers of cocaine and some related compounds as substrates and inhibitors of plasma butyrylcholinesterase. Biochem Pharmacol 41(8):1249–1254
196. Pancook JD, Pecht G, Ader M, Mosko M, Lockridge O, Watkins JD (2003) Application of directed evolution technology to optimize the cocaine hydrolase activity of human butyrylcholinesterase. FASEB J 17(4):A565–A565
197. Pan Y, Gao D, Yang W, Cho H, Yang G, Tai HH, Zhan CG (2005) Computational redesign of human butyrylcholinesterase for anticocaine medication. Proc Natl Acad Sci U S A 102(46):16656–16661
198. Cohen-Barak O, Wildeman J, van de Wetering J, Hettinga J, Schuilenga-Hut P, Gross A, Clark S, Bassan M, Gilgun-Sherki Y, Mendzelevski B, Spiegelstein O (2015) Safety, pharmacokinetics, and pharmacodynamics of TV-1380, a novel mutated butyrylcholinesterase treatment for cocaine addiction, after single and multiple intramuscular injections in healthy subjects. J Clin Pharmacol 55(5):573–583
199. Brimijoin S, Gao Y, Anker JJ, Gliddon LA, Lafleur D, Shah R, Zhao Q, Singh M, Carroll ME (2008) A cocaine hydrolase engineered from human butyrylcholinesterase selectively blocks cocaine toxicity and reinstatement of drug seeking in rats. Neuropsychopharmacology 33(11):2715–2725
200. Chen X, Zheng X, Zhou Z, Zhan CG, Zheng F (2016) Effects of a cocaine hydrolase engineered from human butyrylcholinesterase on metabolic profile of cocaine in rats. Chem Biol Interact. doi:10.1016/j.cbi.2016.05.003
201. Kim K, Tsay OG, Atwood DA, Churchill DG (2011) Destruction and detection of chemical warfare agents. Chem Rev 111(9):5345–5403
202. Alder L, Greulich K, Kempe G, Vieth B (2006) Residue analysis of 500 high priority pesticides: better by GC-MS or LC-MS/MS? Mass Spectrom Rev 25(6):838–865
203. Bhattacharya S, Alsen C, Kruse H, Valentin P (1981) Detection of organo-phosphate insecticide by an immobilized-enzyme system. Environ Sci Technol 15(11):1352–1355
204. Razumas VJ, Kulys JJ, Malinauskas AA (1981) High-sensitivity bioamperometric determination of organo-phosphate insecticides. Environ Sci Technol 15(3):360–361
205. Villatte F, Marcel V, Estrada-Mondaca S, Fournier D (1998) Engineering sensitive acetylcholinesterase for detection of organophosphate and carbamate insecticides. Biosens Bioelectron 13(2):157–164
206. Boublik Y, Saint-Aguet P, Lougarre A, Arnaud M, Villatte F, Estrada-Mondaca S, Fournier D (2002) Acetylcholinesterase engineering for detection of insecticide residues. Protein Eng 15(1):43–50
207. Fremaux I, Mazeres S, Brisson-Lougarre A, Arnaud M, Ladurantie C, Fournier D (2002) Improvement of Drosophila acetylcholinesterase stability by elimination of a free cysteine. BMC Biochem 3:21
208. Schulze H, Muench SB, Villatte F, Schmid RD, Bachmann TT (2005) Insecticide detection through protein engineering of Nippostrongylus brasiliensis acetylcholinesterase B. Anal Chem 77(18):5823–5830

209. Hussein AS, Chacon MR, Smith AM, Tosado-Acevedo R, Selkirk ME (1999) Cloning, expression, and properties of a nonneuronal secreted acetylcholinesterase from the parasitic nematode *Nippostrongylus brasiliensis*. J Biol Chem 274(14):9312–9319
210. Bachmann TT, Schmid RD (1999) A disposable multielectrode biosensor for rapid simultaneous detection of the insecticides paraoxon and carbofuran at high resolution. Anal Chim Acta 401(1–2):95–103
211. Bachmann TT, Leca B, Vilatte F, Marty JL, Fournier D, Schmid RD (2000) Improved multianalyte detection of organophosphates and carbamates with disposable multielectrode biosensors using recombinant mutants of *Drosophila* acetylcholinesterase and artificial neural networks. Biosens Bioelectron 15(3–4):193–201
212. Vanhooke JL, Benning MM, Raushel FM, Holden HM (1996) Three-dimensional structure of the zinc-containing phosphotriesterase with the bound substrate analog diethyl 4-methylbenzylphosphonate. Biochemistry 35(19):6020–6025
213. Serdar CM, Gibson DT, Munnecke DM, Lancaster JH (1982) Plasmid involvement in parathion hydrolysis by *Pseudomonas diminuta*. Appl Environ Microbiol 44(1):246–249
214. Scanlan TS, Reid RC (1995) Evolution in action. Chem Biol 2(2):71–75
215. Iyer R, Iken B (2015) Protein engineering of representative hydrolytic enzymes for remediation of organophosphates. Biochem Eng J 94:134–144
216. Dumas DP, Durst HD, Landis WG, Raushel FM, Wild JR (1990) Inactivation of organophosphorus nerve agents by the phosphotriesterase from *Pseudomonas diminuta*. Arch Biochem Biophys 277(1):155–159
217. Watkins LM, Mahoney HJ, McCulloch JK, Raushel FM (1997) Augmented hydrolysis of diisopropyl fluorophosphate in engineered mutants of phosphotriesterase. J Biol Chem 272(41):25596–25601
218. Chen-Goodspeed M, Sogorb MA, Wu F, Raushel FM (2001) Enhancement, relaxation, and reversal of the stereoselectivity for phosphotriesterase by rational evolution of active site residues. Biochemistry 40(5):1332–1339
219. Cho CM, Mulchandani A, Chen W (2002) Bacterial cell surface display of organophosphorus hydrolase for selective screening of improved hydrolysis of organophosphate nerve agents. Appl Environ Microbiol 68(4):2026–2030
220. Griffiths AD, Tawfik DS (2003) Directed evolution of an extremely fast phosphotriesterase by in vitro compartmentalization. EMBO J 22(1):24–35
221. Hill CM, Li WS, Thoden JB, Holden HM, Raushel FM (2003) Enhanced degradation of chemical warfare agents through molecular engineering of the phosphotriesterase active site. J Am Chem Soc 125(30):8990–8991
222. Cho CM, Mulchandani A, Chen W (2004) Altering the substrate specificity of organophosphorus hydrolase for enhanced hydrolysis of chlorpyrifos. Appl Environ Microbiol 70(8):4681–4685
223. Tsai PC, Fan Y, Kim J, Yang L, Almo SC, Gao YQ, Raushel FM (2010) Structural determinants for the stereoselective hydrolysis of chiral substrates by phosphotriesterase. Biochemistry 49(37):7988–7997
224. Bigley AN, Xu C, Henderson TJ, Harvey SP, Raushel FM (2013) Enzymatic neutralization of the chemical warfare agent VX: evolution of phosphotriesterase for phosphorothiolate hydrolysis. J Am Chem Soc 135(28):10426–10432
225. Cherny I, Greisen P Jr, Ashani Y, Khare SD, Oberdorfer G, Leader H, Baker D, Tawfik DS (2013) Engineering V-type nerve agents detoxifying enzymes using computationally focused libraries. ACS Chem Biol 8(11):2394–2403
226. Jackson CJ, Weir K, Herlt A, Khurana J, Sutherland TD, Horne I, Easton C, Russell RJ, Scott C, Oakeshott JG (2009) Structure-based rational design of a phosphotriesterase. Appl Environ Microbiol 75(15):5153–5156
227. Naqvi T, Warden AC, French N, Sugrue E, Carr PD, Jackson CJ, Scott C (2014) A 5000-fold increase in the specificity of a bacterial phosphotriesterase for malathion through combinatorial active site mutagenesis. PLoS One 9(4):e94177

228. Aharoni A, Gaidukov L, Yagur S, Toker L, Silman I, Tawfik DS (2004) Directed evolution of mammalian paraoxonases PON1 and PON3 for bacterial expression and catalytic specialization. Proc Natl Acad Sci U S A 101(2):482–487
229. Amitai G, Gaidukov L, Adani R, Yishay S, Yacov G, Kushnir M, Teitlboim S, Lindenbaum M, Bel P, Khersonsky O, Tawfik DS, Meshulam H (2006) Enhanced stereoselective hydrolysis of toxic organophosphates by directly evolved variants of mammalian serum paraoxonase. FEBS J 273(9):1906–1919
230. Gupta RD, Goldsmith M, Ashani Y, Simo Y, Mullokandov G, Bar H, Ben-David M, Leader H, Margalit R, Silman I, Sussman JL, Tawfik DS (2011) Directed evolution of hydrolases for prevention of G-type nerve agent intoxication. Nat Chem Biol 7(2):120–125
231. Tsai PC, Bigley A, Li Y, Ghanem E, Cadieux CL, Kasten SA, Reeves TE, Cerasoli DM, Raushel FM (2010) Stereoselective hydrolysis of organophosphate nerve agents by the bacterial phosphotriesterase. Biochemistry 49(37):7978–7987
232. Sepp A, Tawfik DS, Griffiths AD (2002) Microbead display by in vitro compartmentalisation: selection for binding using flow cytometry. FEBS Lett 532(3):455–458
233. Tsai PC, Fox N, Bigley AN, Harvey SP, Barondeau DP, Raushel FM (2012) Enzymes for the homeland defense: optimizing phosphotriesterase for the hydrolysis of organophosphate nerve agents. Biochemistry 51(32):6463–6475
234. Bigley AN, Mabanglo MF, Harvey SP, Raushel FM (2015) Variants of phosphotriesterase for the enhanced detoxification of the chemical warfare sgent VR. Biochemistry 54(35): 5502–5512
235. Horne I, Sutherland TD, Harcourt RL, Russell RJ, Oakeshott JG (2002) Identification of an opd (organophosphate degradation) gene in an *Agrobacterium* isolate. Appl Environ Microbiol 68(7):3371–3376

Recent Advances in Directed Evolution of Stereoselective Enzymes

3

Manfred T. Reetz

Abstract

Directed evolution of enzymes provides a prolific source of biocatalysts for asymmetric reactions in organic chemistry and biotechnology. Nowadays, the real challenge in this research area is the development of mutagenesis methods and strategies which ensure the formation of small and highest-quality mutant libraries requiring minimal screening effort. This chapter constitutes a critical analysis of recent developments. Saturation mutagenesis at sites lining the binding pocket using highly reduced amino acid alphabets has emerged as the superior approach for evolving stereoselectivity.

3.1 Introduction

Asymmetric catalysis stands at the heart of modern synthetic organic chemistry. The three options are chiral transition metal catalysts, organocatalysts, and enzymes. The latter have been applied in synthetic organic chemistry for a century, but biocatalysis has not been accepted as a routine technique for several reasons. However, during the last two decades, notable progress has been made in bioprocess development, reactor design, downstream processing, immobilization, improved expression systems, and genome mining for identifying new enzymes [1]. Nevertheless, serious problems persisted due to the following often observed limitations:

- Poor or wrong stereoselectivity
- Limited substrate scope
- Insufficient activity

M.T. Reetz
Department of Chemistry, Philipps-Universität, Marburg, Germany
e-mail: reetz@mpi-muelheim.mpg.de

M. Alcalde (ed.), *Directed Enzyme Evolution: Advances and Applications*,
DOI 10.1007/978-3-319-50413-1_3

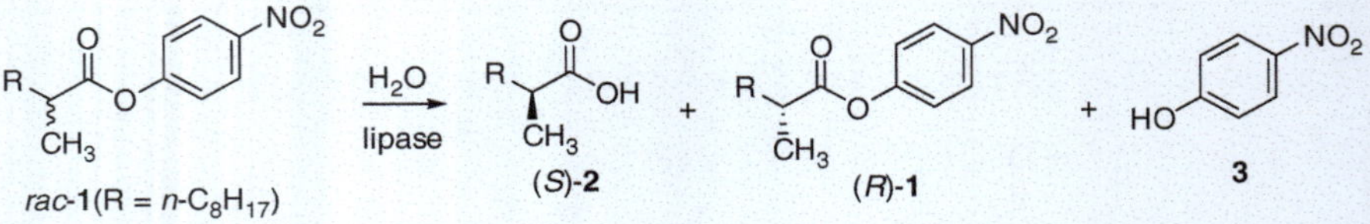

Scheme 3.1 Model reaction used in the first study of directed evolution of a stereoselective enzyme [6]

The so-called rational design based on appropriate site-specific mutagenesis has been shown to be successful in some cases [2], but directed evolution has clearly emerged as the more general and reliable approach [3]. As in genetic optimization of thermostability [4], three basically different gene mutagenesis methods can be applied in order to enhance or invert stereoselectivity: error-prone polymerase chain reaction (epPCR), saturation mutagenesis, and/or DNA shuffling [3, 5]. The first example of directed evolution of an enantioselective enzyme concerned the hydrolytic kinetic resolution of *rac*-**1**, catalyzed by the lipase from *Pseudomonas aeruginosa* (PAL) (Scheme 3.1) [6]. WT PAL leads to low enantioselectivity slightly favoring the formation of (*S*)-**2**, the selectivity factor amounting to only $E=1.4$.

Four cycles of epPCR at low mutation rate with introduction of a single point mutation in each round enhanced enantioselectivity to $E=11$ (*S*). Since the fifth cycle resulted in marginal improvement ($E=15$), which is far from practical application, different mutagenesis strategies were developed. The combination of epPCR, saturation mutagenesis, and DNA shuffling afforded a variant characterized by six point mutations, showing a selectivity factor of $E=51$ and a 250% increase in activity [7]. Only one of the mutations occurred near the active site, five being remote. A theoretical analysis based on QM/MM showed a relay mechanism to be operating. More importantly, it was predicted that only two of the six mutations are necessary for high enantioselectivity. Indeed, the respective double mutant proved to be even more effective ($E=62$) [8].

These observations demonstrated that the genetic approach utilizing epPCR, saturation mutagenesis, and DNA shuffling is successful, but not efficient, a great deal of time-consuming screening being necessary (50,000 transformants). It was also possible to invert enantioselectivity in favor of (*R*)-**2**, but this also involved excessive screening [9]. At the time, several other groups joined efforts in generalizing directed evolution of stereoselectivity using the same strategies, as summarized in a 2004 review [10]. However, efficacy was not a focal point of research. Since the screening step is the bottleneck in the overall directed evolution process, methods and strategies for generating smaller and smarter libraries had to be developed.

After several years of research using PAL and other enzymes, saturation mutagenesis at sites lining the binding pocket as part of the **c**ombinatorial **a**ctive-site **s**aturation **t**est (CAST) emerged as the optimal strategy (Scheme 3.2a) [11].

CAST is a convenient acronym to distinguish it from saturation mutagenesis at other (remote) sites for different purposes. When the "hits" in initial CAST libraries

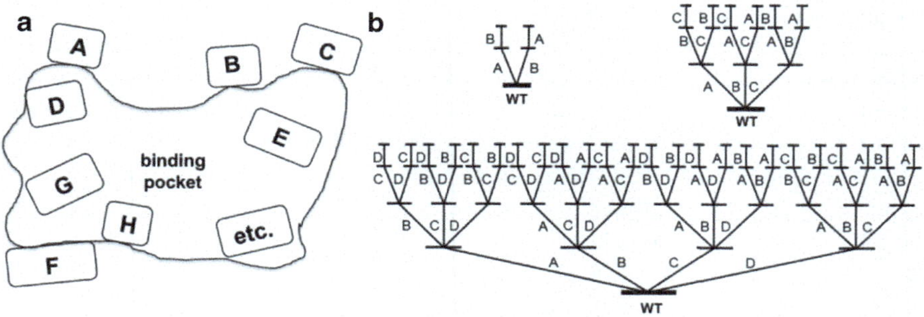

Scheme 3.2 (**a**) Systematization of CASTing; *A*, *B*, *C*, *etc.* denote potential randomization sites, each comprising one, two, or more residues lining the binding pocket. (**b**) Two-, three-, and four-site ISM schemes

Table 3.1 Difference in screening effort when applying NNK (encoding all 20 canonical amino acids) versus NDT (encoding 12 amino acids: Phe, Leu, Ile, Val, Tyr, His, Asn, Asp, Cys, Arg, Ser, Gly) for 95% library coverage [5]

NNK			NDT	
Nr. of amino acid positions at one site	Codons	Transformants needed	Codons	Transformants needed
1	32	94	12	34
2	1028	3066	144	430
3	32,768	98,163	1728	5175
4	$>1.0\times10^{6}$	$>3.1\times10^{6}$	$>2.0\times10^{4}$	$>6.2\times10^{4}$
5	$>3.3\times10^{7}$	$>1.0\times10^{8}$	$>2.5\times10^{5}$	$>5.5\times10^{5}$
6	$>1.0\times10^{9}$	$>3.2\times10^{9}$	$>2.9\times10^{6}$	$>8.9\times10^{6}$
7	$>3.4\times10^{10}$	$>1.\times10^{11}$	$>3.5\times10^{7}$	$>1.1\times10^{8}$
8	$>1.0\times10^{12}$	$>3.3\times10^{12}$	$>4.2\times10^{8}$	$>1.3\times10^{9}$
9	$>3.5\times10^{13}$	$>1.0\times10^{14}$	$>5.1\times10^{9}$	$>1.5\times10^{10}$
10	$>1.1\times10^{15}$	$>3.4\times10^{15}$	$>6.1\times10^{10}$	$>1.9\times10^{11}$

still display insufficient enantioselectivity, a recursive process is recommended: iterative saturation mutagenesis (ISM) [12]. Scheme 3.2b shows the case of two-, three-, and four-site ISM systems involving two, six, and 24 pathways. It is not necessary to explore all theoretically possible pathways, but some may be more productive than others. Since saturation mutagenesis at large randomization sites requires excessive screening for 95% library coverage, reduced amino acid alphabets were introduced [13]. In order to remind the reader of the relationship between the size of a randomization site, the nature of the amino acid alphabet, and the screening effort for 95% library coverage, Table 3.1 is included here which illustrates the difference between NNK and, e.g., NDT codon degeneracy, which encode 20 and 12 canonical amino acids, respectively [5]. Any other reduced amino acid alphabet that the researcher may want to use can be analyzed statistically in the same manner using the CASTER computer aid [5], which is based on the Patrick/Firth algorithm [14]. The Nov metric can also be used, in this case identifying the nth best mutant [15]. CAST/ISM should be guided by X-ray structures (or homology models) and sequence data. Initial NNK-based saturation mutagenesis at individual CAST positions, requiring the screening of only one 96-format microtiter plate, also provides information for choosing a reduced amino acid alphabet in subsequent mutagenesis experiments.

The lessons learned from these methodological developments, reported in a series of studies using different enzyme types [3f, 5], were then used as a guide in the final directed evolution study of PAL for comparison purposes [16]. It was discovered that CAST/ISM provides a notably improved triple mutant showing $E=594$ (*S*) in the model reaction involving the kinetic resolution of *rac*-**1**, while requiring the screening of less than 10,000 transformants. This highlights the progress in methodology development. In the ISM study, three CAST sites A, B, and C, were designed, and NDT codon degeneracy was applied, leading to the highly improved triple mutant [16]. This variant has no superfluous mutations. The origin of the unusually high degree of enantioselectivity was traced on a molecular level to strong cooperative mutational effects. Such synergistic effects (more than additivity) were later found in other ISM-based studies as well [17]. In an independent study utilizing a galactosidase as the enzyme, saturation mutagenesis was likewise shown to be more efficient than DNA shuffling [18].

Since the publication of the best results concerning PAL, further progress in methodology development has been achieved [5, 19, 20]. The primary focus was placed on utilizing the smallest possible reduced amino acid alphabets, again for the purpose of minimizing screening while maximizing library quality. In doing so, two different strategies can be applied (Scheme 3.3). According to strategy 1, one and the same reduced amino acid alphabet is used for the whole CAST randomization site in a single experiment. In contrast, strategy 2 calls for a different reduced amino

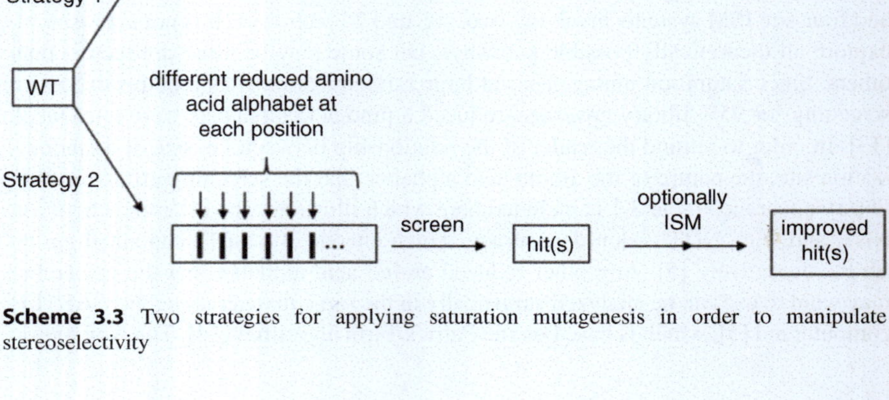

Scheme 3.3 Two strategies for applying saturation mutagenesis in order to manipulate stereoselectivity

acid alphabet for each position of a multi-residue site, a single experiment also being involved. In both cases, ISM can be applied for further optimization.

Nowadays, automated GC or HPLC can handle typically 2000–3000 transformants within a few days. An on-plate pretest for activity is nevertheless recommended, the much smaller number of hits then being analyzed for stereoselectivity by chiral GC or HPLC. The question whether to choose a reduced amino acid alphabet such as NDT in combination with, e.g., two-residue randomization sites followed by ISM, or to opt for a much smaller reduced amino acid alphabet encoding only one, two, or three amino acids in combination with larger randomization sites, e.g., four to ten residues, has been addressed [20–22]. Triple code saturation mutagenesis (TCSM) using three- or four-residue CAST randomization sites appears to be the strategy of choice as shown by several recent studies [22]. In these cases, the initial CAST libraries often harbor stereoselective variants that fulfill all requirements for practical applications or require only one ISM step for final fine-tuning. Nevertheless, more experience is needed for final assessments. When applying CAST/ISM, several guidelines are recommended:

- Library design by the CASTER computer aid (http://www.kofo.mpg.de/en/research/biocatalysis) or the GLUE-IT metric (http://guinevere.otago.ac.nz/cgi-bin/aef/glue-IT.pl), both available free of charge
- Guidance by structural, mechanistic, and (consensus) sequence data [20–22]
- Use of an on-plate pretest for activity followed by chiral GC or HPLC analysis for enantioselectivity [21]
- Application of the quick quality control [23a] or quantitative Q-values [23b] in order to avoid screening something that does not exist
- Application of pooling techniques for reducing the screening effort [23a, 24]
- Techno-economical analysis which considers, inter alia, the number, quality, and cost of primers used in designing and generating mutant libraries [25]

The CAST/ISM-based approach has proven to be particularly efficient, fast, and reliable [5, 19, 20], but other gene mutagenesis methods continue to be used. Unfortunately, comparative experiments are generally not made. In some ISM-based studies, a final round of epPCR was added for further (small) improvements, as in the case of directed evolution of a glycosidase [26]. In other investigations, only epPCR and/or DNA shuffling were employed. In the sections that follow, selected recent examples of directed evolution of stereoselectivity using different gene mutagenesis techniques and strategies are critically analyzed.

3.2 Cytochrome P450 Monooxygenases

Cytochrome P450 monooxygenases (CYPs) catalyze several synthetically useful reaction types, including asymmetric oxidative hydroxylation R-H → R-OH, olefin epoxidation, and sulfoxidation. Several reviews of CYP protein engineering have appeared [27]. The first study featuring the manipulation of enantioselectivity of a

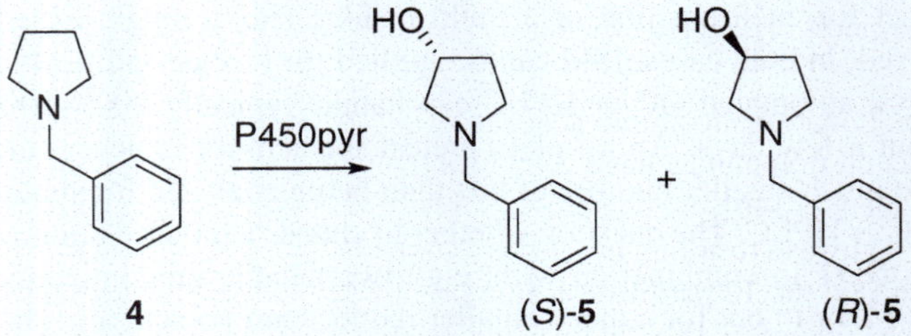

Scheme 3.4 Model reaction used in the first case of directed evolution of a P450 monooxygenase leading to high regio- and stereoselectivity [28]

cytochrome P450 monooxygenase while maintaining essentially complete regioselectivity concerned the P450-pyr catalyzed oxidative hydroxylation of *N*-benzylpyrrolidine (**4**) with formation of product **5** (Scheme 3.4) [28]. WT P450-pyr shows complete regioselectivity in favor of the 3-hydroxy product, but with a mere 43% ee (*S*). The substrate appears to be bound in two energetically similar poses within reach of the catalytically active high spin heme-Fe=O (Cpd I) in the large binding pocket. The mechanism is known to involve a radical abstraction of the H-atom with formation of a carbon-centered radical followed by rapid C-O bond formation. The ideal O-H-C angle has been calculated to be about 130° [29].

Using a homology model, 17 residues were identified within 5 Å of the heme-docked substrate [28a]. With the exception of residues C366 and G256, they were subjected individually to saturation mutagenesis using NNK codon degeneracy. This would require for 95% library coverage the screening of only 94 transformants in each case. In fact, an excess of 180 transformants were screened to ensure even higher coverage [28a]. Whereas variant F403L improves (*S*)-selectivity to 65% ee, a single point mutation (N100S) induces the reversal of enantioselectivity. In order to enhance (*R*)-selectivity, a simplified version of ISM was applied by employing mutant N100S as the template in saturation mutagenesis at other CAST sites. The best double mutant proved to be N100S/T186I with an enantioselectivity of 83% ee (*R*) and no trade-off in regioselectivity.

In a subsequent study, (*S*)-selectivity was improved by exploring more of protein sequence space [28b]. First, the crystal structure of WT P450-pyr was determined and used as a guide in choosing CAST sites. Twenty residues, A77, I82, I83, L98, P99, N100, I102, A103, S182, D183, T185, T186, L251, V254, G255, D258, T259, L302, M305, and F403, were targeted using the same simplified ISM-based strategy. Only nine initial libraries had to be generated, since the remaining 11 had already been obtained in the previous study. The best variant I83H/M305Q/A77S was obtained in three rounds of iterative saturation mutagenesis (ISM), showing an ee-value of 98% (*S*) at fully maintained regioselectivity and little trade-off in activity [28b]. It would be interesting to test TCSM [22] in this reaction.

Particularly challenging goals arise when regio- *and* enantioselectivity need to be evolved. In a study dedicated to solving this problem, P450-BM3 was employed [30]. It was found that WT P450-BM3 catalyzes the hydroxylation of cyclohexene-1-carboxylic

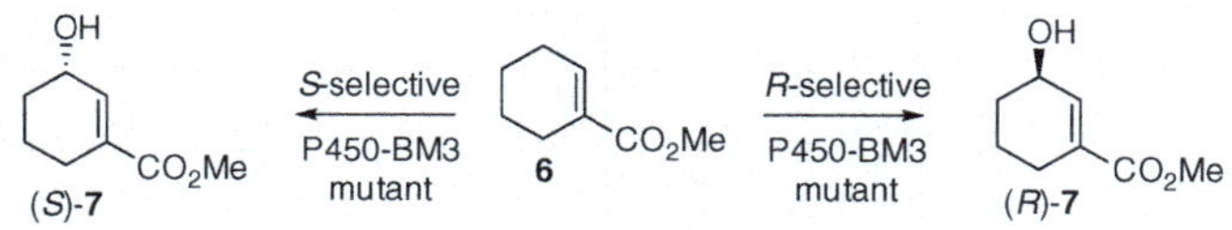

Scheme 3.5 Model reaction used in the first case of directed evolution of a P450 monooxygenase leading to high regio- and stereoselectivity with optional formation of either enantiomeric product [30]

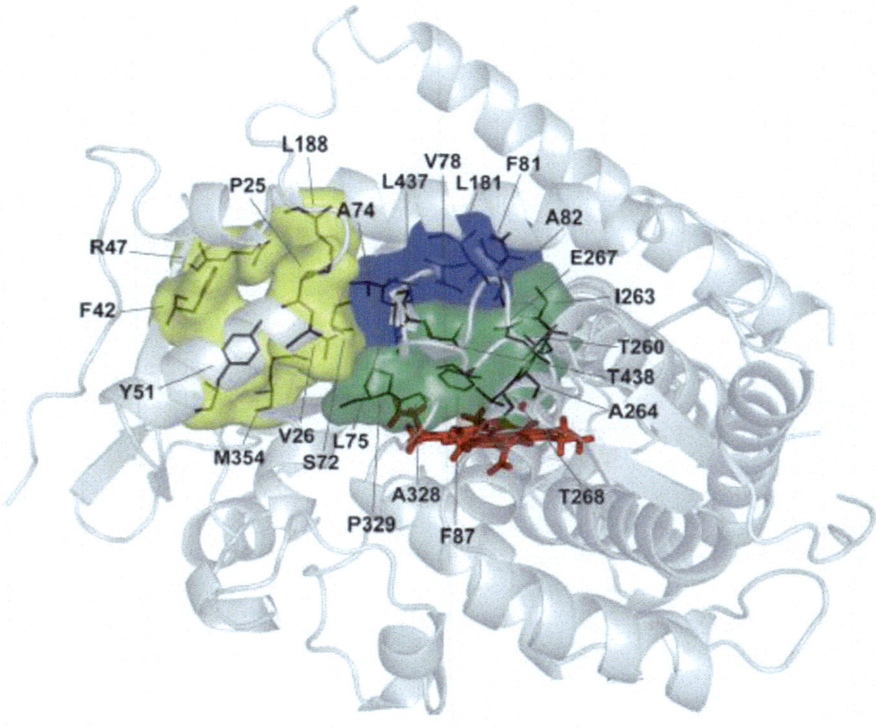

Fig. 3.1 The 24 P450-BM3 residues considered for saturation mutagenesis. They were assigned to three categories marked in *green* (residues closest to the heme), *blue* (residues further away but still lining the binding pocket), and *yellow* (residues at entrance to the large binding pocket) [30]

acid methyl ester (**6**) with insufficient regioselectivity (84%) and poor enantioselectivity (34%) ee in slight favor of (*R*)-**7**) (Scheme 3.5).

In order to achieve practical levels of regioselectivity as well as >95% (*R*)- and (*S*)-selectivity on an optional basis, ISM was applied. On the basis of the crystal structure of P450-BM3, a total of 24 CAST residues were identified (first- and second-sphere residues) (Fig. 3.1) [30].

NNK-based saturation mutagenesis was initially applied at 23 of the 24 positions, leading to a limited set of improved mutants with enhanced and reversed

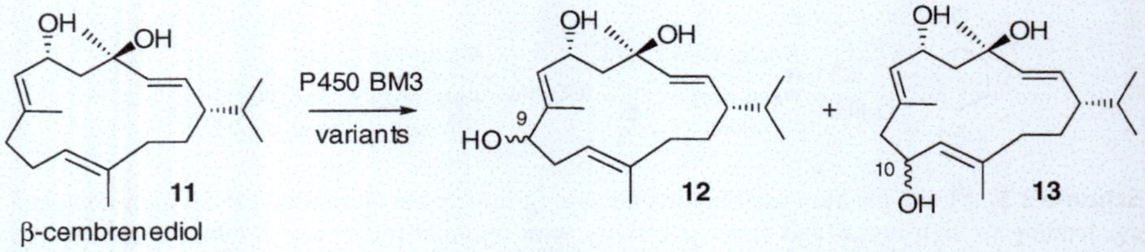

Scheme 3.6 P450-BM3 catalyzed oxidative hydroxylation of the diterpenoid ß-cembrenediol [32]

enantioselectivity. This information formed the basis of ISM optimization. For example, the gene of the best (*R*)-selective variant was chosen as the template for saturation mutagenesis at other sites which had also favored (*R*)-selectivity, and the analogous procedure was applied for reversing (*S*)-selectivity. This strategy provided selective variants (97–98% ee) favoring (*R*)- and (*S*)-**7**, respectively, and displaying regioselectivities in the range of 93–98% [30].

ISM was also applied to P450-BM3 catalyzed oxidative hydroxylation of steroids as part of "late-stage oxidation" [31]. For example, testosterone was used as the substrate, for which 2ß- and 15ß-selective variants were evolved showing >95% regio- and diastereoselectivity.

Yet another example of late-stage oxidative hydroxylation concerns P450-BM3 catalyzed reactions of the monocyclic diterpenoid ß-cembrenediol (**11**), which is essentially not accepted by the WT (2% conversion) (Scheme 3.6) [32]. At the time of this project, a number of other protein engineering studies utilizing various structurally different substrates were available, the mutational data being of significant help in the new endeavor [27]. This is an important point, because it is logical to learn from past experience and not to start from "scratch" in each new project. For example, it was known that residue F87 is a sensitive position, smaller amino acids generally being necessary to enable substrate acceptance of sterically demanding compounds because the phenyl group of F87 prevents complete access to the catalytically active heme-Fe=O (Cpd I). Therefore, known variants F87A and F87G [27] were tested first, which indeed led to acceptable levels of activity. These were then used as templates for site-specific mutagenesis at selected first-sphere CAST residues A74, L75, V78, I263, A264, and L437. Iterative site-specific mutagenesis was performed in several rounds, in each case a new point mutation being introduced [32]. A total of 29 variants were produced by this technique, which resembles ISM without the need to screen libraries. Structure-guided successive site-specific mutagenesis with the creation of minimally sized libraries constitutes a fusion of rational design and directed evolution. Whereas the particular choice of the mutations limits structural diversity drastically, the results proved to be of practical interest in this project. Several variants were generated in this way, F87A/I263L catalyzing complete regioselective hydroxylation at position C-9 with an 89:11 diastereomeric ratio. The triple mutant L75A/V78A/F87G enables hydroxylation at position C-10 (97% regioselectivity) with a 74:26 diastereomeric ratio. Assignment of the absolute configuration at the new stereogenic centers was not reported [32].

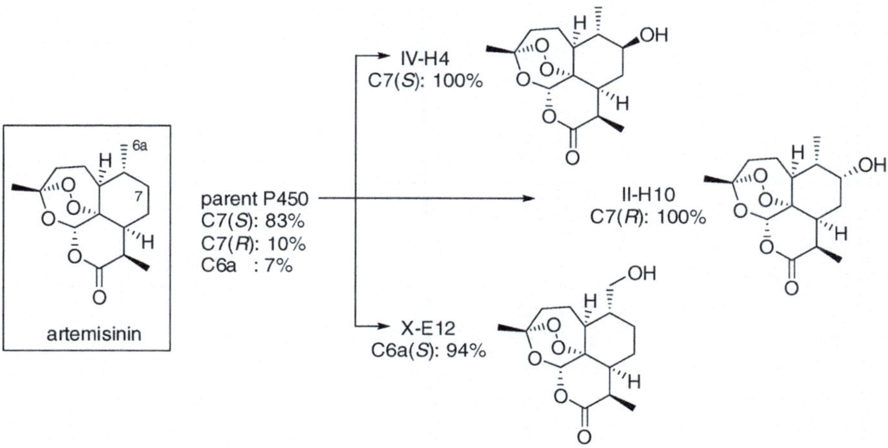

Scheme 3.7 Oxidative hydroxylation of artemisinin by P450-BM3 mutants [33]

This study nicely shows that very small semi-rationally designed mutant libraries may well suffice, provided sufficient previous knowledge of mutational effects is available. Indeed, after years of protein engineering of P450-BM3, the huge mutational data serves as a convenient guide when targeting new substrates using rational design or directed evolution.

Another interesting example of directed evolution of a CYP concerns the late-stage regio- and stereoselective hydroxylation of the antimalaria drug artemisinin catalyzed by P450-BM3 (Scheme 3.7) [33]. Here again a semi-rational approach was implemented using saturation mutagenesis at first-sphere CAST sites, this time based on initial P450 fingerprinting followed by fingerprint-driven reactivity predictions and final ISM experiments. WT P450-BM3 does not accept the substrate. First, 125,000 transformants were screened for activity using not the actual substrate artemisinin, but five semisynthetic chromogenic probes, which gave rise to 1950 active variants that accept such sterically demanding substrates (criterion: ">10% of parent enzyme activity on at least one of the fingerprint probes"), and 522 functionally unique variants (criterion: "larger than 20% variation on at least one of the five fingerprint components compared to the parent enzyme and any other member of the library") [33]. After correlating the P450-BM3 fingerprints with the actual artemisinin reactivity, 75 variants remained and were tested as catalysts for oxidation of the real substrate by HPLC analysis. The best hit was variant FL#62, which was found to have 16 point mutations. This work was then followed by several ISM experiments. The result of all of these efforts is summarized in Scheme 3.7 [33]. Whereas the "parent" enzyme showed 83% C7(*S*)-, 10% C7(*R*)-, and 7% C6a-selectivities, notable improvements were achieved in the final saturation mutagenesis experiments. The tendency to hydroxylate at C6 and C7 was maximized by directed evolution. If for some reason a different position were to be the target, then the challenging question of how to achieve such regioselectivity would arise.

In the directed evolution of CYPs, as in other enzyme types, a clear trend to utilize saturation mutagenesis has emerged [27, 34]. However, some researchers in recent CYP studies rely on epPCR and site-specific mutagenesis. An example is the evolution of P450-BM3 mutants that hydroxylate the contraceptive drug norethisterone at the 15ß- or 16ß-position [35a]. The combination of epPCR and DNA shuffling has also been used, as in activity optimization of the 13-hydroperoxide lyase CYP74B; the products are used for the production of C6-aldehydes [35b].

To date, the concept of P450-catalyzed late-stage oxidative hydroxylation of natural products or synthetic compounds suffers from the lack of predictability as to where hydroxylation will occur. When designing a synthetic pathway in natural product synthesis, it is generally unclear whether P450-catalyzed hydroxylation will occur at the desired position. In such a situation, the initial library must be diverse enough to harbor variants that provide many different regioisomers. If the desired product is formed to a small extent, then the respective mutant can be chosen for further mutational improvements with the aim of turning the minor into the major product.

3.3 Baeyer-Villiger Monooxygenases

The first directed evolution study of a Baeyer-Villiger monooxygenase (BVMO) concerned the optimization of cyclohexanone monooxygenase (CHMO) as the catalyst in the oxidative desymmetrization of 4-hydroxycyclohexanone [36]. One of the best variants evolved by epPCR, F345S, was then used in the desymmetrization of structurally different ketones (Table 3.2) [37]. It can be seen that in essentially all

Table 3.2 CHMO variant Phe432Ser as the catalyst in oxidative desymmetrization [37]

Substrate	ee (%)	Substrate	ee (%)
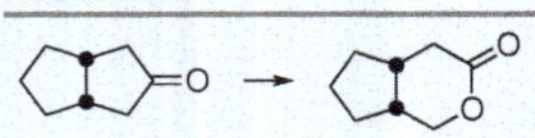	94	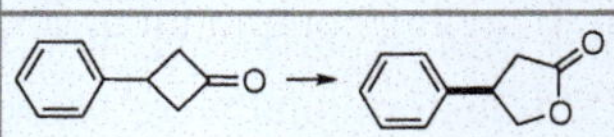	96
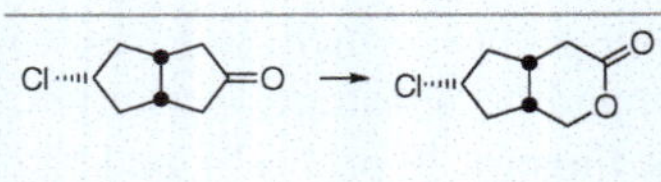	99	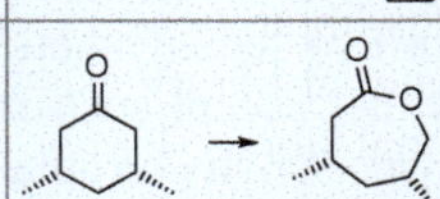	>99
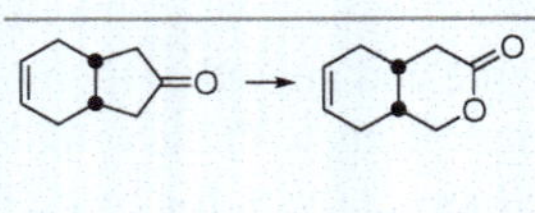	91	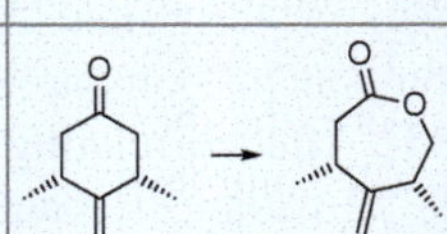	>99
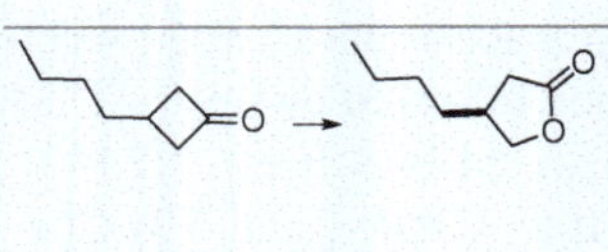	97	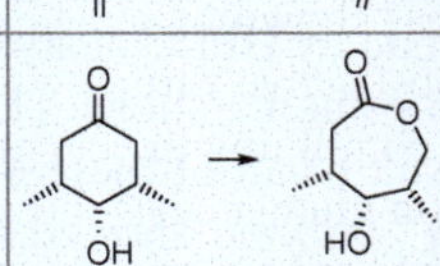	>99
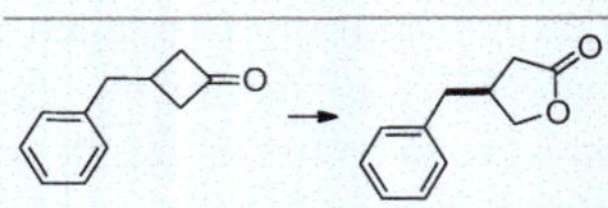	78	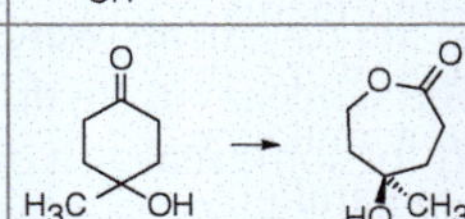	99

Scheme 3.8 Oxidative kinetic resolution catalyzed by mutants of the Baeyer-Villiger monooxygenase PAMO [40]

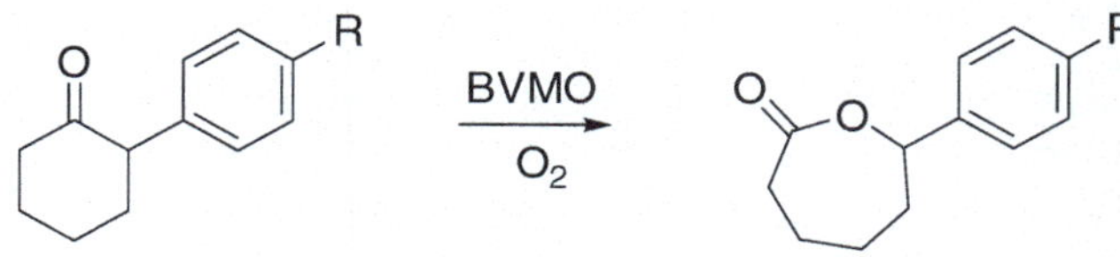

rac-**14a** R = H
b R = Cl

(*R*)- or (*S*)-**15a** R = H
b R = Cl

Scheme 3.9 Sequence alignment of eight BVMOs (loop 441–444 in gray box) [40]

cases, high enantioselectivity as well as chemoselectivity was achieved without having to perform additional mutagenesis experiments.

These results are impressive when utilizing whole cells, but CHMO suffers from insufficient thermostability. The discovery of the unusually robust phenyl acetone monooxygenase (PAMO) aroused a great deal of interest, although its substrate scope was shown to be very narrow [38]. Such substrates as cyclohexanone or derivatives thereof are not accepted by this BVMO, apparently because the binding pocket is too small to accommodate such compounds. In order to solve this problem, a series of CAST-based directed evolution studies have appeared [39].

A bioinformatics-based study of PAMO as the catalyst in oxidative kinetic resolution of compounds **14a-b** deserves special attention for three reasons (Scheme 3.8) [40]: (1) It not only utilized structural data for choosing appropriate CAST sites, but also (2) sequence alignment information in order to derive optimal reduced amino acid alphabets, and (3) a different codon degeneracy according to strategy 2 as outlined in Scheme 3.3 (Introduction). Ketone **14a** was used in all screening assays, the best evolved variant then being tested in the oxidative kinetic resolution of **14b** as well [40].

Guided by the PAMO crystal structure [38b], four residues in loop 441–444 next to the binding pocket were chosen as CAST sites. NNK-based randomization of a four-residue CAST site would require the screening of 3.1 million transformants for 95% library coverage, while NDT codon degeneracy would still call for excessive screening($\approx$ 62,000 transformants) (Table 3.1, Introduction). Therefore, eight Baeyer-Villiger monooxygenases were aligned, the loop region 441–444 being of interest (Scheme 3.9) [40]. A limited number of amino acids are conserved at the

Table 3.3 Codon degeneracies chosen at each position in the PAMO loop 441–444. Degenerate codons: A (adenine), B (cytosine/guanine/thymine), C (cytosine), G (guanine), S (cytosine/guanine), K (guanine/thymine), N (adenine/cytosine/guanine/thymine. WT amino acids are shown in parentheses [40]

Amino acid positions	Codon degeneracy	Encoded amino acids	Codons	Oversampling for 95% coverage
441	KCA	A, (S)	864	2587
442	KBG	S, (A), L, V, W, G		
443	BGC	F, H, (L), V, Y, G, D, R, C		
444	NSC	(S), A, P, T, R, G, C		

four positions: Ser and Ala (position 441); Ala, Val, Gly, and Leu (position 442); Leu, Phe, Gly, and Tyr (position 443); and Ser, Ala, Cys, and Thr (position 444). Consequently, these amino acids were used as building blocks at the respective positions of the four-residue randomization site, these reduced amino acid alphabets minimizing the screening effort drastically.

Codon degeneracies were then designed for matching the amino acids occurring at these four positions while also introducing a limited number of additional amino acids for randomization experiments in order to minimize primer costs and enhance diversity (Table 3.3) [40]. WT amino acid is maintained throughout. Interestingly, at position 441, KCA codon degeneracy correlates with the introduction of only one new amino acid (Ala). At positions 442–444, structural diversity was designed to be higher.

After screening a mere 1700 transformants (95% library coverage would require 2587), several active variants were discovered, PAMO mutant Ser441Ala/Ala442Trp/Leu443Tyr/Ser444Thr displaying the highest activity and enantioselectivity ($E=70$ in favor of (*R*)-**15a**) (Scheme 3.8) [40]. This variant proved to be an even better catalyst for the oxidation of the *p*-chloro-derivative **14b** ($E>200$). As in the case of **14a**, WT PAMO does not accept this substrate. It was concluded that bioinformatics can be used to define a reduced amino acid alphabet and that a different designed reduced amino acid alphabet can well be effective at each position within a multi-residue randomization site in a single saturation mutagenesis experiment (strategy 2 in Scheme 3.3). This approach was later employed in the directed evolution of other enzyme types [21b, 22].

It should be mentioned that following these and other CAST-based studies of BVMOs [39], examples of epPCR as an alternative were reported. For example, BmoFI from *Pseudomonas fluorescens* DSM 50106 was used as the catalyst in the oxidative kinetic resolution of *rac*-**16** with preferential formation of (*S*)-**17** (Scheme 3.10) [41]. Enantioselectivity of the WT ranges between $E=55$ (at small scale) and $E=71$ (growing *E. coli* cells in shake flasks). Application of epPCR and screening 3500 transformants provided several improved mutants displaying $E=77$–92. Upon combining the respective point mutations, an excellent variant was identified, His51Leu/Ser136Leu displaying a selectivity factor of $E=86$ [41].

In a very different approach not relying on CASTing, epPCR, or DNA shuffling, a theoretically predicted remote two-residue site comprising positions 93 and 94 in

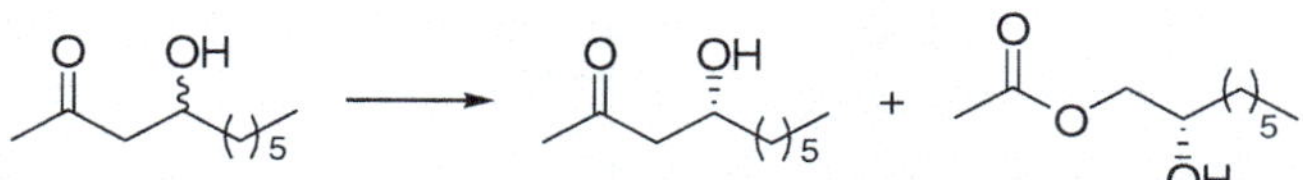

rac-**16** (*R*)-**16** (*S*)-**17**

Scheme 3.10 Oxidative kinetic resolution catalyzed by mutants of the Baeyer-Villiger monooxygenase BmoFI [41]

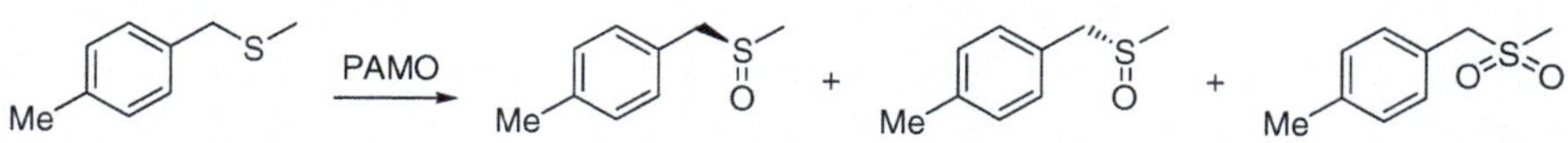

18 (*R*)-**19** (*S*)-**19** **20**

Scheme 3.11 Asymmetric sulfoxidation catalyzed by mutants of the Baeyer-Villiger monooxygenase PAMO [22a]

PAMO was subjected to saturation mutagenesis with the expectation of an allosteric effect in the absence of an effector molecule [42]. Several four-substituted cyclohexanone derivatives (methyl, ethyl, *n*-butyl, *tert*-butyl) were chosen as substrates which are not accepted by WT PAMO. This procedure proved to be surprisingly successful, the double mutant Gln93Asn/Pro94Asp showing 97–98% (*R*)-selectivity for the methyl, ethyl, and *n*-butyl derivatives in desymmetrization reactions. The *tert*-butyl derivative was not accepted due to steric factors. Instead of an effector molecule inducing allostery, it is the mutational change that is causing a conformational reorganization at the binding site. Calculations of Karplus-type covariance maps [43] of the double mutant versus WT PAMO showed that the expected conformational motions were indeed occurring [42]. Deconvolution of Gln93Asn/Pro94Asp showed that the single mutants Gln93Asn and Pro94Asp are inactive, signaling pronounced cooperative mutational effects.

To date, the concept of focusing on mutationally induced remote allosteric effects has not been applied to other substrates. In further optimization, the double mutant Gln93Asn/Pro94Asp could be used as a template for CASTing. However, researchers have preferred to concentrate from the very beginning on CAST-based ISM when new projects are initiated. For example, using this strategy, PAMO was evolved as the catalyst in asymmetric sulfoxidation reactions using prochiral thioether **18** as the substrate (Scheme 3.11) [22a]. WT PAMO leads to the preferential formation of (*S*)-**19** with notable selectivity (90% ee). The primary goal was to evolve inverted enantioselectivity in favor of (*R*)-**19**.

Six potential randomization residues were first identified on the basis of the PAMO crystal structure [38b] and past studies which identified "hot spots": P440, A442, and L443 (as CAST loop residues) and V54, I67, and Q152 (as traditional CAST sites). One possible strategy would be to group them into three two-residue randomization sites and to apply ISM. In this study, a different procedure was chosen [22a]. In exploratory experiments, residues V54, I67, Q152, and A442 were chosen for individual saturation mutagenesis. Instead of applying traditional NNK codon degeneracy, the "smart-intelligent" library construction according to Tang [44] was

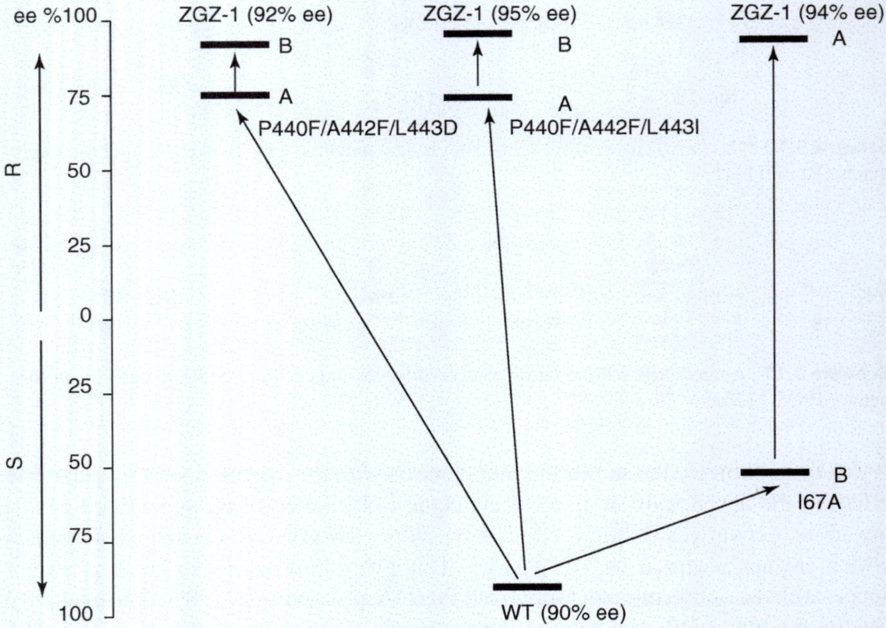

Scheme 3.12 Two ISM pathways in the directed evolution of PAMO as catalyst in the asymmetric sulfoxidation of **18**, WT → A → B providing two (*R*)-selective variants ZGZ-1 and ZGZ-2 and WT → B → A leading to an equally (*R*)-selective variant ZGZ-3 [22a]

used. Four pairs of primers with degenerate codons of NDT (encoding 12 amino acids N, S, I, H, R, L, Y, C, F, D, G, and V), codon degeneracy VMA (encoding six amino acids E, A, Q, P, K, T), codon degeneracy ATG (encoding M), and codon degeneracy TGG (encoding W) were considered at the target sites [22a]. They were mixed in a ratio of 12:6:1:1, and for each library creation, only 60 transformants had to be screened for 95% coverage. Whereas saturation mutagenesis at V54 failed to provide any hits, the other libraries harbored improved hits.

Following these initial experiments, a two-site ISM scheme starting from WT PAMO was designed involving site A (P440/A442/L443) with reduced amino acid alphabets and site B (I67 as a hot spot identified earlier) with 20 amino acids. The reduced amino acid alphabet applied at P440 involved three building blocks as the components of a triple code in addition to the WT amino acid, Tyr, Leu, and Phe. This is an example of triple code saturation mutagenesis (TCSM) [22a] as part of strategy 2 (Scheme 3.3). The results of this minimal search in protein sequence space are summarized in Scheme 3.12 [22a]. It can be seen that pathway WT → A → B provided two variants showing highly reversed enantioselectivity: ZGZ-1 (I67C/P440F/A442F/L443D (92% ee) and ZGZ-2 (I67Q/P440F/A442N/L443I (95% ee) in favor of (*R*)-**19**. The opposite pathway WT → B → A was also successful, leading to variant I67A/P440Y/A442V/L443I with 94% ee (*R*). Overoxidation with formation of the sulfone **20** occurred to only a minimal degree.

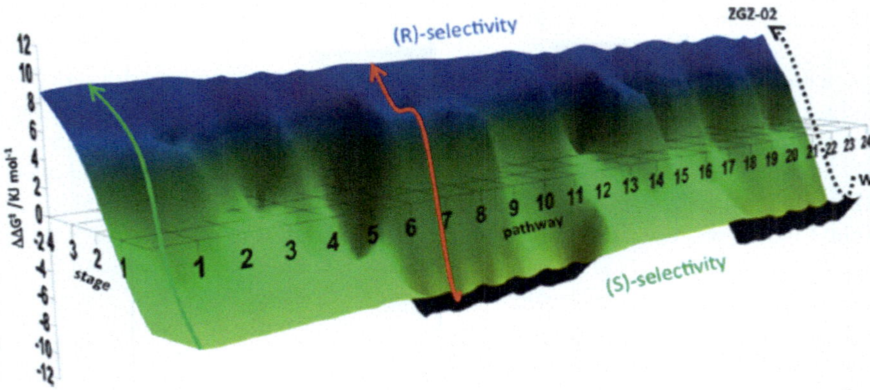

Fig. 3.2 Fitness pathway landscape featuring 24 pathways which by necessity lead from the *(S)-selective* WT PAMO to the best *(R)-selective* mutant *ZGZ-2* allowing the formation of *(R)-**19*** [22a]. The *green line* denotes a typical trajectory lacking any local minima, and the *red* one a trajectory characterized by at least one local minimum. The upward climb is associated with 3.9 kcal/mol

Complete reversal of enantioselectivity is impressive, because the change in energy $\Delta\Delta G^{\#}$ in going from WT PAMO (90% ee, *S*) to variant ZGZ-2 (95% ee, *R*) amounts to 3.9 Kcal/mol. Deconvolution of this highly (*R*)-selective quadruple mutant ZGZ-2 (I67Q/P440F/A442N/L443I) led to the surprising discovery that the respective single mutants are all (*S*)-selective: I67Q (69% ee), P440F (97% ee), A442N (69% ee), and L443I (98% ee) [22a]. If these four (*S*)-selective single mutants had been generated separately by other means, few researchers would combine them in order to generate the opposite (*R*)-selectivity! This kind of synergistic nonadditive effects continues to be observed whenever time and effort are invested in deconvolution. It is an indication of the efficacy of ISM [17].

Exhaustive deconvolution was performed, meaning the generation of variants formed by combining the four point mutations in the form of all theoretically possible double and triple mutants. It allowed a fitness pathway landscape to be constructed, comprising 4! = 24 experimental pathways which lead from (*S*)-selective WT PAMO to (*R*)-selective variant ZGZ-2 (Fig. 3.2) [22a]. Six of the 24 pathways have no local minima along the respective trajectories, meaning the absence of libraries which contain no variants with improved enantioselectivity. Eighteen pathways are characterized by at least one local minimum along the evolutionary trajectory. Analysis of the intermediate stages of all 24 pathways revealed strong cooperative mutational effects. Thus, much can be learned from deconvolution studies of this kind. In combination with MD/docking computations, they throw light on the origin of stereoselectivity while also illuminating the efficacy of ISM. It should be noted that this type of "constrained" fitness landscape [22a, 45] is different from constructing all theoretically possible pathways of an ISM scheme, as was implemented experimentally in the case of a four-site ISM system with 24 pathways [46]. The latter has been termed "unconstrained" fitness pathway landscape [17, 46], in which every trajectory leads to a different result (see Sect. 3.5).

3.4 Lipases and Esterases

Directed evolution has been extensively applied to lipases and esterases in order to manipulate stereoselectivity, substrate scope, and activity [3]. As delineated in the Introduction, the lipase from *Pseudomonas aeruginosa* (PAL) is the most systematically studied stereoselective enzyme in the field of directed evolution. In the final PAL study [16], which included deconvolution experiments, ISM was applied to the original model reaction involving the hydrolytic kinetic resolution of *rac*-**1** with preferential formation of (*S*)-**2** (Scheme 3.1). Based on the crystal structure of PAL, a three-site ISM scheme comprising two-residue CAST sites A, B, and C was considered. Following exploratory mutagenesis experiments, the best pathway proved to be WT → B → A which provided the triple mutant 1B2 (Leu162Asn/Met16Ala/Leu17Phe) showing a selectivity factor of $E=94$ (*S*) and notably enhanced activity: WT PAL ($k_{cat}=37\times10^{-3}$ s^{-1}; $k_{cat}/K_m=43.5$ s^{-1} M^{-1}) versus variant 1B2 ($k_{cat}=1374\times10^{-3}$ s^{-1}; $k_{cat}/K_m=4041$ s^{-1} M^{-1}) [16].

This result was achieved in the following way: In the first ISM step, site B (Leu159/Leu162) was randomized using NNK codon degeneracy, leading to single mutant Leu162Asn with $E=8$ (*S*). It was employed as the template for DNT-based saturation mutagenesis at site A (Met16/Leu17), which like NDT involves 12 amino acids as building blocks. The ISM pathway is illustrated in Scheme 3.13 [16].

Scheme 3.13 seems to suggest that the second mutational introduction, Met16Ala/Leu17Phe, contributes most to the overall result. However, this assumes classical mutational additivity [17]. Therefore, deconvolution was performed by generating

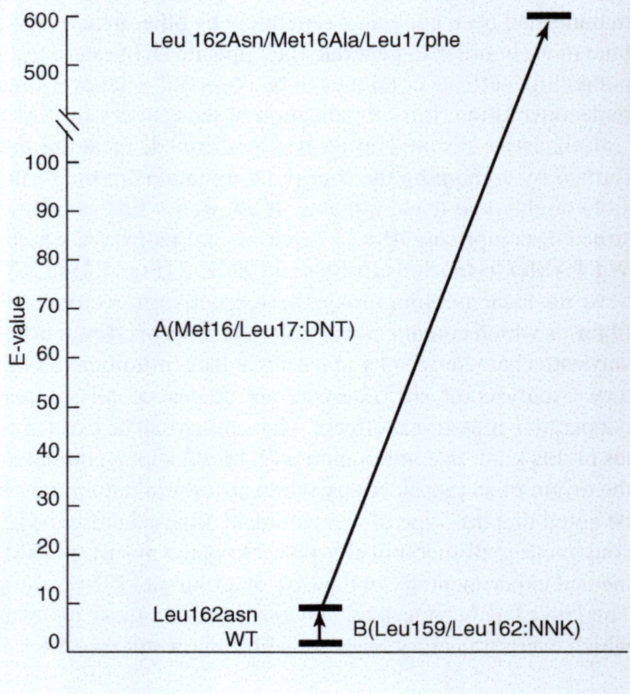

Scheme 3.13 ISM pathway WT → B → A in the directed evolution of PAL as the catalyst in the hydrolytic kinetic resolution of *rac*-**1** [16]

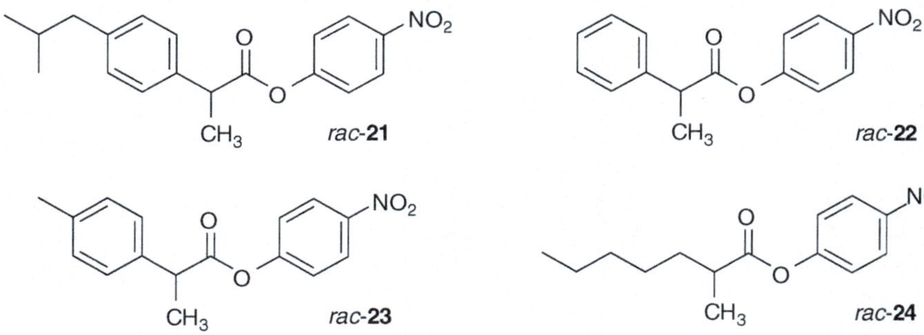

Scheme 3.14 Model compounds used in the directed evolution study of CALA-catalyzed hydrolytic kinetic resolution [21b]

the double mutant Met16Ala/Leu17Phe and testing it in the model reaction. Surprisingly, it proved to be a poor catalyst with $E = 2.6$ (*S*). Thus, a strong cooperative mutational effect is operating. An MD/docking analysis uncovered the molecular basis of this epistatic synergism. It involves the creation of an extended H-bond network as a consequence of the interaction of Leu162Asn in concert with Met16Ala/Leu17Phe [16].

This study required the screening of less than 10,000 transformants at a time when subsequent optimization of saturation mutagenesis strategies and techniques such as triple code saturation mutagenesis (TCSM) [22b, c] were not yet available. It is likely that TCSM as applied to PAL would require even less screening.

PAL served as a useful model system to test various mutagenesis strategies in a comparative manner. However, for several reasons, it is not likely to become a lipase of practical utility in organic chemistry. The situation is very different in the case of *Candida antarctica* lipase B (CALB), one of the most popular enzymes in biocatalysis [1]. It has been employed in ISM-based directed evolution in order to expand substrate scope and to invert enantioselectivity [47].

The homolog CALA has also been subjected to ISM-based directed evolution in order to accept chiral phenyl-substituted carboxylic acid esters [48], but the bulkier analogs of the ibuprofen type were not accepted. Therefore, a different strategy was tested following strategy 2 (Scheme 3.3). The plan was to use a single amino acid as building block (in addition to WT) at most of the positions of a nine-residue site in a single mutagenesis experiment [21b]. The hydrolytic kinetic resolution of *rac*-**21** was chosen for screening, and racemic esters **22–24** were also tested following the mutagenesis experiments (Scheme 3.14). The goal was enhancing activity and enantioselectivity while minimizing screening.

All experiments began with the use of the triple mutant F149Y/I150N/F233G as template, obtained in the earlier ISM-based study [48]. Substrate **21** was docked into the CALA binding pocket in the oxyanion form (tetrahedral intermediate at Ser184), leading to the identification of nine residues at the acyl binding region: positions 149, 150, 215, 221, 225, 233, 234, 237, and 431 (Fig. 3.3) [21b]. Other residues near this large CAST site were not considered because they are highly conserved as shown by

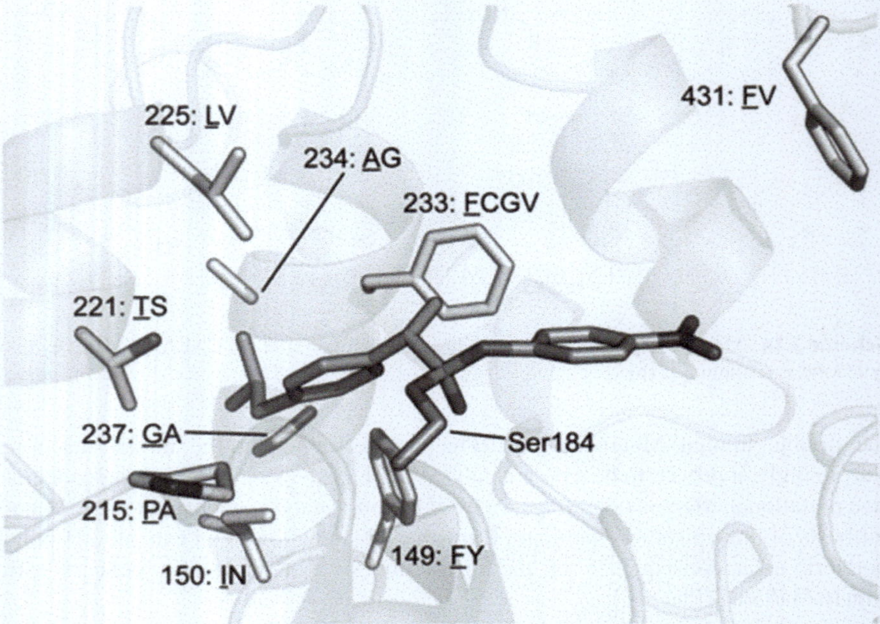

Fig. 3.3 View of CALA binding pocket harboring covalently bound substrate **21** as an oxyanion [21b]. Nine CAST residues with the respective reduced amino acid alphabet(s) used in saturation mutagenesis are shown, the original WT amino acids being underlined

an alignment analysis. In view of the PAMO study [40], this would not necessarily be mandatory. Structure-based decisions were made regarding the different reduced amino acid alphabets at the nine randomization positions.

Substrate **21** is sterically so demanding that it is not accepted by the triple mutant with acceptable rate. Thus, small amino acids were mostly chosen for saturation mutagenesis. In the earlier CALA study [48], mutations Phe149Tyr and Ile150Asn had been shown to be important for high enantioselectivity toward similar substrates. Therefore, at these positions, Tyr and Asn were chosen as the respective building blocks (Table 3.4) [21b]. At position 233, three amino acids were employed (in addition to WT). A certain degree of intuition was involved in some of the decisions.

A single highly condensed library was created using appropriately designed primers, followed by screening about 2400 transformants in the model reaction of *rac*-**21** (≈90% library coverage). Only a few variants proved to be active toward substrate *rac*-**21** [21b]. The best hit was a penta-substituted variant Thr221Ser/Leu225Val/Phe233Cys/Gly237Ala/Phe431Val in which four different amino acids were introduced at five different positions. It shows high (*S*)-stereoselectivity ($E = 100$). This and other CALA variants were tested in the hydrolytic kinetic resolution of the substrates **21–24**, which likewise resulted in acceptable levels of enantioselectivity [21b]. The alternative of applying conventional NNK codon degeneracy

Table 3.4 Reduced amino acid alphabets used in simultaneous saturation mutagenesis of a nine-residue randomization site of CALA [21b]

Position	WT residue	Alternative residue(s)
149	Phe	Tyr
150	Ile	Asn
215	Pro	Ala
221	Thr	Ser
225	Leu	Val
233	Phe	Cys/Gly/Val
234	Ala	Gly
237	Gly	Ala
431	Phe	Val

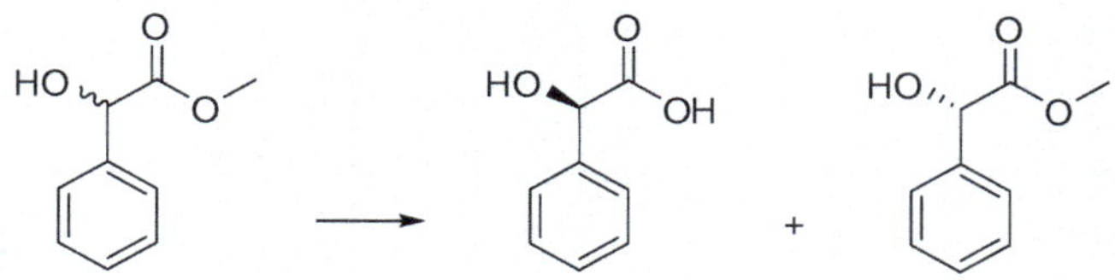

rac-**25** (*R*)-**26** (*S*)-**25**

Scheme 3.15 Model hydrolytic kinetic resolution catalyzed by mutants of the esterase RspE [50a]

encoding all 20 canonical amino acids would have required for 95% library coverage the screening of 10^{14} potentially enantioselective transformants. A limited number of deconvolution experiments revealed cooperative mutational effects. This suggested that the particular variant would not be accessible by ISM. The same strategy was later successfully applied to CALA as the catalyst in acylating kinetic resolution of chiral alcohols [49].

It can be concluded that both approaches to the use of reduced amino acid alphabets are successful, strategy 1 as well as strategy 2 as delineated in Scheme 3.3. It would be interesting to test strategy 1 using triple code saturation mutagenesis. Whether such a procedure would also allow reversal of enantioselectivity is currently a matter of speculation.

Esterases have also been targeted by directed evolution for improving stereoselectivity [3], recent studies utilizing various approaches including epPCR alone [50a], epPCR in combination with saturation mutagenesis [50b], saturation mutagenesis alone [50c, d], or site-specific mutagenesis in combination with saturation mutagenesis [50e]. One study is featured here which relied solely on epPCR. The carboxyl esterase from *Rhodobacter sphaeroides* (RspE) was subjected to three recursive rounds of epPCR at low mutation rate, the model reaction being the hydrolytic kinetic resolution of methyl mandelate (*rac*-**25**) with preferential formation of the carboxylic acid (*R*)-**26** (Scheme 3.15) [50a]. WT RspE shows a selectivity factor of $E=3.1(R)$.

In each epPCR cycle, 4000–6000 transformants were screened for activity using an on-plate pH-dependent color test followed by conventional ee-determination. In this way, it was possible to boost (*R*)-selectivity to $E=30.3$ (Scheme 3.16) [50a]. This study is a new example of iterative epPCR in the successful attempt to evolve

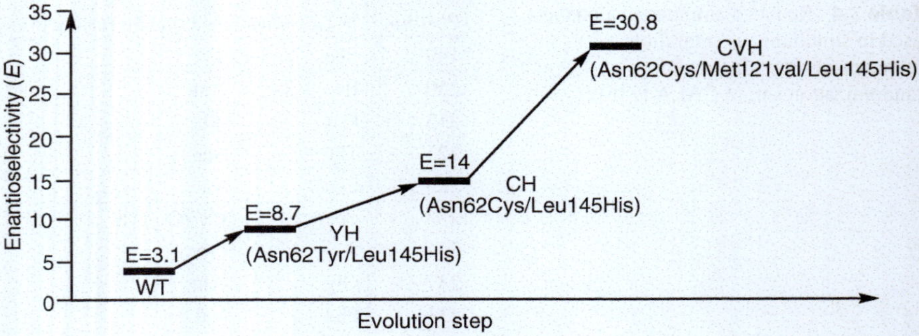

Scheme 3.16 Hydrolytic kinetic resolution of *rac*-**25** with preferential formation of (*R*)-**26** (Scheme 3.15), catalyzed by RspE mutants which were evolved by recursive epPCR [50a]

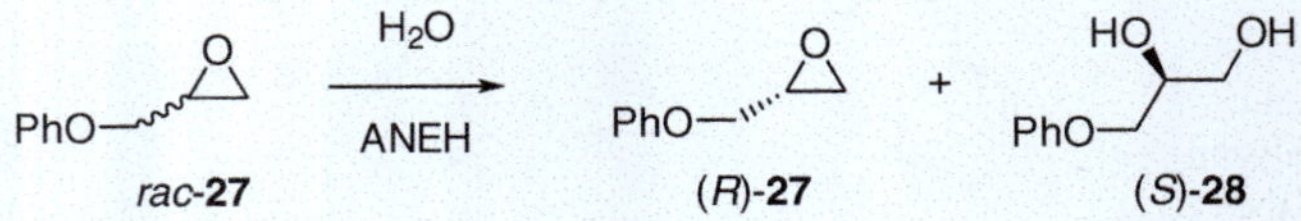

Scheme 3.17 ANEH-catalyzed hydrolytic kinetic resolution used in the construction of a complete four-site ISM system featuring 24 evolutionary pathways [46]

enhanced stereoselectivity of an esterase. Reversal of enantioselectivity was not reported, but the best mutant in the model reaction showing $E=30.3$ was used in the kinetic resolution of structurally related substrates including acetates of chiral alcohols with selectivity factors of up to $E=92$. It would be interesting to see how well CASTing/ISM would perform in this enzyme system.

3.5 Epoxide Hydrolases

Several recent protein engineering studies of epoxide hydrolases have contributed to methodology development in laboratory evolution [3f, 5]. The goal of one of them was the exploration of all 24 pathways of a four-site ISM scheme [46]. To date, it is the only case of complete exploration of such an ISM system. The hydrolytic kinetic resolution of *rac*-**27** was chosen as the model reaction, catalyzed by mutants of the epoxide hydrolase from *Aspergillus niger* (ANEH) (Scheme 3.17). WT is a poor catalyst in this reaction ($E=4.6$ in slight favor of (*S*)-**28**).

Four CAST randomization sites were considered on the basis of the WT ANEH crystal structure, each comprising two residues. Subsequently all saturation mutagenesis libraries were constructed using NDT codon degeneracy encoding 12 amino acids (Phe, Leu, Ile, Val, Tyr, His, Asn, Asp, Cys, Arg, Ser, and Gly), which is a balanced "cocktail" of polar/nonpolar, charged/non-charged, hydrophobic/non-hydrophobic, and aromatic/nonaromatic amino acids as building blocks. All 24 pathways provided in the final saturation mutagenesis rounds notably improved variants characterized by different sequences [46]. Some pathways proved to be more productive than others. The 12 best pathways resulting in selectivity factors in the range $E = 78$–159 are shown in Scheme 3.18a; the 12 pathways leading to the least improved variants ($E = 28$–78) are pictured in Scheme 3.18b. These results suggest that if the researcher is faced with the question of choosing an appropriate ISM pathway, an arbitrary choice has a high probability of providing improved variants. This explains why in essentially all ISM studies reported thus far arbitrarily chosen pathways led to notably improved variants [5]. Nevertheless, superior variants may have been missed. It can therefore be concluded that ISM systems should be designed in a way that involves less pathways and therefore less mutant libraries, which correlates with a lower number of decisions for the researcher to make. Indeed, methodology development since the publication of this study has focused, inter alia, on step-economy [21, 22].

Noteworthy is another feature of the experimental results collected in Scheme 3.18. In several ISM pathways, libraries occurred which failed to harbor any improved variants. This phenomenon signals a local minimum in the fitness landscape. Such "dead ends" may occur in any directed evolution project irrespective of the mutagenesis method [3]. In the present case, ISM was not abandoned, but an inferior mutant was used as the template in the subsequent saturation mutagenesis experiment at the next randomization site. This trick led to notably improved mutants. This unique way of escaping from a local minimum is reminiscent of neutral drift [4d, 51] or the Eigen concept of quasi-species [52a]. The latter has been invoked occasionally in directed evolution studies [52b].

Maximal step-economy would involve the generation of a single saturation mutagenesis library, which should be small and ideally harbor both (*R*)- and (*S*)-selective mutants. If the degree of stereoselectivity should not be fully satisfactory, one ISM round could then be undertaken for fine-tuning. Along these lines, the directed evolution of a different epoxide hydrolase was reported recently, namely, limonene epoxide hydrolase (LEH) as the catalyst in the model hydrolytic desymmetrization of cyclohexene oxide (**29**) with formation of (*R*,*R*)- and (*S*,*S*)-**30** (Scheme 3.19) [21a]. WT LEH shows poor enantioselectivity with minimal preference for (*S*,*S*)-**30**, the enantiomeric ratio (er) amounting to a mere 48:52 (4% ee).

Based on the crystal structure of WT LEH, ten CAST residues were identified for saturation mutagenesis (Leu74, Phe75, Met78, Ile80, Leu103, Leu114, Ile116, Phe134, Phe139, and Leu147) [21a]. Tyr53 activates the substrate by forming an H-bond to the epoxide O-atom, Asp101 being mainly responsible for positioning water which initiates the rate-determining S_N2 reaction with ring opening.

Saturation mutagenesis using NNK codon degeneracy (20-amino-acid alphabet) or NDT codon degeneracy (12-amino-acid alphabet) would require the screening of about 10^{15} or 10^{11} transformants for 95% coverage, respectively. The use the smallest amino acid alphabet, a single amino acid, would call for only ≈3000 transformants. This would reduce structural diversity dramatically, suggesting that such a strategy would fail. Nevertheless, it could be successful if the right decision is made concerning the choice of the amino acid. Such an approach constitutes single codon

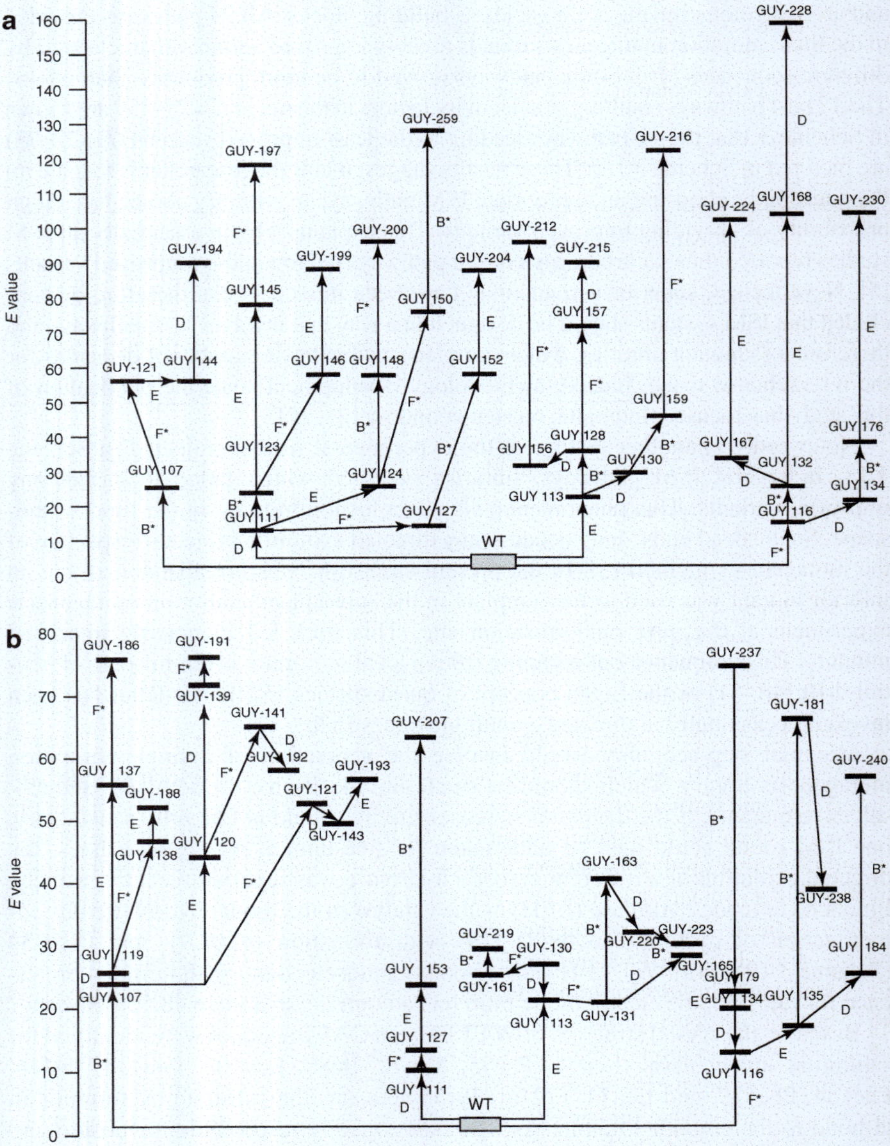

Scheme 3.18 Experimental exploration of a complete 24-pathway ISM system involving the ANEH-catalyzed hydrolytic kinetic resolution of *rac*-**27** (Scheme 3.17) [46]. (**a**) Portion of the 24-pathway ISM scheme featuring the 12 most productive pathways leading to ANEH variants displaying $E=78–159$ (*S*); (**b**) portion of the 24-pathway ISM scheme showing the 12 least productive pathways providing ANEH variants with $E=28–78$ (*S*)

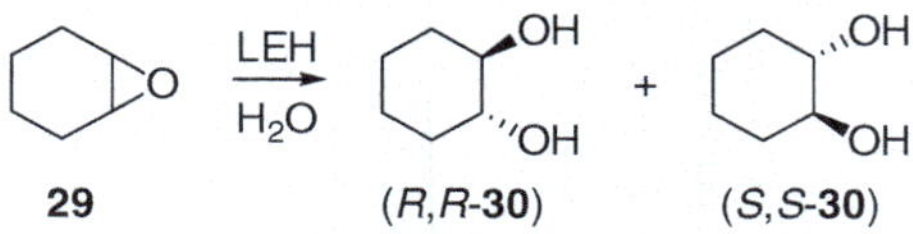

Scheme 3.19 Hydrolytic desymmetrization catalyzed by LEH mutants evolved by applying single codon saturation mutagenesis (SCSM) [21a] and more recently using triple codon saturation mutagenesis (TCSM) [22b]

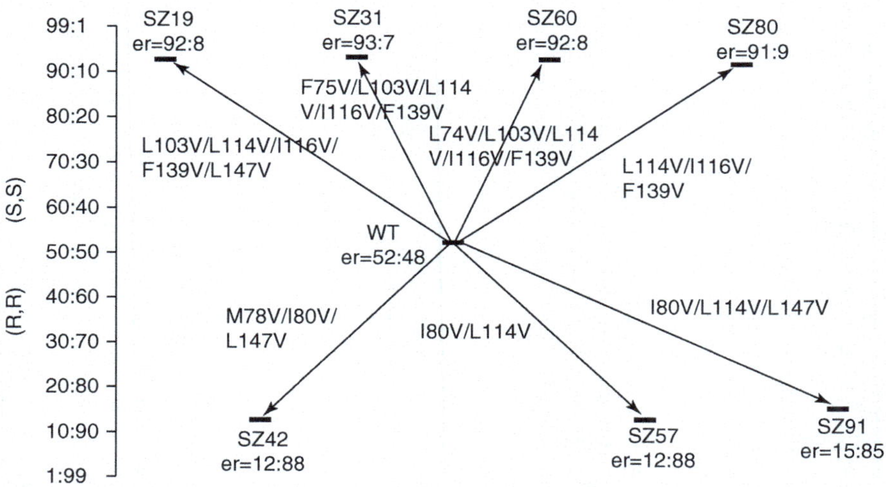

Scheme 3.20 Results of generating one single codon saturation mutagenesis (SCSM) library on the basis of valine as the sole building block, hydrolytic desymmetrization of cyclohexene oxide (**29**) serving as the model reaction [21a]

saturation mutagenesis (SCSM) as part of strategy 1 in Scheme 3.3 and was first tested in the directed evolution of LEH [21a]. In doing so, the choice of the single building block was crucial. The crystal structure of LEH reveals that most of the amino acids surrounding the binding pocket are hydrophobic. Therefore, valine, having a hydrophobic and sterically demanding side chain, was chosen as the smallest reduced amino acid alphabet for SCSM in a single saturation mutagenesis experiment at the ten-residue site. The results following the screening of about 3200 transformants are shown in Scheme 3.20.

It can be seen that both (*R*,*R*)- and (*S*,*S*)-selective occur in one and the same small but high-quality library. The number of introduced valines ranges between two and five, depending upon the particular variant. In a single ISM step, fine-tuning was performed, boosting the enantiomeric ratio to 98:2 (96% ee). Crystal structures of the variants with and without product in combination with MD/docking computations uncovered the origin of stereoselectivity [21a]. In a control experiment, the use

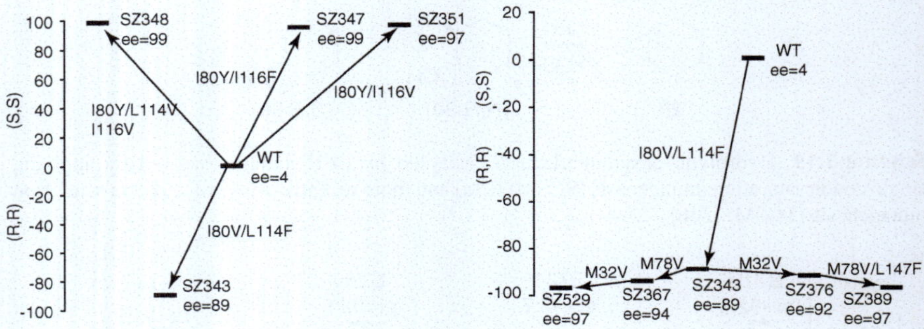

Scheme 3.21 The results of triple code saturation mutagenesis (TCSM) when applied to LEH as the catalyst in the hydrolytic desymmetrization of cyclohexene oxide (**29**) [22b]

of serine failed completely, which supports the original hypothesis regarding the successful choice of valine.

Single code saturation mutagenesis cannot be expected to be general. Therefore, triple code saturation mutagenesis (TCSM) was developed [22b, c]. A reduced amino acid alphabet of three members at a ten-residue randomization site would require for 95% library coverage excessive screening. Therefore, if so many CAST residues are chosen, they need to be grouped into smaller randomization sites. As part of strategy 1 (Scheme 3.3), this approach was first tested with LEH as the catalyst in the same model reaction involving the desymmetrization of epoxide **29** (Scheme 3.19). The ten previously identified CAST residues were grouped into three randomization sites: (A) (V83/L114/I116), (B) (L74/M78/L147), and (C) (M32/L35/L103) [22b].

By considering the crystal structure of LEH, it was logical to test valine, phenylalanine, and tyrosine as the reduced amino acid alphabet in TCSM, even if the previous saturation mutagenesis experiments using these amino acids as building blocks had not been performed. The V-F-Y triple code was applied to the three randomization sites A, B, and C in separate experiments, requiring for 95% library coverage the screening of only 576, 192, and 192 transformants, respectively. The best results were obtained in library A (Scheme 3.21). As before, one and the same library harbors both (*R*,*R*)- and (*S*,*S*)-selective variants, but this time much better results were obtained. (*S*,*S*)-selectivity amounts to 96–99% ee in three different variants, while the best (*R*,*R*)-mutant results in 89% ee which was boosted to 97% ee by ISM. X-ray structures of selected mutants, MD/docking computations, and kinetic data were reported, which shed light on the origin of mutational effects [22b].

3.6 Transaminase

Transaminases are enzymes that catalyze the reductive amination of prochiral ketones with formation of the respective chiral primary amines [1]. An impressive industrial example of directed evolution was reported by Codexis, specifically in the

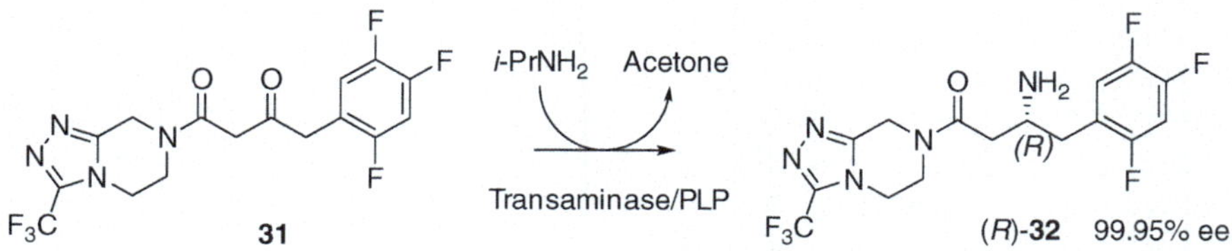

Scheme 3.22 Asymmetric reductive amination with formation of the antidiabetic therapeutic drug sitagliptin (*R*)-**32**, catalyzed by mutants of transaminase ATA-117 [53]

asymmetric reductive amination of ketone **31** with formation of the antidiabetic drug sitagliptin (**32**) (Scheme 3.22) [53].

The transaminase ATA-117 was chosen as the enzyme, which is related to the structurally well-characterized homolog from *Arthrobacter* sp. Both were known to be (*R*)-selective in the reductive amination of methyl ketones and small cyclic ketones. ATA-117 proved to be (*R*)-selective in the desired reaction as well, but activity was extremely low [53]. Thus, the goal was to enhance activity while maintaining stereoselectivity. At the beginning of the project, the industrial researchers did not use the "real" substrate **31**, but first resorted to in vitro coevolution which means testing simpler but still structurally related compounds (substrate walking) [54]. With the help of a homology model of ATA-117, docking computations were carried out which allowed reasonable choices for randomizing sites lining the binding pocket (CAST sites). NNK-based saturation mutagenesis led to variant S223P with an 11-fold increase in activity in the reaction of a simplified model ketone. The mutant was then employed as a template for ISM experiments using the "real" substrate **31** [53]. Docking computations indicated that the trifluoromethyl group interacts with residues V69, F122, T283, and A284. Therefore, four NNK-based saturation mutagenesis libraries were generated separately at these four positions. Moreover, a combinatorial library using several residues simultaneously was created. The combinatorial library harbored an active variant characterized by four point mutations lining "small" and "large" parts of the binding pocket. Double mutants F122I/V69G, F122I/A284G, F122V/V69G, F122V/A284G, F122L/V69G, and F122L/A284G were the best hits. They all contain the parent mutation S223P. Activity was still quite low, yet without point mutation S223P, no activity whatsoever resulted, as shown by a deconvolution experiment.

The gene of the most active variant was then used as the parent for the next round of ISM, followed by combining the beneficial mutations from the small-pocket and large-pocket saturation mutagenesis libraries. This provided a variant with 12 point mutations and a 75-fold increase in activity. Subsequently, 11 additional rounds of mutagenesis/screening were performed using DNA shuffling, epPCR, rational design, and saturation mutagenesis at second-sphere sites. Process development was also performed in parallel. A total of 36,480 transformants were screened using an LC/MS/MS screen. The best variant was shown to have 27 point mutations. In 50% DMSO, this catalyst converts 200 g/L of the prositagliptin ketone **31** to sitagliptin (**32**) with >99.95% ee (*R*) [53].

The catalytic performance of the best ATA-117 variant under operating conditions underscores the success of the project. However, it is difficult to assess the efficacy of the applied mutagenesis approach. It is not clear whether the order of the mutagenesis cycles in the overall multistep process was actually planned or whether corrections in the strategy had to be undertaken. Why were the particular mutagenesis events chosen in the reported order?

3.7 Alternative Mutagenesis Approaches

As noted in the Introduction, such mutagenesis techniques such as epPCR or DNA shuffling can also be applied in order to manipulate the stereoselectivity of enzymes [3, 5, 10], but in several comparative studies, these have been shown to be less efficient [16, 18]. Nevertheless, following several cycles of saturation mutagenesis, it may be useful to add one round of epPCR for activity enhancement [26]. Other approaches such as neutral drift [4d, 51], domain swapping [55], and circular permutation [56] likewise deserve mention, but these strategies have not been applied very often to the evolution of stereoselectivity. Neutral drift can be used for identifying superior starting points for protein engineering by exploring accessible sequence space on the basis of recursive cycles of (random) mutagenesis and screening or selection. The technique identifies accumulating mutations which are neutral for the native function but which may prove to be useful for novel catalytic profiles. An example is the evolution of promiscuity by turning a ß-glucuronidase into a ß-galactosidase [51c].

Domain swapping was originally used in the study of natural evolution and for addressing mechanistic questions in protein science, but it has also been applied occasionally in directed evolution [3, 55]. For example, a glycosyltransferase was engineered for different substrate specificity [55b]. However, it is not well suited for enhancing or inverting stereoselectivity. A special form of this technique is circular permutation in which the N- and C-termini of an enzyme are relocated [56]. A seminal example concerns the engineering of enhanced activity of the lipase from *Candida antarctica* (CALB) [56a, b]. New locations of the N- and C-termini in WT CALB were designed to occur at positions 282 and 283 in hope of influencing local backbone flexibility and perhaps active site accessibility. Indeed, this led to higher activity [56b], but enantioselectivity was not addressed at the time nor in a subsequent review [56a].

Conclusions and Perspectives

This chapter provides a summary of the most important recent methodological developments in the directed evolution of stereoselective enzymes as catalysts in synthetic organic chemistry and biotechnology. Rather than being comprehensive, five different types of enzymes were chosen which illustrate important recent advances in strategies and methods.

It can safely be concluded that structure-guided saturation mutagenesis at sites in vicinity of the active site (CASTing) is a distinctly logical approach to reshape the bonding pockets of enzymes in the quest to manipulate stereoselectivity, substrate scope, and activity. The use of reduced amino acid alphabets at relatively large CAST randomization sites has emerged as the preferred strategy, allowing for the creation of small yet high-quality mutant libraries requiring less screening than in the past. In this respect, triple code saturation mutagenesis (TCSM) appears to be an optimal compromise between limited structural diversity and screening effort (bottleneck of directed evolution). TCSM allows for step-economy, since initial mutant libraries already contain notably improved (*R*)- as well as (*S*)-variants. If needed, further tuning is possible by iterative saturation mutagenesis (ISM). It has been demonstrated that it is better to strive for higher library coverage at reduced structural diversity regarding the chosen amino acid alphabet rather to maintain maximum structural diversity using NNK codon degeneracy at the same screening effort correlating with considerably lower library coverage [13a].

One of the remaining fundamental challenges in directed evolution is the development of a general guide for optimizing more than one or two enzyme parameters, e.g., stereo-/regioselectivity, substrate scope, rate, and thermostability. Efforts are underway in several laboratories.

References

1. (a) Drauz K, Gröger H, May O (eds) (2012) Enzyme catalysis in organic synthesis, 3rd edn. Wiley-VCH, Weinheim; (b) Faber K (2011) Biotransformations in organic chemistry, 6th edn. Springer, Heidelberg; (c) Liese A, Seelbach K, Wandrey C (eds) (2006) Industrial biotransformations, Wiley-VCH, Weinheim.
2. (a) Pleiss J (2012) Rational design of enzymes. In: Drauz K, Gröger H, May O (eds) Enzyme catalysis in organic synthesis, 3rd edn. Wiley-VCH, Weinheim, p 89–117; (b) Ema T, Nakano Y, Yoshida D, Kamata S, Sakai T (2012) Redesign of enzyme for improving catalytic activity and enantioselectivity toward poor substrates: manipulation of the transition state. Org Biomol Chem 10:6299–6308; (c) Steiner K, Schwab H (2012) Recent advances in rational approaches for enzyme engineering. Comput Struct Biotechnol J 2:e201209010.
3. Reviews of directed evolution: (a) Bommarius AS (2015) Biocatalysis, a status report. Annu Rev Chem Biomol Eng 6:319–345; (b) Denard CA, Ren H, Zhao H (2015) Improving and repurposing biocatalysts via directed evolution. Curr Opin Chem Biol 25:55–64; (c) Currin A, Swainston N, Day PJ, Kell DB (2015) Synthetic biology for the directed evolution of protein biocatalysts: navigating sequence space intelligently. Chem Soc Rev 44:1172–1239; (d) Gillam EMJ, Copp JN, Ackerley DF (eds) (2014) Directed evolution library creation. In: Methods in molecular biology, Humana Press, Totowa; (e) Widersten M (2014) Protein engineering for development of new hydrolytic biocatalysts. Curr Opin Chem Biol 21:42–47; (f) Reetz MT (2012) Directed evolution of enzymes. In Drauz K, Gröger H, May O (eds) Enzyme catalysis in organic synthesis, 3rd edn. Wiley-VCH, Weinheim, p 119–190
4. Review of directed evolution of protein thermostability: (a) Arnold FH (1998) Design by directed evolution. Acc Chem Res 31:125–131; (b) Petrounia IP, Arnold FH (2000) Designed evolution of enzymatic properties. Curr Opin Biotechnol 11: 325–330; (c) Polizzi KM, Bommarius AS, Broering JM, Chaparro-Riggers JF (2007) Stability of biocatalysts. Curr Opin Chem Biol 11:220–225; (d) Tokuriki N, Tawfik DS (2009) Stability effects of mutations and protein evolvability. Curr Opin Struct Biol 19:596–604

5. Reviews of directed evolution of stereoselectivity[3f]: (a) Reetz MT (2011) Laboratory evolution of stereoselective enzymes: a prolific source of catalysts for asymmetric reactions. Angew Chem Int Ed 50:138–174; (b) Reetz MT (2013) Biocatalysis in organic chemistry and biotechnology: past, present, and future. J Am Chem Soc 135:12480–12496
6. Reetz MT, Zonta A, Schimossek K, Liebeton K, Jaeger KE (1997) Creation of enantioselective biocatalysts for organic chemistry by in vitro evolution. Angew Chem Int Ed Eng 36:2830–2832
7. Reetz MT, Wilensek S, Zha D, Jaeger KE (2001) Directed evolution of an enantioselective enzyme through combinatorial multiple-cassette mutagenesis. Angew Chem Int Ed 40:3589–3591
8. Reetz MT, Puls M, Carballeira JD, Vogel A, Jaeger KE, Eggert T, Thiel W, Bocola M, Otte N (2007) Learning from directed evolution: further lessons from theoretical investigations into cooperative mutations in lipase enantioselectivity. Chembiochem 8:106–112
9. Zha DX, Wilensek S, Hermes M, Jaeger KE, Reetz MT (2001) Complete reversal of enantioselectivity of an enzyme-catalyzed reaction by directed evolution. Chem Commun 24:2664–2665
10. Reetz MT (2004) Controlling the enantioselectivity of enzymes by directed evolution: practical and theoretical ramifications. Proc Natl Acad Sci U S A 101:5716–5722
11. Reetz MT, Bocola M, Carballeira JD, Zha DX, Vogel A (2005) Expanding the range of substrate acceptance of enzymes: combinatorial active-site saturation test. Angew Chem Int Ed 44:4192–4196
12. (a) Reetz MT, Wang LW, Bocola M (2006) Directed evolution of enantioselective enzymes: iterative cycles of CASTing for probing protein-sequence space. Angew Chem Int Ed 45:1236–1241; (b) Reetz MT (2005) Evolution im Reagenzglas: Neue Perspektiven für die Weiße Biotechnologie. Tätigkeitsberichte der Max-Planck-Gesellschaft 327–331
13. (a) Reetz MT, Kahakeaw D, Lohmer R (2008) Addressing the numbers problem in directed evolution. Chembiochem 9:1797–1804; (b) Clouthier CM, Kayser MM, Reetz MT (2006) Designing new Baeyer-Villiger monooxygenases using restricted CASTing. J Org Chem 71:8431–8437
14. (a) Firth AE, Patrick WM (2008) GLUE-IT and PEDEL-AA: new programmes for analyzing protein diversity in randomized libraries. Nucleic Acids Res 36 (Web Server issue):W281–285; (b) Denault M, Pelletier JN (2007) Protein library design and screening: working out the probabilities. In: Arndt KM, Müller KM (eds) Protein engineering protocols, Humana Press, Totowa, p 127–154
15. Nov Y (2012) When second best is good enough: another probabilistic look at saturation mutagenesis. Appl Environ Microbiol 78:258–262
16. Reetz MT, Prasad S, Carballeira JD, Gumulya Y, Bocola M (2010) Iterative saturation mutagenesis accelerates laboratory evolution of enzyme stereoselectivity: rigorous comparison with traditional methods. J Am Chem Soc 132:9144–9152
17. Reetz MT (2013) The importance of additive and non-additive mutational effects in protein engineering. Angew Chem Int Ed 52:2658–2666
18. Parikh MR, Matsumura I (2005) Site-saturation mutagenesis is more efficient than DNA shuffling for the directed evolution of beta-fucosidase from beta-galactosidase. J Mol Biol 352:621–628
19. Acevedo-Rocha CG, Hoebenreich S, Reetz MT (2014) Iterative saturation mutagenesis: a powerful approach to engineer proteins by systematically simulating Darwinian evolution. Methods Mol Biol 1179:103–128
20. Sun Z, Wikmark Y, Bäckvall J-E, Reetz MT (2016) New concepts for increasing the efficiency in directed evolution of stereoselective enzymes. Chem Eur J 22:5046–5054
21. (a) Sun Z, Lonsdale R, Kong XD, Xu JH, Zhou J, Reetz MT (2015) Reshaping an enzyme binding pocket for enhanced and inverted stereoselectivity: use of smallest amino acid alphabets in directed evolution. Angew Chem Int Ed 54:12410–12415; (b) Sandström AG, Wikmark Y, Engström K, Nyhlen J, Bäckvall JE (2012) Combinatorial reshaping of the Candida antarctica lipase A substrate pocket for enantioselectivity using an extremely condensed library. Proc Natl Acad Sci U S A 109:78–83

22. (a) Zhang ZG, Lonsdale R, Sanchis J, Reetz MT (2014) Extreme synergistic mutational effects in the directed evolution of a baeyer-villiger monooxygenase as catalyst for asymmetric sulfoxidation. J Am Chem Soc 136:17262–17272; (b) Sun Z, Lonsdale R, Wu L, Li G, Li A, Wang J, Zhou J, Reetz MT (2016) Structure-guided triple code saturation mutagenesis: efficient tuning of the stereoselectivity of an epoxide hydrolase. ACS Catal 6:1590–1597; (c) Sun Z, Lonsdale R, Ilie A, Li G, Zhou J, Reetz MT (2016) Catalytic asymmetric reduction of difficult-to-reduce ketones: triple code saturation mutagenesis of an alcohol dehydrogenase. ACS Catal 6:1598–1605
23. (a) Bougioukou DJ, Kille S, Taglieber A, Reetz MT (2009) Directed Evolution of an enantioselective enoate-reductase: testing the utility of iterative saturation mutagenesis. Adv Synth Catal 351:3287–3305; (b) Sullivan B, Walton AZ, Stewart JD (2013) Library construction and evaluation for site saturation mutagenesis. Enzym Microb Technol 53:70–77
24. Polizzi KM, Parikh M, Spencer CU, Matsumura I, Lee JH, Realff MJ, Bommarius AS (2006) Pooling for improved screening of combinatorial libraries for directed evolution. Biotechnol Prog 22:961–967
25. Acevedo-Rocha CG, Reetz MT, Nov Y (2015) Economical analysis of saturation mutagenesis experiments. Sci Rep 5:10654
26. Kwan DH, Constantinescu I, Chapanian R, Higgins MA, Kotzler MP, Samain E, Boraston AB, Kizhakkedathu JN, Withers SG (2015) Toward efficient enzymes for the generation of universal blood through structure-guided directed evolution. J Am Chem Soc 137:5695–5705
27. (a) Whitehouse CJ, Bell SG, Wong LL (2012) P450(BM3) (CYP102A1): connecting the dots. Chem Soc Rev 41:1218–1260; (b) Fasan R (2012) Tuning P450 enzymes as oxidation catalysts. ACS Catal 2:647–666; (c) Bernhardt R, Urlacher VB (2014) Cytochromes P450 as promising catalysts for biotechnological application: chances and limitations. Appl. Microbiol Biotechnol 98:6185–6203; (d) Holtmann D, Fraaije MW, Arends IW, Opperman DJ, Hollmann F (2014) The taming of oxygen: biocatalytic oxyfunctionalisations. Chem Commun 50:13180–13200; (e) Roiban GD, Reetz MT (2015) Expanding the toolbox of organic chemists: directed evolution of P450 monooxygenases as catalysts in regio- and stereoselective oxidative hydroxylation. Chem Commun 51:2208–2224; (f) Girvan HM, Munro AW (2016) Applications of microbiol cytochrome P450 enzymes in biotechnology and synthetic biology. Curr Opin Chem Biol 31:136–145
28. (a) Tang WL, Li Z, Zhao H (2010) Inverting the enantioselectivity of P450pyr monooxygenase by directed evolution. Chem Commun 46:5461–5463; (b) Pham SQ, Pompidor G, Liu J, Li XD, Li Z (2012) Evolving P450pyr hydroxylase for highly enantioselective hydroxylation at non-activated carbon atom. Chem Commun 48:4618–4620
29. Lonsdale R, Harvey JN, Mulholland AJ (2010) Inclusion of dispersion effects significantly improves accuracy of calculated reaction barriers for cytochrome P450 catalyzed reactions. J Phys Chem Lett 1:3232–3237
30. Agudo R, Roiban GD, Reetz MT (2012) Achieving regio- and enantioselectivity of P450-catalyzed oxidative CH activation of small functionalized molecules by structure-guided directed evolution. Chembiochem 13:1465–1473
31. Kille S, Zilly FE, Acevedo JP, Reetz MT (2011) Regio- and stereoselectivity of P450-catalysed hydroxylation of steroids controlled by laboratory evolution. Nat Chem 3:738–743
32. Le-Huu P, Heidt T, Claasen B, Laschat S, Urlacher VB (2015) Chemo-, regio-, and stereoselective oxidation of the monocyclic diterpenoid beta-cembrenediol by P450 BM3. ACS Catal 5:1772–1780
33. Zhang K, Shafer BM, Demars MD 2nd, Stern HA, Fasan R (2012) Controlled oxidation of remote sp3 C-H bonds in artemisinin via P450 catalysts with fine-tuned regio- and stereoselectivity. J Am Chem Soc 134:18695–18704
34. (a) Chen MM, Snow CD, Vizcarra CL, Mayo SL, Arnold FH (2012) Comparison of random mutagenesis and semi-rational designed libraries for improved cytochrome P450 BM3-catalyzed hydroxylation of small alkanes. Prot Eng Des Sel 25:171–178; (b) Ritter C, Nett N, Acevedo-Rocha CG, Lonsdale R, Kraling K, Dempwolff F, Hoebenreich S, Graumann PL, Reetz MT, Meggers E (2015) Bioorthogonal enzymatic activation of caged compounds. Angew

Chem Int Ed 54:13440–13443; (c) Dennig A, Lulsdorf N, Liu H, Schwaneberg U (2013) Regioselective o-hydroxylation of monosubstituted benzenes by P450 BM3. Angew Chem Int Ed 52:8459–8462; (d) Agudo R, Roiban GD, Lonsdale R, Ilie A, Reetz MT (2015) Biocatalytic route to chiral acyloins: P450-catalyzed regio- and enantioselective alpha-hydroxylation of ketones. J Org Chem 80:950–956; (e) Hoebenreich S, Zilly FE, Acevedo-Rocha CG, Zilly M, Reetz MT (2015) Speeding up directed evolution: combining the advantages of solid-phase combinatorial gene synthesis with statistically guided reduction of screening effort. ACS Synth Biol 4:317–331; (f) Roiban GD, Agudo R, Reetz MT (2014) Cytochrome P450 catalyzed oxidative hydroxylation of achiral organic compounds with simultaneous creation of two chirality centers in a single C-H activation step. Angew Chem Int Ed 53:8659–8663; (g) Hu S, Huang J, Mei LH, Yu Q, Yao SJ, Jin ZH (2010) Altering the regioselectivity of cytochrome P450 BM-3 by saturation mutagenesis for the biosynthesis of indirubin. J Mol Catal B-Enzym 67:29–35; (h) Nguyen KT, Virus C, Gunnewich N, Hannemann F, Bernhardt R (2012) Changing the regioselectivity of a P450 from C15 to C11 hydroxylation of progesterone. Chembiochem 13:1161–1166; (i) Yang Y, Liu J, Li Z (2014) Engineering of P450pyr hydroxylase for the highly regio- and enantioselective subterminal hydroxylation of alkanes. Angew Chem Int Ed 53:3120–3124

35. (a) Reinen J, Vredenburg G, Klaering K, Vermeulen NPE, Commandeur JNM, Honing M, Vos JC (2015) Selective whole-cell biosynthesis of the designer drug metabolites 15-or 16-ß-hydroxynorethisterone by engineered Cytochrome P450 BM3 mutants. J Mol Catal B-Enzym 121:64–74; (b) Bruhlmann F, Bosijokovic B, Ullmann C, Auffray P, Fourage L, Wahler D (2013) Directed evolution of a 13-hydroperoxide lyase (CYP74B) for improved process performance. J Biotechnol 163:339–345
36. Reetz MT, Brunner B, Schneider T, Schulz F, Clouthier CM, Kayser MM (2004) Directed evolution as a method to create enantioselective cyclohexanone monooxygenases for catalysis in Baeyer-Villiger reactions. Angew Chem Int Ed 43:4075–4078
37. Mihovilovic MD, Rudroff F, Winninger A, Schneider T, Schulz F, Reetz MT (2006) Microbial Baeyer-Villiger oxidation: stereopreference and substrate acceptance of cyclohexanone monooxygenase mutants prepared by directed evolution. Org Lett 8:1221–1224
38. (a) Fraaije MW, Wu J, Heuts DP, van Hellemond EW, Spelberg JH, Janssen DB (2005) Discovery of a thermostable Baeyer-Villiger monooxygenase by genome mining. Appl Microbiol Biotechnol 66:393–400; (b) Malito E, Alfieri A, Fraaije MW, Mattevi A (2004) Crystal structure of a Baeyer-Villiger monooxygenase. Prod Natl Acad Sci 101:13157–13162
39. Reviews of directed evolution of Baeyer-Villiger monooxygenases: (a) Zhang ZG, Parra LP, Reetz MT (2012) Protein engineering of stereoselective Baeyer-Villiger monooxygenases. Chem Eur J 18:10160–10172; (b) Balke K, Kadow M, Mallin H, Sass S, Bornscheuer UT (2012) Discovery, application and protein engineering of Baeyer-Villiger monooxygenases for organic synthesis. Org Biomol Chem 10:6249–6265
40. Reetz MT, Wu S (2008) Greatly reduced amino acid alphabets in directed evolution: making the right choice for saturation mutagenesis at homologous enzyme positions. Chem Commun 43:5499–5501
41. Kirschner A, Bornscheuer UT (2006) Kinetic resolution of 4-hydroxy-2-ketones catalyzed by a Baeyer-Villiger monooxygenase. Angew Chem 45:7004–7006
42. Wu S, Acevedo JP, Reetz MT (2010) Induced allostery in the directed evolution of an enantioselective Baeyer-Villiger monooxygenase. Prod Natl Acad Sci USA 107:2775–2780
43. Ichiye T, Karplus M (1991) Collective motions in proteins – a covariance analysis of atomic fluctuations in molecular-dynamics and normal mode simulations. Proteins 11:205–217
44. Tang L, Gao H, Zhu X, Wang X, Zhou M, Jiang R (2012) Construction of "small-intelligent" focused mutagenesis libraries using well-designed combinatorial degenerate primers. BioTechniques 52:149–158
45. Reetz MT, Sanchis J (2008) Constructing and analyzing the fitness landscape of an experimental evolutionary process. Chembiochem 9:2260–2267
46. Gumulya Y, Sanchis J, Reetz MT (2012) Many pathways in laboratory evolution can lead to improved enzymes: how to escape from local minima. Chembiochem 13:1060–1066

47. Wu Q, Soni P, Reetz MT (2013) Laboratory evolution of enantiocomplementary Candida antarctica lipase B mutants with broad substrate scope. J Am Chem Soc 135:1872–1881
48. Engström K, Nyhlen J, Sandström AG, Bäckvall JE (2010) Directed evolution of an enantioselective lipase with broad substrate scope for hydrolysis of alpha-substituted esters. J Am Chem Soc 132:7038–7042
49. Wikmark Y, Svedendahl Humble M, Bäckvall JE (2015) Combinatorial library based engineering of Candida antarctica lipase A for enantioselective transacylation of sec-alcohols in organic solvent. Angew Chem Int Ed 54:4284–4288
50. (a) Ma J, Wu L, Guo F, Gu J, Tang X, Jiang L, Liu J, Zhou J, Yu H (2013) Enhanced enantioselectivity of a carboxyl esterase from Rhodobacter sphaeroides by directed evolution. Appl Microbiol Biotechnol 97:4897–4906; (b) Luan ZJ, Li FL, Dou S, Chen Q, Kong XD, Zhou JH, Yu HL, Xu JH (2015) Substrate channel evolution of an esterase for the synthesis of cilastatin. Catal Sci Technol 5:2622–2629; (c) Godinho LF, Reis CR, van Merkerk R, Poelarends GJ, Quax WJ (2012) An esterase with superior activity and enantioselectivity towards 1,2-o-isopropylideneglycerol esters obtained by protein design. Adv Synth Catal 354:3009–3015; (d) Nobili A, Tao Y, Pavlidis IV, van den Bergh T, Joosten HJ, Tan T, Bornscheuer UT (2015) Simultaneous use of in silico design and a correlated mutation network as a tool to efficiently guide enzyme engineering. Chembiochem 16:805–810
51. (a) Gupta RD, Tawfik DS (2008) Directed enzyme evolution via small and effective neutral drift libraries. Nat Methods 5:939–942; (b) Kaltenbach M, Tokuriki, N (2014) Generation of effective libraries by neutral drift. Methods Molec Biol 1179:69–81; (c) Smith WS, Hale JR, Neylon C (2011) Applying neutral drift to the directed molecular evolution of a ß-glucuronidase into a ß-galactosidase: two different evolutionary pathways lead to the same variant. BMC Res Notes 4:138; (d) Bernath-Levin K, Shainsky J, Sigawi L, Fishman A (2014) Directed evolution of nitrobenzene dioxygenase for the synthesis of the antioxidant hydroxytyrosol. Appl Microbiol Biotechnol 98:4975–4985
52. (a) Eigen M, McCaskill J, Schuster P (1988) Molecular quasi-species. J Phys Chem 92:6881–6891; (b) Kurtovic S, Mannervik B (2009) Identification of emerging quasi-species in directed enzyme evolution. Biochemistry 48:9330–9339
53. Savile CK, Janey JM, Mundorff EC, Moore JC, Tam S, Jarvis WR, Colbeck JC, Krebber A, Fleitz FJ, Brands J, Devine PN, Huisman GW, Hughes GJ (2010) Biocatalytic asymmetric synthesis of chiral amines from ketones applied to sitagliptin manufacture. Science 329:305–309
54. Chen Z, Zhao H (2005) Rapid creation of a novel protein function by in vitro coevolution. J Mol Biol 348:1273–1282
55. (a) Ostermeier M, Benkovic SJ (2000) Evolution of protein function by domain swapping. Adv Protein Chem 55:29–77; (b) Park AH, Park HY, Sohng JK, Lee HC, Liou K, Yoon YJ, Km BG (2009) Expanding substrate specificity of GT-B fold glycosyltransferase via domain swapping and high-throughput screening. Biotechnol Bioeng 102:988–994
56. (a) Yu Y, Lutz S (2011) Circular permutation: a different way to engineer enzyme structure and function. Trends Biotechnol 29:18–25; (b) Qian Z, Lutz S (2005) Improving the catalytic activity of Candida antarctica lipase B by circular permutation. J Am Chem Soc 127:13466–13467; (c) Yu Y, Lutz S (2010) Improved triglyceride transesterification by circular permutated Candida antarctica lipase B. Biotechnol Bioeng 105:44–50.

Improving CO_2 Fixation by Enhancing Rubisco Performance

4

Robert H. Wilson and Spencer M. Whitney

Abstract

The photosynthetic enzyme linking the inorganic and organic phases of the biosphere is ribulose-1,5-bisphosphate [RuBP] carboxylase/oxygenase (Rubisco). The complicated catalytic chemistry of Rubisco slows its CO_2 fixation rate, allows for competitive inhibition by oxygen and permits the production of misfire products that can self-inhibit activity. Significant effort has been invested into better understanding the structure-function details of Rubisco as improving its performance is recognised as a viable means to enhance the photosynthetic efficiency and yield potential of crops. While rational design approaches have still been unable to provide catalysis enhancing solutions, modern directed evolution tools are posing a promising conduit to improving Rubisco. Advances in the design of effective selection systems for mutagenic Rubisco library screening have strategically increased their focus on using *Escherichia coli*. The inherent sensitivity of *E. coli* viability to the pentose sugar substrate of Rubisco, RuBP, is being exploited in an increasingly effective manner to select for Rubisco mutants with increased activity. Here we review the differing directed evolution technologies used to evolve Rubisco, examine the merits of available high-throughput Rubisco-dependent *E. coli* (RDE) selection systems and postulate approaches for improving their functionality.

R.H. Wilson • S.M. Whitney (✉)
Research School of Biology, The Australian National University,
Acton, Australian Capital Territory 2601, Australia
e-mail: spencer.whitney@anu.eud.au

M. Alcalde (ed.), *Directed Enzyme Evolution: Advances and Applications*,
DOI 10.1007/978-3-319-50413-1_4

4.1 Rubisco: A Target for Improvement

4.1.1 A Rate-Limiting Enzyme in Photosynthesis

The CO_2-fixing enzyme ribulose-1,5-bisphosphate carboxylase/oxygenase (Rubisco) is the catalytic gatekeeper linking the inorganic and organic phases of the biosphere's carbon cycle. Rubisco initiates carbon assimilation in the Calvin cycle by fixing CO_2 to the pentose sugar substrate ribulose-1,5-bisphosphate (RuBP) and cleaving it into two molecules of 3-phosphoglycerate (3-PGA, Fig. 4.1a). The 3-PGA is the precursor to carbohydrate synthesis that is required for energy and growth. In plants and algae the carboxylation reaction of Rubisco occurs at a slow pace (~1–5 cycles per second) resulting in its catalytic properties often limiting the rate of photosynthesis and growth in these organisms [9, 45]. To compensate for this shortcoming, many photosynthetic organisms require high amounts of Rubisco to meet their metabolic needs. For example, Rubisco can comprise up to 50% of the soluble protein in rice and wheat leaves [9, 68]. This high investment in Rubisco is therefore critical to supporting primary productivity in the global food chain which results in it being the most abundant enzyme on Earth [16].

Rubisco is a bifunctional enzyme as it also catalyses RuBP oxygenation to produce 3-PGA and 2-phosphoglycolate (2-PG). As 2-PG is toxic to growth, organisms have evolved metabolic mechanisms to recycle it [15]. In plants this occurs via photorespiration, a multi-organelle pathway requiring energy and emitting CO_2 – resulting in the loss of fixed carbon (Fig. 4.1a). In C_3 crops like rice and wheat, this carbon loss typically amounts to 25% of the CO_2 assimilated by photosynthesis [45]. As a consequence of these catalytic inefficiencies, improving the performance of Rubisco has been an objective spanning more than three decades [17, 53, 85].

Fig. 4.1 Photosynthesis, Rubisco and Rubisco activase. (**a**) Summary of oxygenic photosynthesis where the light reactions produce O_2 from H_2O photolysis and the derived energy is used in anabolic reactions such as the Calvin cycle in chloroplasts. The cycle begins with the carboxylation of ribulose-1,5-bisphosphate (RuBP) to produce 3-phosphoglycerate (3-PGA) that is reduced to glyceraldehyde-3-phosphate, the precursor for carbohydrate (CHO) synthesis or RuBP regeneration. Oxygenation of RuBP by Rubisco produces 2-phosphoglycolate (2-PG) whose recycling to 3-PGA by photorespiration occurs at the cost of energy and release of fixed CO_2 (shown in grey). (**b**) Distribution of the differing Rubisco forms in nature, their subunit structure, gene composition (*rbc*L ± *rbc*S), genome location and copy number. The multiplicity of ancillary proteins that influence Rubisco biogenesis varies throughout nature. Examples include the protein folding GroEL-GroES or corresponding Cpn60/Cpn20-23/Cpn10 (CPN, chloroplasts) and thermosome (TSM) chaperonin complexes, Rubisco accumulation factors 1 and 2 (Raf1, Raf2), RbcX and the stromal protein bundle sheath defective2 protein (BSD2) (Reviewed in [30]). (**c**) Carbamylation of each catalytic site via the binding of CO_2 (C) and then Mg^{2+} (M) is required to form an active ternary complex (ECM) that can bind RuBP and initiate its carboxylation or oxygenation. Inactive sites form when RuBP binds before carbamylation (ER) or when carbamylated catalytic sites are occupied by inhibitory sugar phosphate ligands (ECMI). Example inhibitors are xylulose-1,5-bisphosphate (XuBP, a catalytic misfire product) and carboxyarabinitol 1-phosphate (CA1P, regulatory "shade" inhibitor in plants) whose removal is facilitated by Rubisco activase via transitory intermolecular interactions powered by ATP hydrolysis

4.1.2 Natural Diversity in Rubisco Form and Function

Over 50 years of research has provided valuable knowledge regarding Rubisco's genetic and structural diversity in nature. Confidence that improving Rubisco for agricultural applications is a feasible objective stems from the discovery of more

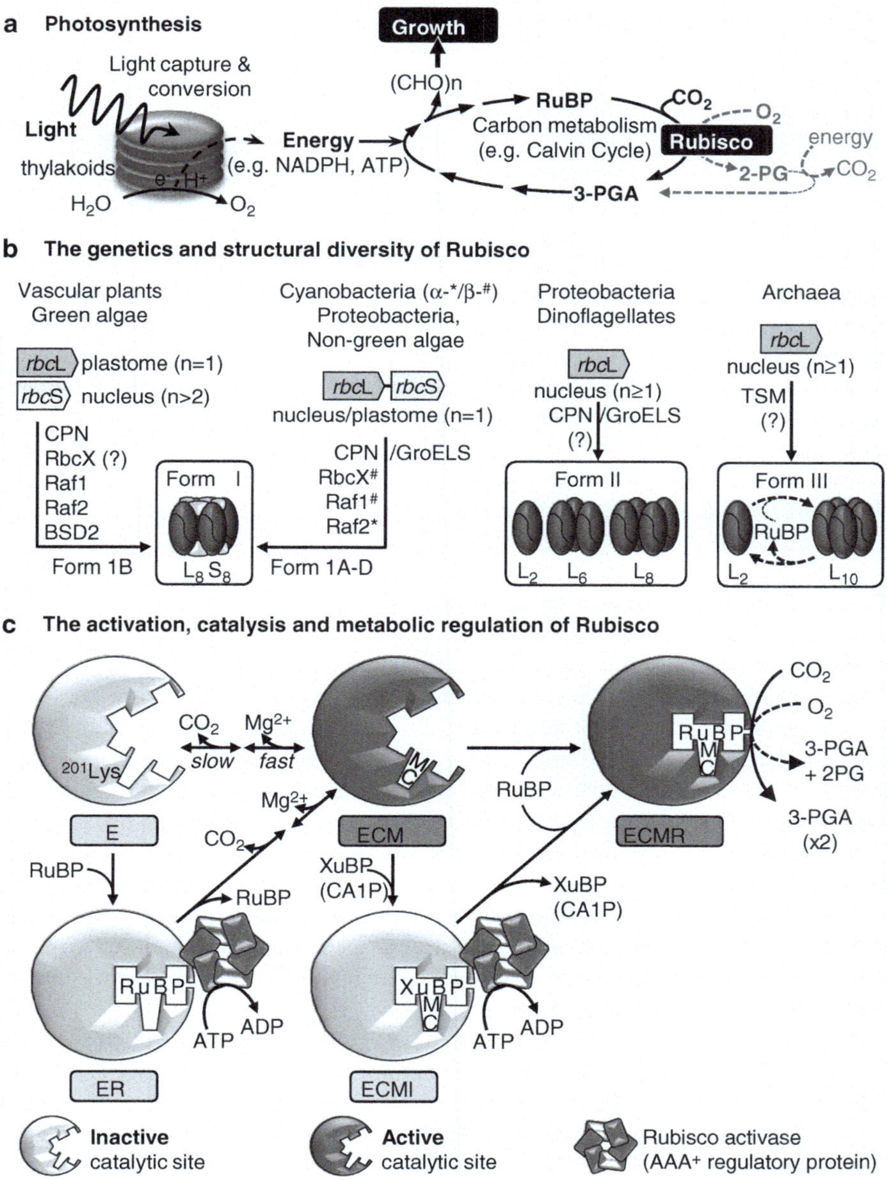

efficient versions of Rubisco found naturally in some red algae [6, 83]. Of benefit is the oligomeric structure of Rubisco in plants and algae, as well as many photosynthetic prokaryotes (e.g. cyanobacteria and many photolithotrophic bacteria), which share the same subunit complement – eight ~50 kDa large (L) subunits and eight ~14–17 kDa small (S) subunits [4]. The L_8S_8 hexameric holoenzyme is termed form I Rubisco (Fig. 4.1b) and constitutes the most abundant Rubisco isoform in the biosphere. In plants and green algae, the S-subunits are coded by multiple *rbc*S genes and the L-subunit by a single *rbc*L gene in the plastid genome (plastome) [85]. Each S-subunit is transported into the chloroplast for assembly with L_8 cores in a highly complex process that requires many known and suspected molecular partners [30, 88] (Fig. 4.1b). While some of these assembly requirements appear conserved for some L_8S_8 prokaryotic Rubiscos, the requirements for Rubisco biogenesis in nongreen algae are unclear and are not met by leaf chloroplasts [30, 83]. Associated with their different assembly requirements are likely the preserved coding of Rubisco as a *rbc*L – *rbc*S operon in the nongreen algae plastome (Fig. 4.1b) [1].

While the S-subunit of L_8S_8 Rubisco is critical for catalysis and enzyme stability, other forms of Rubisco found in nature comprise only L-subunits [4]. Form II Rubisco exists as L_2 dimers or oligomers of dimers (Fig. 4.1b). The best studied form II Rubisco is the L_2 Rubisco from the bacterium *Rhodospirillum rubrum*. Form III Rubisco can be found in Archaea and have no role in CO_2 assimilation but are capable of sustaining photosynthetic growth when transformed into leaf chloroplasts [2, 89]. Form III Rubisco functions to metabolise the RuBP produced in some archaea during the nucleotide salvaging pathways [63] and can exist as L_2 or as RuBP-induced tetrameric/pentameric assemblages of L_2 (Fig. 4.1b) [2].

4.1.3 A Highly Conserved Catalytic Site and Mechanism

Crystal structure examples for all three Rubisco forms have been solved [4, 9]. Consistent with a conserved catalytic mechanism, all forms of Rubisco show a striking conservation in L-subunit tertiary structure and amino acid arrangement in the catalytic site. This conservation occurs despite the L-subunit primary sequence differing up to 70%. The structural conservation of the catalytic site suggests that the basic scaffold of Rubisco is essential for function, although attributing what specific sequences elicit the vast catalytic variation found in the Rubisco superfamily continues to prove challenging (Table 4.1). The catalytic sites form at the interfaces between adjoining pairs of L-subunits that orient head to tail to form a L_2 – the minimal functional unit for Rubisco. Conserved amino acid residues from the C-terminal domain of one L-subunit and the N-terminal domain of the other L-subunit contribute to the two catalytic sites in each L_2 unit [4].

Prior to catalysis Rubisco must first be activated (Fig. 4.1c). Activation requires the binding of a CO_2 molecule to the ε-amino group of a conserved L-subunit ^{201}lysine (K201) in the catalytic site [10, 28]. Carbamylation of K201 produces an anionic carbamate that is stabilised by Mg^{2+} binding resulting in an activated (carbamylated) ternary "ECM" complex (E for enzyme catalytic site, C for

Table 4.1 Catalytic variation at 25 °C among the Rubisco superfamily

Rubisco form	Organism	k_{cat}^{C} (s^{-1})	K_C (µM)	K_O (µM)	$S_{C/O}$ (mol.mol^{-1})	k_{cat}^{O} (s^{-1})
IB	C_3-plants[a]	2.2–3.6	8–16	230–600	80–100	1.1–2.1
	C_4-NAD ME[a]	2.1–3.4	7–13	270–445	80–85	0.9–1.6
	C_4-NADP ME[b]	3.9–6.0	18–19	470–620	70–80	2.0–2.8
	C_4-PCK[a]	5.0–5.7	14–16	265–470	78–85	1.3–2.2
	β-cyanobacteria[a]	11.6–14.3	250–340	200–440	41–52	<0.5
IC	Proteobacteria [a]	4.1	58	1130	60	1.4
ID	Red algae[a]	1.2–2.6	29.9	306	79	1.0
	Diatoms[a]	2.1–3.7	23–65	420–1200	57–120	0.4–1.3
II	Proteobacteria[1]	6–9	40–150	160–300	9–41	<1.5
	Dinoflagellates[c,d]	1.2	n.d	n.d	37[f]	n.d
III	Archaea[e]	<2	52–130	2–100	1–11	<0.4

Organisms with a CCM are shaded in grey
n.d not determined
[a]Young et al. [92]
[b]Sharwood et al. [67]
[c]Leggat et al. [42]
[d]Whitney and Andrews [81]
[e]Alonso et al. [2] and Wilson et al. [89]
[f]Values measured at 10 °C

carbamylated K201, M for Mg^{2+}, Fig. 4.1c). In this active state, the catalytic site can bind RuBP and form an enediol intermediate that provides a nucleophilic site at the C2 carbon of RuBP for binding substrate CO_2 (carboxylation) or O_2 (oxygenation). A series of hydration, protonation and cleavage reactions follow that result in formation of the respective 3-PGA or 2-PG products (Fig. 4.1a). The complexity of this catalytic chemistry, that is compounded by the inability of Rubisco to bind either gas substrate, has posed a significant challenge to both nature and scientific research in identifying solutions for improving Rubisco performance [53]. Nevertheless, the large natural diversity in the catalytic properties of Rubisco (Table 4.1) confirms evolutionary adaptation of the enzyme. For example, Rubisco from organisms with CO_2 concentrating mechanisms (CCM) that elevate CO_2 concentrations around the enzyme typically have higher CO_2 fixation rates (*kcatC*) but lower CO_2 affinities (i.e. a higher K_m for CO_2, K_C) [67], although this does not appear to be the case for diatom (nongreen microalgae) Rubisco (Table 4.1) [92].

4.1.4 What Constitutes a "Better" Rubisco?

Throughout the literature are misconceptions about what constitutes a better Rubisco. Improving *kcatC* is a desired change in organisms containing a CCM able to provide saturating levels of CO_2 that can compensate for any accompanying reductions in CO_2 affinity (i.e. increasing K_C) or decline in the enzyme specificity for CO_2 over O_2 ($S_{c/o}$) [67]. In organisms lacking a CCM (e.g. C_3 plants), improving *kcatC* does not necessarily constitute a better Rubisco. For example, the photosynthesis and growth of C_3 plants producing the high *kcatC* Rubisco from a cyanobacterium (Table 4.1) will not exceed the endogenous plant Rubisco unless the multiple components of a functional CCM are co-introduced [57]. Therefore, according to the C3 photosynthesis models of Farquhar et al. (1980) [20], a better Rubisco within the context of a C_3 plant is one with an improvement in $S_{c/o}$ that is accompanied by an increase in carboxylation efficiency under ambient O_2 (defined as $kcatC/K_C^{21\%O2}$) [6, 67]. Importantly, modest declines in *kcatC* can be tolerated if sufficiently offset by substantial improvement in CO_2 affinity (i.e. low $K_C^{21\%O2}$).

For many Rubisco isoforms the values for $S_{C/O}$ tend to be inversely correlated with *kCcat* suggesting a catalytic trade-off between these two parameters. This feature has questioned the feasibility of engineering a faster and more CO_2-specific enzyme [64, 76]. However, kinetic surveys indicate this relationship significantly diverges when considering Rubisco isoforms other than those from vascular plants and a small sampling of algae and proteobacteria. For example, the $S_{C/O}$-*kCcat* inverse correlation strongly diverges when Rubisco from cyanobacteria, Archaea and nongreen algae are considered [89, 92]. This is best illustrated by the slightly lower *kCcat*, but vastly higher $S_{C/O}$ of Rubisco from the red algae *Griffithsia monilis* that has the potential to enhance the photosynthetic efficiency of plants with a C_3 physiology like wheat and rice by up to 30% [45, 83]. The feasibility of uncoupling this relationship by directed evolution has also been demonstrated by the identification of point mutations in archaeal Rubisco from *Methanococcoides burtonii* that improve both *kcatC* and $S_{c/o}$ [89].

4.1.5 Factors Limiting Mutagenic Study of Rubisco

The discovery that vascular plant Rubisco is not the pinnacle of evolution has engendered confidence that the challenge of improving the enzyme for agricultural applications is not insurmountable. Efforts to improve form I Rubisco function have been hindered by the inherent complexity of eukaryote L_8S_8 Rubisco biogenesis [30]. In particular the L-subunit folding and assembly needs are not met in *E. coli* [33], even when co-expressed with cognate chaperonin [11] or the Rubisco-specific assembly chaperones RbcX and Rubisco accumulation factor 1 (Raf1) [21]. Most Rubisco mutagenic studies have therefore focused on Rubisco from cyanobacteria and proteobacteria that can be functionally expressed in *E. coli*, albeit to varying degrees of competency [51]. Direct mutagenic testing of Rubisco in photosynthetic hosts such as *Rhodobacter sphaeroides* (proteobacterium, [19]), *Synechococcus* sp. PCC6803 (cyanobacteria, [3]) and *Chlamydomonas reinhardtii* (green algae, [61]) has been facilitated by the availability of *rbc*L ± *rbc*S-deficient mutants (Fig. 4.2a). Unfortunately the viability of these photosynthetic hosts for screening Rubisco mutagenic libraries is primarily confined by their low transformation efficiency (Fig. 4.2a). Similarly, while leaf chloroplast transformation provides the only system for study of recombinant plant Rubiscos, the high cost and slowness of this technology limit its usefulness for random mutagenesis studies of Rubisco [50, 85, 86].

4.2 Directed Evolution of Rubisco

Exploration of Rubisco sequence space to identify solutions that improve performance has been a long-standing challenge. Our extensive knowledge of Rubisco structure and function still remains insufficient to allow improved performance by rational design. The success of directed evolution in numerous enzyme bioengineering endeavours has lured researchers in both academia and industry into applying these tools to improve Rubisco performance [8, 23, 51, 89, 93]. Of particular appeal is the level of control available through both unbiased mutagenesis and curated selection using directed evolution and the capacity to successively sample sequence space to make incremental improvements towards a desired function [12, 13]. The last two decades has seen increasing success in these endeavours with examples of forms I, II and III Rubisco having been subject to directed evolution. As summarised in Table 4.2, most studies have focused on cyanobacteria L_8S_8 Rubisco with particular attention made on mutating the *rbc*L gene and only in a few cases including *rbc*S [8, 50, 69]. The general outcomes for some key mutagenic screening studies are described in Table 4.2.

4.2.1 The Challenges of Identifying an "Improved Rubisco"

A key consideration in any directed evolution application is identifying a suitable screening system. This requires consideration on whether the screening system meets the desired throughput, has sufficient fidelity to avoid selection of "false positives" and is suitably sensitive to identify minor and incremental improvements in a desired trait (i.e. successive increases in fitness). With regard to Rubisco, selection systems incorporating an in vitro catalytic screening component are typically compromised by throughput and fidelity. The challenges are primarily associated with being able to accurately measure Rubisco content and quantify the multiplicity of kinetic parameters that equate to an improved enzyme. In organisms with a CCM, this might equate to simply increasing *kcatC*; in non-CCM-containing hosts, increases in $S_{c/o}$ and $kcatC/K_C^{21\%O2}$ without extensive undermining of *kcatC* are typically required (Sect. 4.1.4).

A lack of suitable infrastructure and experimental familiarity in how to correctly measure Rubisco content and kinetics continues to confuse the kinetic literature. As a consequence, large differences in the kinetics for a Rubisco are often reported between studies. For example, large errors in the reported kinetic properties of a hybrid plant Rubisco comprising the sunflower L-subunit and tobacco S-subunits [36] and for *Thermococcus kodakaraensis* form III Rubisco [18] have been corrected in subsequent studies [40, 66]. With regard to Rubisco directed evolution studies, some mutants have been incorrectly interpreted as catalytically enhanced [52] rather than "solubility" mutants with enhanced L-subunit folding and assembly [26]. These mistakes highlight how accurately measuring changes in Rubisco synthesis can be problematic, especially using polyacrylamide gel electrophoresis methods that can be misleading without appropriate controls and experimental rigour [87]. Potential problems also arise measuring Rubisco kinetics using commercial sources of RuBP contaminated with inhibitory sugar phosphates that reduce the reliability of the catalytic measurements [68].

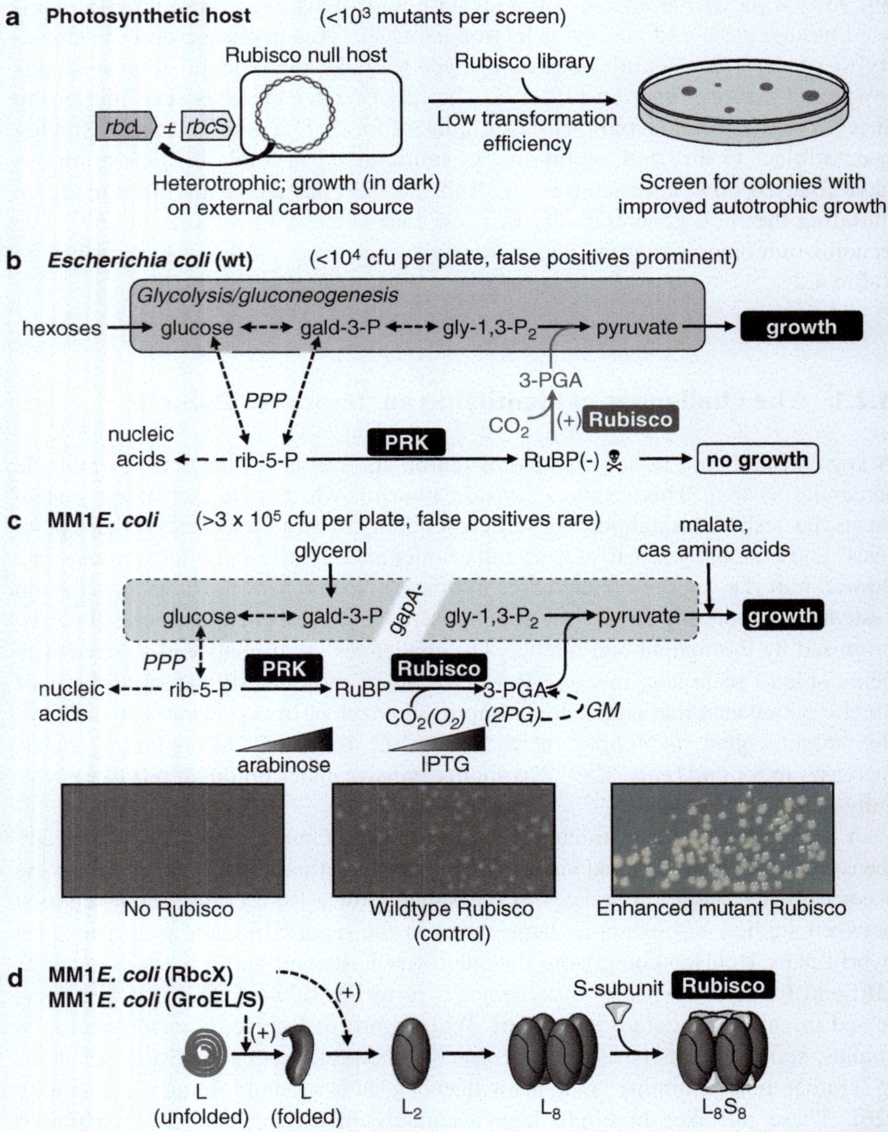

Table 4.2 The biochemical properties of Rubisco mutants selected by directed evolution

Rubisco origin	Main mutations identified and effect on Rubisco	References
Selection host: *Rhodobacter capsulatus*		
Synechococcus PCC6301 (L_8S_8)	L-subunit – F342V: twofold increase in host doubling time, 50% reduction to RuBP affinity (K_mRuBP)	[69]
Synechococcus PCC6301 (L_8S_8)	L-subunit – A375V: improved CO_2 and reduced O_2 affinities, carboxylation rate 12% of wild type	[60]
Selection host: *Chlamydomonas reinhardtii*		
Chlamydomonas reinhardtii (L_8S_8)	L-subunit – A99T, A281S, D352G: improved specificity, carboxylation rate and affinity for CO_2	[93]
Nicotiana tabacum (L_8S_8)	L-subunit – A99T, A281S, D352G: little influence on catalytic properties of tobacco Rubisco	[25]
Selection host: *Escherichia coli* (RDE)		
Synechococcus PCC6301 (L_8S_8)	L-subunit – M259T, F342S: four- to ninefold increase in Rubisco assembly	[26]
Rhodospirillum rubrum (L_2)	H44N, D117V: altered affinity for CO_2 and O_2	[47]
Synechococcus PCC6301 (L_8S_8)	L-subunit – L161M, M169L: up to 11-fold increase in Rubisco assembly	[51]
Synechococcus PCC7002 (L_8S_8)	S-subunit – E49V, D82G: increased carboxylation efficiency by 45%	[8]
Synechococcus PCC6301 (L_8S_8)	L-subunit – F140I: 2.9-fold increase in carboxylation efficiency, 9% decrease in specificity	[14]
Methanococcoides burtonii (L_2–L_{10})	E138V, K332E: 2.8–3.6-fold improvement to carboxylation efficiency in air	[89]

Fig. 4.2 Directed evolution selection screens for Rubisco, past and present. (**a**) Low transformation efficiencies have limited the versatility, and success, of using Rubisco-deficient photosynthetic host mutants to screen Rubisco libraries. (**b**) By exploiting the sensitivity of *E. coli* to RuBP toxicity (−), expressing phosphoribulokinase (PRK, converts ribulose 5-phosphate, rib-5-P, from the pentose phosphate pathway, *PPP*, to RuBP) can be used to visually select for increased Rubisco activity (+, faster RuBP turnover) via improved cell viability (faster colony growth). A high proportion of false positives reduce the sensitivity and throughput of this selection approach. Maximum practical colony-forming units (cfu) able to be screened per plate are indicated in brackets. (**c**) Interruption of glycolysis/gluconeogenesis in *E. coli* via glyceraldehyde-3-phosphate dehydrogenase deletion (*gapA*$^-$) produced a higher fidelity Rubisco-dependent *E. coli* (RDE) selection system (MM1) that requires a functional PRK (arabinose inducible) and Rubisco (IPTG inducible) shunt to bypass glycolysis and allow carbon metabolism to flow from supplied hexose substrate through to the citric acid cycle for energy and growth [47]. Trace levels of malate, casamino acids and glycerol are needed to support early cell division upon induction of the PRK-Rubisco shunt. The 3-PGA or 2-PG produced by Rubisco catalysis is metabolised in *E. coli* by glycolysis or glycolate metabolism (GM; see biocyc.org/ECOLI), respectively [54]. Shown are the colony phenotypes of RDE-MM1 cells producing no Rubisco (left), *Synechococcus* PCC6301 Rubisco (middle) and the higher solubility M259T PCC6301 mutant (right) [26]. (**d**) Alternative selection screens using MM1 co-expressing either additional GroEL-GroES chaperonin complexes (which stimulate (+) L-subunit folding) or the Rubisco chaperone RbcX (which "staple" post-chaperonin folded L-subunits into stable L_2 units to facilitate L_8 formation) altered the selectivity for novel *Synechococcus* PCC6803 Rubisco mutants with differing catalytic and biophysical properties [14]

4.2.2 Rubisco Mutant Selection Using a Photosynthesis Deficient Host

Directed evolution selection systems require the target molecule under selection to be identifiable via a screen that can report a desired functional change. As Rubisco is essential for photosynthesis, initial directed evolution selection systems utilised an available Rubisco deletion mutant of the photoheterotrophic bacterium *Rhodobacter capsulatus* (strain SBI-II) [19]. Although non-photosynthetic, the Rubisco-deficient SBI-II *R. capsulatus* cells can grow heterotrophically on media such as peptone yeast extract medium [69]. Upon transforming with mutagenic cyanobacteria Rubisco libraries, photosynthetic potential can be screened through growth on Ormerod's minimal medium and by varying the growth CO_2 levels (Fig. 4.2a). While able to effectively identify L-subunit residues that influence the catalytic properties of cyanobacteria (Table 4.2), the approach was limited in throughput by the very low transformation competency of the SBI-II cells [59], an impediment that also limits the versatility of a new Rubisco-deletion strain of the bacterium *Ralstonia eutropha* for directed evolution studies [62] due to low (4×10^3 cfu/μg DNA) electroporation efficiency [70].

The availability of a non-photosynthetic *rbc*L deletion strain of *Chlamydomonas reinhardtii* (strain MX3312) has also been exploited for use in a directed evolution study of Rubisco from this green algae [93]. Like *R. capsulatus*, heterotrophic growth of the Rubisco null MX3312 strain can be sustained in the dark on acetate-containing medium. The practicality of the system to screen mutagenic Rubisco libraries is again limited by the low chloroplast transformation efficiency of *C. reinhardtii* (Fig. 4.2a). Nevertheless, the approach proved successful in identifying Rubisco mutants with improvements in carboxylation efficiency (Table 4.2). Unfortunately, independent testing showed one mutant comprising three amino acid mutations was not translatable to influencing the kinetic properties of Rubisco in tobacco, the model C_3 plant primarily used for Rubisco mutagenic testing [25].

4.2.3 Selection Using Rubisco-Dependent *E. coli*

The high transformation efficiency and fast growth rate of *E. coli* make it a favoured host for many directed evolution applications. The value of a high-throughput bacterial selection system for thoroughly exploring the adaptive potential of Rubisco has been a long-standing objective [46]. A potential caveat is that the range of Rubisco types for experimentation may be limited to the prokaryotic and archaeal isoforms capable of functional expression in *E. coli* [50].

Current Rubisco selection systems in *E. coli* require phosphoribulokinase (PRK) co-expression. The PRK catalyses the production of RuBP via ATP-dependent phosphorylation of the ribulose-5-phosphate (R5P) produced in the pentose phosphate pathway (Fig. 4.2b). In *E. coli* RuBP cannot be metabolised and its accumulation is toxic to growth [32]. The cause of this toxicity remains unclear, although it does not appear due to reductions in R5P availability since deletion of 6-phosphogluconate dehydrogenase in the pentose phosphate pathway needed to produce R5P has no discernible effect on *E. coli* growth [35]. Expression of Rubisco in *E.coli*-PRK-expressing cells can alleviate RuBP toxicity by conversion into 3-PGA, a natural metabolite of the glycolysis/gluconeogenesis pathway, or 2-PG that can be converted into glycolate, a usable carbon source for *E. coli* growth [54] (Fig. 4.2c). The dependence of *E.coli*-PRK-expressing cells on Rubisco activity for survival forms the basis of all Rubisco-dependent *E. coli* (RDE) selection systems.

The initial application of RDE utilised a cyanobacterial PRK under the control of an arabinose-inducible BAD promoter in plasmid pACYC184 [52]. When expressed in wild-type *E. coli*, this simple RDE screen isolated a small range of *Synechococcus* PCC6301 (cyanobacteria) L-subunit mutants that all shared a Met-259-Thr (M259 T) substitution [52]. The M259T mutation was shown to enhance L_8S_8 Rubisco biogenesis ~fivefold in *E. coli* and modestly improve the overall kinetic properties of *Synechococcus* PCC6301 Rubisco [26] (Table 4.2). More recent success with this RDE system found E49V and D82G mutations in the *Synechococcus* PCC6301 S-subunit improved carboxylation efficiency by 50% [8]. A continuing limitation of these studies is the low fidelity of the RDE system due to >100:1 ratio of false to bona fide Rubisco mutants produced. This necessitates lower cell plating densities to avoid excess colony formation thereby significantly increasing the processing time and cost of analysis (Fig. 4.2b). The foremost cause of false positives appears to be silencing of PRK synthesis via mutations or transposon insertion in the *prk*A gene – thus reverting the cells to a wild-type growth rate [8, 50, 52].

A higher fidelity RDE with increased throughput was developed by Mueller-Cajar et al. (2007) that incorporates a shunt comprising PRK and Rubisco where both enzymes are necessary for carbon metabolism and cell growth (Fig. 4.2c). The RR1 *E. coli* strain in this alternative RDE system (called RDE-MM1) contains a deletion in glyceraldehyde-3-phosphate dehydrogenase (*gap*A$^-$) that inhibits carbon flux through the glycolysis/gluconeogenesis cycle. The PRK-Rubisco shunt bridges the *gap*A$^-$ metabolic gap to enable carbon flux from hexose carbon sources in the growth media through to the citric acid cycle for energy and growth ([47], Fig. 4.2c). An arabinose-inducible BAD promoter is used to provide stringent regulation of PRK expression [37], while Rubisco production is regulated by an IPTG-inducible *lac* promoter (Fig. 4.2c). As PRK expression is directly linked to MM1 survival in this RDE system, the production of false positives derived from "PRK silencing" is largely avoided [50]. The improved fidelity of the MM1-RDE system has successfully isolated form II *R. rubrum* L_2 Rubisco mutants that unveiled conserved $S_{c/o}$ determining amino acids [47], identified novel L-subunit mutations that improve cyanobacterial L_8S_8 assembly ("solubility") and affinity for RuBP [51] and improved carboxylation efficiencies [14] (Table 4.2). More recent use of the MM1-RDE in directed evolution of the archaeal *Methanococcoides burtonii* L_{10} Rubisco (MbR) successfully selected two MbR mutants with significant improvements in *kcatC*, $S_{c/o}$ and *kcatC*/$K_C^{21\%O2}$ (Table 4.2) [89].

4.2.4 Expressing Molecular Partners of Rubisco Can Alter the Kinetic Outcome

The majority of attempts to evolve *Synechococcus* PCC6301 Rubisco in *E. coli* have identified L-subunit mutations that improve cellular Rubisco levels (Table 4.2). These mutations included M259T [52], I174V, Q212L and F345I/L [51] and C172G, V189I, S398C and I465V (and various substitutions at amino acid F345, [14]). The mechanism behind the increased production of assembled L_8S_8 Rubisco is uncertain. It is hypothesised they enhance the interaction of the nascent L-subunit chains with the *E. coli* GroEL-GroES chaperonin complex [14, 26] or provide the monomers more time in vivo to form the more stable L_2 to L_8 complexes to which the soluble S-subunits can rapidly self-associate and form functional L_8S_8 complexes. A recent study highlighted the differential influence *E. coli* chaperonin (GroEL-GroES) and the Rubisco-specific assembly chaperone RbcX have on *Synechococcus* PCC6301 L-subunit evolution in RDE-MM1 cells [14]. As summarised in Fig. 4.2d, both ancillary proteins improve rates of cyanobacterial L_8S_8 biogenesis, the GroEL-GroES facilitate L-subunit folding [24] and the RbcX stabilise post-chaperonin folded L-subunits into L_2-$(RbcX_2)_2$ units that form L_8 cores for S-subunit binding [30]. In RDE-MM1, over-expression of GroEL-GroES had little influence on the Rubisco mutational range selected despite stimulating cell growth through increased Rubisco biogenesis. In contrast, selection of some Rubisco mutants was prevented in RDE-MM1 cells producing RbcX due to apparently constraining the selection of mutations to solvent exposed surface amino acids. These findings support the notion that the requirement of form I Rubisco for a range of assembly factors (Fig. 4.1b) with suitable compatibility [84] has had a pervasive influence on Rubisco evolution, possibly constraining its potential to enhance catalysis [14, 30, 50, 53].

4.2.5 Limitations with Using Fusion Marker Selection

Efforts to mutate eukaryotic form I Rubisco to facilitate its assembly in *E. coli* have yet to yield success [11, 47]. This apparent impasse is most likely due to the complementarity requirements of eukaryotic Rubisco with its multiple assembly factors for subunit folding, stabilisation and assembly into L_8S_8 holoenzyme (Fig. 4.1b, [30, 88]). This potential hurdle has however not averted attempts to develop approaches for identifying solubility enabling eukaryotic L-subunit mutants. In theory this could be achieved using RDE-MM1 to select for mutations that generate Rubisco activity. The feasibility of this approach depends on modulating the RDE-MM1 sensitivity to detect very low levels of L_8S_8 assembly – a possibility still untested. Instead we have trialled the viability of using reporter protein fusions to screen for scalable changes in Rubisco solubility. Past success has demonstrated the potential to detect improved protein assembly (solubility) using reporter fusions that increase antibiotic resistance [72], enhance fluorescence [39] or alter the amount of dye-based reporter produced [71, 80]. In the case of Rubisco, our research has shown L-subunit fusions comprising a truncated N- or C-terminal 85-amino-acid (9 kDa) beta-galactosidase alpha fragment proved insensitive to reporting variations in

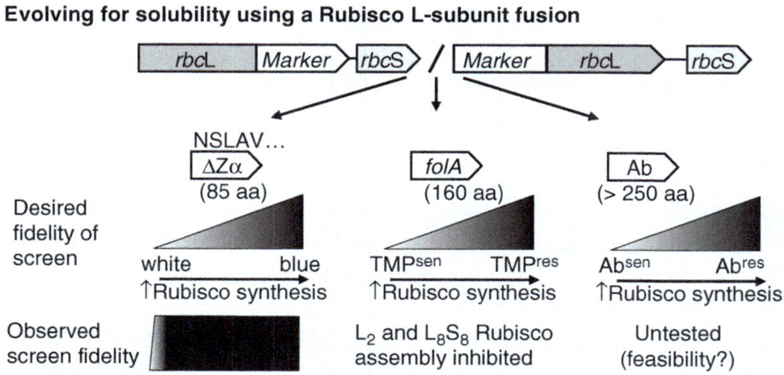

Fig. 4.3 Limitations in using L-subunit fusions for selecting improved Rubisco solubility. Studies in our laboratory have tested fusion constructs using various selective markers fused in frame to the N- or C-termini of *R. rubrum* or *Synechococcus* PCC6301 L-subunits, the latter comprising a *rbc*L-*rbc*S operon as shown. Fusions with the alpha peptide of *E. coli* β-galactosidase (Zα) performed poorly as a scalable screen of Rubisco assembly as colonies producing little or no assembled Rubisco were *vivid blue*. Similarly, fusions with *E. coli* dihydrofolate reductase (*fol*A) conferred resistance to trimethoprim (TMP) that did not correlate with the levels of L_2 or L_8S_8 Rubisco produced. The viability of using L-subunit fusions comprising antibiotic (Ab) markers has not been tested

solubility by blue-white screening (Fig. 4.3). Even Rubisco mutants lacking assembly capabilities produced vivid blue colonies through X-gal conversion making the fidelity of the screen untenable for visually detecting changes in Rubisco solubility. The use of a 160-amino-acid (18 kDa) dihydrofolate reductase (DHFR) fusion in selecting for trimethoprim resistance linked changes in enzyme solubility has been successful in identifying acetyltransferase mutants with improved solubility [44]. Such an approach proved unsuccessful with *R. rubrum* L_2 and cyanobacterial L_8S_8 Rubisco as the additional DHFR sequence prevented L-subunit folding and holoenzyme assembly (Fig. 4.3). In hindsight this strategy was flawed from inception due to the GroEL-GroES folding requirements of both Rubisco L-subunits [43] and DHFR [27] and that the 71 kDa size of the L-DHFR and DHFR-L fusions exceed the ~60 kDa protein folding limit of the *E. coli* chaperonin cage [31, 58]. The prohibitive influences fusion reporter proteins have on L-subunit folding and assembly make this a challenging, if not untenable, approach for detecting changes in Rubisco solubility in *E. coli*.

4.3 Evolutionary Outcomes, Applications and Limitations

A major caveat with Rubisco directed evolution studies to date is correlating changes in catalysis with changes in the growth performance of the host. While the RDE systems provide the potential for higher throughput of mutagenic library screening, its reliance on RuBP detoxification to report Rubisco mutants with improved catalytic potential remains inexact.

4.3.1 A Need to Improve RDE Selection

The low levels of false positives selected using the RDE-MM1 system appear linked to a heightened sensitivity of MM1 to PRK expression [47]. This phenotype may stem from the diminished viability of the *gapA*⁻ MM1 strain that requires slow growth (23 °C) on minimal media for prolonged periods (6–16 days) which dramatically impedes experimental throughput [50, 65]. The MM1 strain also suffers from a low transformation efficiency as growth in LB, or media with a high sugar content, causes cell lysis [34]. It is also unclear why RDE-MM1 requires CO_2 concentrations 50–65 times atmospheric levels for Rubisco selection. Possibly the high CO_2 may help maintain the activation status of the Rubisco catalytic sites (Fig. 4.1c) and/or facilitate permissible rates of RuBP carboxylation in the bacterial cytoplasm [50]. Somewhat paradoxically even under such high CO_2 conditions, there remains the potential for the selecting Rubisco mutants with increased RuBP oxygenation, not carboxylation, properties (Table 4.2). It is presumed this can arise as the 2PG produced can be metabolised into glycolate and used by *E. coli* for growth [47, 54]. Although increasing oxygenation rates is undesired within the context of improving photosynthesis, such mutagenic outcomes help to better understand Rubisco structure-function.

4.3.2 Selection Using a Photosynthetic Host: Is It Really Advantageous?

An advantage of using a photosynthetic organism in Rubisco directed evolution applications is the selected enzymes are supported by cognate, or near-cognate, chaperoning resources of the host. Theoretically this should allow, for example, the evolution of eukaryote Rubisco in a eukaryote host. As indicated in Sect. 4.2.2, however, the throughput of such selection systems is relatively slow and costly. There also appear a number of unresolved ambiguities in some photosynthetic host studies that likely arise from challenges described in Sect. 4.2.1 associated with accurately measuring Rubisco content and catalysis. For example, the reductions in CO_2 fixation rate, CO_2 affinity and poorer specificity of the F342 V *Synechococcus* PCC6301 Rubisco mutant selected in the *R. capsulatus* SBI-II Rubisco null line are inconsistent with it benefiting photosynthetic growth [69]. The other F342I and M259 T mutants selected in the *R. capsulatus* screen correlate with those repeatedly selected in RDE studies due to enhanced cyanobacteria L_8S_8 assembly (Table 4.2). It therefore seems likely the *R. capsulatus* screen successfully identified *Synechococcus* PCC6301 Rubisco mutants with improved L_8S_8 biogenesis not beneficial changes in catalysis (Table 4.2). A subsequent directed evolution study using *Chlamydomonas* as a selection host identified three L-subunit mutations that improved Rubisco catalysis in the algae with the same substitutions also able to enhance tobacco Rubisco catalysis [93]. This finding was not supported in an independent study where the mutations had no appreciable influence on tobacco Rubisco kinetics, leaf photosynthesis or plant growth [25]. This brings into question the versatility of mutagenic studies of L_8S_8 Rubisco from *Chlamydomonas*, or photosynthetic prokaryotes, with regard to providing solutions of direct translatable benefit to Rubisco in vascular plants.

4.3.3 Translational Success to Improving Photosynthesis

The last few years have seen the successful translational application of RDE-derived Rubisco mutants to enhance the productivity of a photosynthetic host. Traditionally the most direct route for improving RDE fitness using *Synechococcus* PCC6301 Rubisco has been the selection of solubility-enhancing mutants (Sect. 4.2.4). In contrast, using RDE-MM1 Durão et al. [14] identified point mutations in the *Synechococcus* PCC6301 L-subunit that increased carboxylation efficiency ($kcatC/K_C^{21\%O2}$) by either 70% (V189I) or nearly threefold (F140I), the latter mutation not affecting L_8S_8 assembly. When transformed into *Synechocystis* PCC6803 cells, the F140I mutation improved the rate of photosynthesis by 40% and supported wild-type growth rates in cells comprising ~20% less Rubisco [14]. While a highly motivating outcome, the underpinning question is why had the mutation not already been selected during evolution? Possibly the photosynthesis improvements are not sustainable under alternative, nonoptimal, growth conditions – a hypothesis for future experimental scrutiny.

A recent directed evolution study has highlighted the potential of improving the performance of archaeal *Methanococcoides burtonii* L_{10} Rubisco (MbR) in a photosynthetic role [89]. As indicated in Sect. 4.1.2, form III archaeal Rubisco serves a non-photosynthetic function, comprises only L-subunits and is highly soluble when expressed in *E. coli* [2, 38, 91]. The alternative biological function of archaeal Rubisco to metabolise RuBP produced as a by-product during nucleotide metabolism [22, 63] has seen it evolve distinctive, somewhat variable, catalytic properties relative to contemporary Rubisco isoforms [74]. This includes uniquely high affinities for RuBP and atypically low carboxylation properties [2]. These atypical properties relative to photosynthetic Rubisco suggest archaeal Rubisco has been subject to alternative evolutionary trajectories, possibly improving its potential for carboxylation enhancement [2, 50]. Consistent with this hypothesis, two MbR mutants (E138V and K332E) with three- to fourfold improvements in carboxylation efficiency in air ($kcatC/K_C^{21\%O2}$) were isolated from a single round of selection in the RDE-MM1 system ([89]; Table 4.2). When expressed in tobacco chloroplasts, both mutant MbR variants supported higher rates of photosynthesis and much faster plant growth compared to control plants producing wild-type MbR. The efficient expression of archaeal Rubisco in leaves, and overall success with evolving MbR catalysis, inspires further rounds of evolution to improve catalytic parameters along evolutionary trajectories that further enhance its photosynthetic potential.

4.3.4 Rubisco Production Can Be Toxic to *E. coli*

An important consideration of the RDE systems is being able to carefully regulate PRK expression to ensure the intercellular RuBP production meets the desired level of Rubisco activity sought. While the BAD promoter has proven suitable to fine-tuning PRK expression by varying arabinose induction ([47], Fig. 4.2c), little attention has been made to modulating Rubisco expression. This may require closer attention when one considers the expression of form I [52] or form II [82] Rubisco can itself impede, sometimes prevent, *E. coli* growth. As shown in Fig. 4.4, the growth of XL-1Blue *E. coli* cells expressing *R. rubrum* or *Synechococcus* PCC6301 Rubisco is increasingly impeded under rising IPTG induction. In contrast, growth is prevented in cells producing the highly soluble *M. burtonii* L_2 or the F345I mutated *Synechococcus* PCC6301 L_8S_8 Rubiscos. Awareness on how Rubisco toxicity impacts the fidelity of RDE selection remains unexplored. For example, it may preclude selection of activity-enhancing mutants that additionally escalate Rubisco biogenesis. Alternatively it may beneficially focus selection of catalytic enhancements that have little or no influence on Rubisco biogenesis, somewhat akin to the F140I *Synechococcus* PCC6301 mutant identified by Durão et al. (2015) [14].

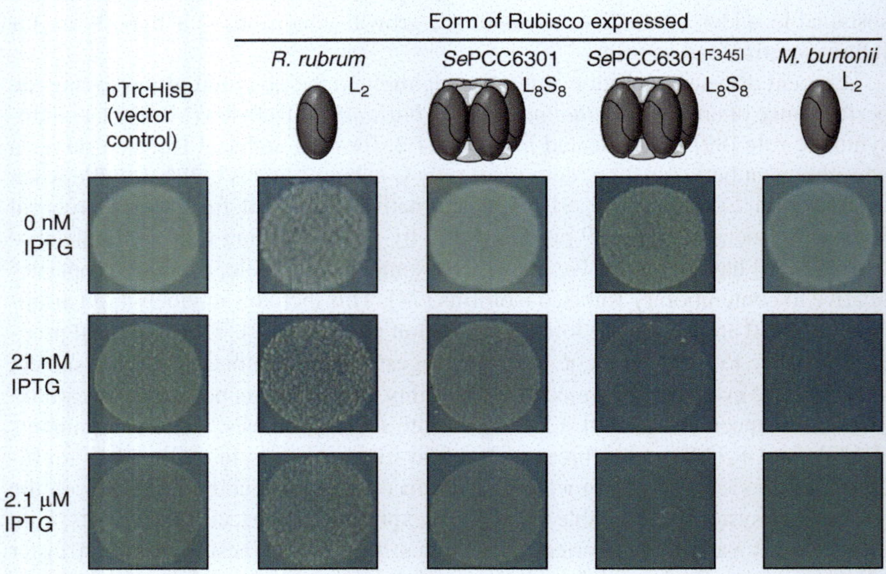

Fig. 4.4 High levels of Rubisco expression can inhibit *E. coli* growth. The differential influence of Rubisco expression on *E. coli* growth was determined by spot plating of XL1-blue cells transformed with pTrcHisB plasmids coding the Rubisco forms shown. Shown is the extent of growth on LB agar containing 200 µg/mL Amp and varying IPTG concentrations (shown) after 5 days in air at room temperature. *Synechococcus* PCC6301 (*Se*PCC6301), mutation Phe415Ile (F345I) improves Rubisco assembly by eight- to 15-fold in *E. coli* [51]

4.4 New In Vivo Screening Strategies for Rubisco Directed Evolution

The above examples of possible limitations in the versatility and sensitivity of RDE-MM1 incite further development of the selection system to increase its efficiency and broaden the scope of suitable Rubisco clients amenable to directed evolution study. In this section we explore possible ways to meet these objectives both in directed evolution studies of Rubisco and its activity-regulating protein Rubisco activase (Fig. 4.1c).

4.4.1 Avoiding PRK Escape Mutants by Dual Selection

Antibiotic resistance markers are common selective agents used in molecular biology to detect and maintain genetically transformed cells. This selective trait can also be employed to detect recombinant protein production by fusing it to proteins coding antibiotic resistance. While there appears little merit in using Rubisco-reporter protein fusions (Sect. 4.2.5), linking an antibiotic resistance onto PRK may avoid selection of false positives that currently plague RDE systems using wild-type *E. coli* (Sect. 4.2.3). These false positives primarily arise from interruption of PRK expression via integration of varying transposon elements into differing regions of the *prk*A gene (Fig. 4.5a–c). By appending an in-frame resistance marker to the C-terminus of PRK (possibly via a flexible peptide linker), transposon interruption to PRK synthesis will relinquish antibiotic resistance, preventing cell growth on selective media (Fig. 4.5d, option 1). Such a strategy requires the PRK, and resistance marker remains active as a PRK-Ab^R fusion peptide (possibly necessitating a flexible peptide linker peptide). Notably the *Synechococcus* PCC6301 PRK typically used in RDE studies is a homodimer [75]. Therefore, the antibiotic marker should ideally function as a monomer to avoid steric hindrance and oligomerisation assembly issues that impede PRK functionality. The feasibility of developing suitable PRK-Ab^R fusion peptides that can improve the fidelity and throughput potential of the RDE systems using wild-type *E. coli* remains to be explored.

4.4.2 Tailoring to the Assembly Needs of Eukaryotic Rubisco

Rubisco isoforms from crop plants – especially grain crops of significant nutritional and agricultural value – pose idyllic targets for catalytic modification by directed evolution tools. However, their extensive requirement for compatible ancillary proteins during L_8S_8 biogenesis, incorporating chaperonin folding and downstream assembling factors (Fig. 4.1b), precludes their functional production in *E. coli* [30, 47]. As efforts so far have been unable to mutate plant L-subunits to meet the folding and assembly machinery of *E. coli* [11, 51], it is likely the inclusion of one or more of Rubisco's chloroplast assembly factors may be needed to facilitate Rubisco solubility in *E. coli*. Fortunately the last decade has seen dramatic advances in our understanding of L_8S_8 Rubisco biogenesis [30, 89]. Unlike the *E. coli* chaperonin GroEL cylindrical ring and heptameric GroES cap, the chaperonins of chloroplasts are structurally more diverse [7, 78, 79]. They comprise heptameric Cpn60 cages comprising α and β subunits capped by hetero-oligomeric complexes of Cpn10 and Cpn20 subunits. While the feasibility of expressing chloroplast chaperonins in *E. coli* has been shown [7, 78], their ability to fold Rubisco in this context remains uncertain. Possibly the expression of chloroplast chaperonin subunits in an RDE may allow directed evolution of eukaryotic Rubisco (Fig. 4.5d). Vascular plant Rubisco may however require expression of additional assembly factors in the

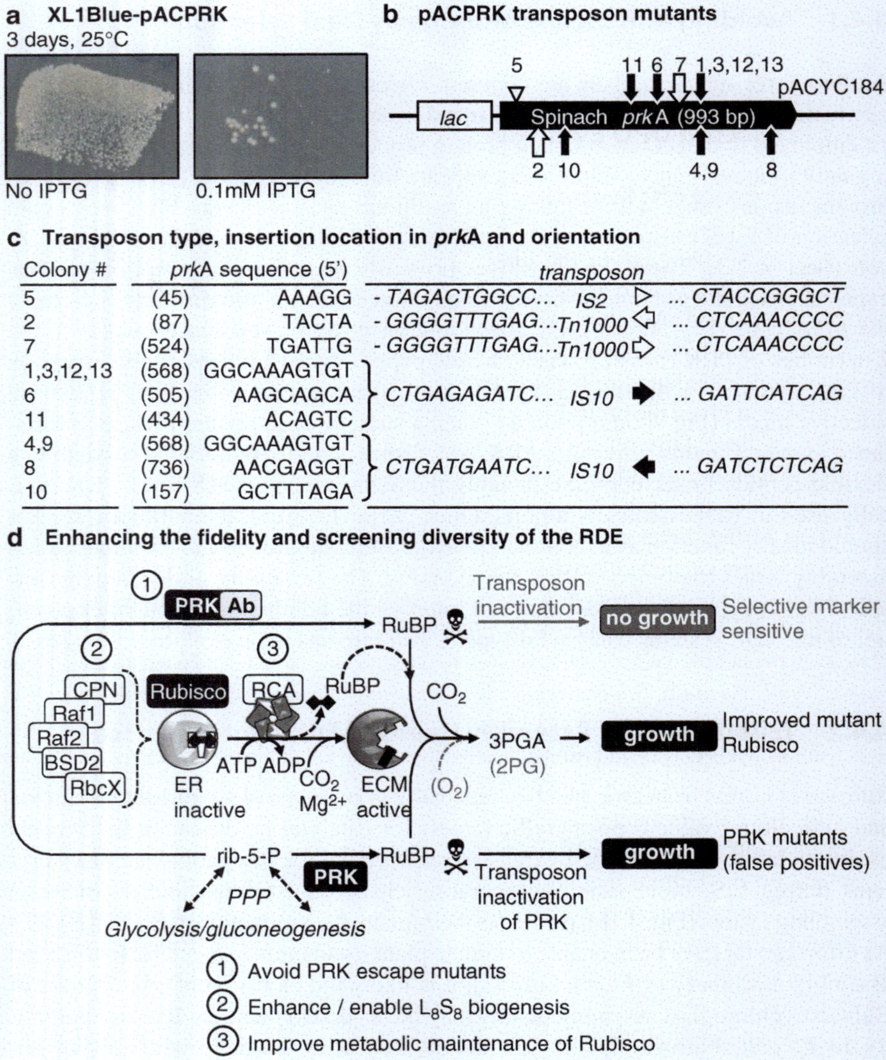

RDE. Examples include Rubisco accumulation factors 1 and 2 (Raf1, Raf2), RbcX and/or the bundle sheath defective2 (BSD2) protein (Figs. 4.1c and 4.5d, option 2) that serve differing, sometimes crucial, roles in L_8S_8 biogenesis [21, 30, 84]. Engineering an RDE with the required multiplicity of assembly factors to facilitate higher plant Rubisco (or algae Rubisco) assembly in *E. coli* or within in vitro reconstitution systems is clearly a significant, and risky, challenge but one of extremely high reward if able to provide a suitable platform for high-throughput mutagenic study of Rubisco from diverse origins.

4.4.3 Is There a Need for Metabolic Regulation of Rubisco in an RDE System?

An outcome yet to be considered for improving Rubisco fitness using an in vivo screen is if the selected mutation(s) influence(s) the extent Rubisco activity is inhibited by pentose phosphate sugars, in particular its own substrate RuBP. As shown in Fig. 4.1c, formation of inactive "ER" Rubisco arises when RuBP binds to non-carbamylated catalytic sites. Recent work has highlighted the pervasive influence ER formation affects Rubisco activity in plant and algal chloroplasts, reducing the proportion of catalytically active Rubisco sites by 10–70% [68, 92]. In the absence of exogenous RuBP, the rate of substrate dissociation from ER occurs within a half-time of 0.5–5 min, depending on Rubisco isoform (Pearce 2006). However, ER formation is sustained under saturating RuBP (Sharwood 2016). Therefore, to counter the formation of ER, and other catalytically inactive sugar phosphate inhibited Rubisco complexes (ECMI, Fig. 4.1c), photosynthetic organisms utilise isoforms of the Rubisco activase (RCA) enzyme family to reverse this inhibition [5, 9, 48, 50]. While having differences in oligomeric structure and mode of action, all RCA homologues are part of the AAA^+ protein superfamily and use the power of ATP

Fig. 4.5 Improving the function of RDE selection. Expanding the potential of RDE selection that suffers from (**a–c**) a high frequency of transposon induced silencing of PRK in wild-type *E. coli* by (d) utilising a PRK-antibiotic fusion (option 1), improving Rubisco activity through co-expression of complimentary biogenesis factors (option 2) and/or Rubisco activase (option 3). (**a**) Comparing growth after 3 days at 23 °C of ~2×10^3 XL1-Blue-pACPRK cells on non-inducing (no IPTG) and spinach PRK-inducing (0.1 mM IPTG) LB media. The large colonies arising under inducing conditions were due to (**b**) alleviation of RuBP toxicity by inactivation of spinach PRK expression via (**c**) transposon insertion at differing positions in the spinach *prk*A gene in pACPRK. Regions of the *prk*A sequences were repeated upstream (5′, as shown) and downstream (3′, not shown) at the transposon insertion sites with the position of the first nucleotide relative to the *prk*A initiator codon shown in parenthesis. The partial flanking sequence (ten nucleotide) of the IS2 (white triangle), Tn1000 (*white arrow*) and IS10 (*black arrow*) transposon elements are shown in italics and their relative orientation (*ori*) indicated by the arrow direction. The selection of these false positives might be avoided by (**d**) fusing an in-frame antibiotic selectable marker to PRK. The fidelity, selection stringency and variety of Rubisco substrates that can be screened by a RDE might be increased through co-expression of ancillary Rubisco folding/assembly factors and/or a compatible Rubisco activase (RCA) (described in Sects. 4.4.1, 4.4.2, and 4.4.3)

hydrolysis to facilitate sugar phosphate release from Rubisco (Fig. 4.1c) via poorly understood transitory interactions [29, 49, 73, 77].

The necessity for RCA to regulate ER (and ECMI) levels in chloroplasts of illuminated leaves arises from the saturating steady-state levels of RuBP (>2 mM in tobacco chloroplasts, [56]) that resemble the concentration of Rubisco catalytic sites (~2.5–3.5 mM). By analogy, RDE-MM1 cells producing abundant levels of Rubisco (>5% of the cellular soluble protein) also retain saturating RuBP levels (0.2–0.9 mM) [47]. Untested is what proportion of the Rubisco pool in the RDE cells is represented by inactive ER complexes. In vascular plants the reduction or elimination of RCA results in a high-CO_2-requiring phenotype where growth is CO_2 assimilation limited due to low levels of activated, catalytically competent, Rubisco within the stroma [48, 90]. Somewhat akin to this is the extraordinarily high CO_2 requirement by all RDE systems for Rubisco activity selection to be undertaken in air supplemented with >1% (v/v) CO_2. As summarised in Fig. 4.5d (option 3), these observations suggest directed evolution studies using RDE may benefit, possibly alter, the evolutionary outcomes, through co-expressing a RCA that is complimentary to the mutated client Rubisco under examination.

Potential examples of how ER formation might influence RDE selection arise from the study of Mueller-Cajar et al. [47]. Here the content and catalytic properties of the D117V/H and H44Q/N *R. rubrum* L_2 Rubisco mutants did not accord with increased rates of RuBP fixation or improved substrate affinities. How the L_2 mutants improved RDE-MM1 fitness has remained an enigma. Both D117 and H44 are conserved among form II Rubisco and form a hydrogen bond that influences the positioning of a conserved E48 (E60 in plant Rubisco). During catalysis E48 forms a salt bridge with the conserved K329 (K334 in plant Rubisco) residue located at the apex of loop 6 that closes over the catalytic site to initiate catalysis. Mutation of D117 and H44 is therefore thought to alter E48 positioning and perturb loop 6 closure [47]. Conceivably, the altered properties to loop 6 closure conferred by the D117V/H and H44Q/N mutations may alter the capability for inactive ER complex formation. This hypothesis, as well as testing how RCA co-expression influences Rubisco mutant selection in an RDE system, poses considerations for future analysis (Fig. 4.5d, option 3).

4.4.4 A Role for RDE-MM1 in Directed Evolution Studies of RCA

The last few years has seen rapid expansion in our appreciation into the molecular diversity of RCA in nature, in regard to variations in quaternary structure, phylogenetic distribution, mechanistic machinery and recognition specificity among both form I and II Rubisco [29, 48, 49, 73, 77]. Should RCA co-expression benefit directed evolution studies of Rubisco using RDE (Sect. 4.4.3), it may be possible to adapt RDE cells to screen mutated *rca* gene libraries for evolved RCA function. Already a directed evolution study has successfully identified more thermostable plant RCA mutants [41]. However, such in vitro activity assay approaches suffer from low throughput (<7 × 10^3 mutants) and are highly labour intensive, limitations that theoretically would be averted via an in vivo RDE Rubisco screen. Potential enzymatic phenotypes to select for are RCA mutations that unveil information about residues that enable (or stimulate) recognition and binding with cognate or heterologous Rubisco and those that improve RCA thermostability. Both objectives need to consider two key limitations using RDE systems for Rubisco bio-selection. Firstly, while all forms of RCA can be produced in *E. coli* [29, 48, 49, 73, 77], functional Rubisco expression is limited to those sourced from Archaea and some bacteria but not plants and algae (Sect. 4.1.5). This limitation might be avertible by using Rubisco chimers comprising regions of eukaryotic Rubisco subunit sequence transposed into cyanobacterial Rubisco. A comparable approach using *Chlamydomonas* Rubisco chimers (produced by chloroplast transformation) has helped identify how compatibility between amino acids located in the equatorial region of plant L_8S_8 (L-subunit residues 89–94) and those in the short "specificity" helix (H9) of RCA confers the ability of a plant RCA to distinguish Rubisco from Solanaceae and non-Solanaceae species [55]. A second limitation of an RDE Rubisco screen is whether the temperature can be modulated to select for improvements in RCA thermostability. Currently RDE selection requires low temperatures to slow growth and optimise the fidelity of mutant selection [50], a limitation potentially overcome using a PRK-Ab^R fusion (Sect. 4.4.1).

4.5 Summary

The last decade has witnessed a wealth of Rubisco engineering studies aimed at improving CO_2 fixation in crops to increase productivity in response to growing concerns of global food security [17, 45, 53, 67, 85]. The projections are dire – an estimated global population of nine billion by 2050 requires we produce more food in the coming four decades than has been produced in mankind's entire history [23]. While modern directed evolution tools are beginning to show increasing promise towards improving Rubisco performance, they need to advance at a faster pace if we are to reach the improvement level required. Here we propose strategies for improving the fidelity, throughput and expansion of client Rubisco diversity amenable to RDE screening using PRK-antibiotic fusions, through co-expression of appropriate factors to facilitate Rubisco assembly in *E. coli* and possibly through expression of the regulatory protein Rubisco activase (Fig. 4.5d). If directed evolution is to play a role in improving Rubisco to enhance crop photosynthesis and yield potential, then there is a need to steer away from the continuing focus on high *kcatC* prokaryotic Rubisco whose CCM requirements defy what is currently achievable in chloroplasts. There is a need to refocus efforts to developing capabilities to screen mutagenic libraries of more efficient Rubiscos from plants and red algae and fully evaluate the evolutionary potential for enhancing the carboxylation properties of non-photosynthetic Rubisco from Archaea.

References

1. Allen JF, de Paula WBM, Puthiyaveetil S, Nield J (2011) A structural phylogenetic map for chloroplast photosynthesis. Trends Plant Sci 16(12):645–655
2. Alonso H, Blayney MJ, Beck JL, Whitney SM (2009) Substrate-induced assembly of *Methanococcoides burtonii* D-ribulose-1, 5-bisphosphate carboxylase/oxygenase dimers into decamers. J Biol Chem 284(49):33876–33882
3. Amichay D, Levitz R, Gurevitz M (1993) Construction of a *Synechocystis* PCC6803 mutant suitable for the study of variant hexadecameric ribulose bisphosphate carboxylase/oxygenase enzymes. Plant Mol Biol 23:465–476
4. Andersson I, Backlund A (2008) Structure and function of Rubisco. Plant Physiol Biochem 46(3):275–291
5. Andralojc PJ, Madgwick PJ, Tao Y, Keys A, Ward JL, Beale MH, Loveland JE, Jackson PJ, Willis AC, Gutteridge S, Parry MAJ (2012) 2-Carboxy-D-arabinitol 1-phosphate (CA1P) phosphatase: evidence for a wider role in plant Rubisco regulation. Biochem J 442:733–742
6. Andrews TJ, Whitney SM (2003) Manipulating ribulose bisphosphate carboxylase/oxygenase in the chloroplasts of higher plants. Arch Biochem Biophys 414(2):159–169
7. Bai C, Guo P, Zhao Q, Lv Z, Zhang S, Gao F, Gao L, Wang Y, Tian Z, Wang J, Yang F, Liu C (2015) Protomer roles in chloroplast chaperonin assembly and function. Mol Plant 8(10): 1478–1492
8. Cai Z, Liu G, Zhang J, Li Y (2014) Development of an activity-directed selection system enabled significant improvement of the carboxylation efficiency of Rubisco. Protein Cell 5(7):552–562
9. Carmo-Silva E, Scales JC, Madgwick PJ, Parry MAJ (2015) Optimizing Rubisco and its regulation for greater resource use efficiency. Plant Cell Environ 38(9):1817–1832
10. Cleland WW, Andrews TJ, Gutteridge S, Hartman FC, Lorimer GH (1998) Mechanism of Rubisco – the carbamate as general base. Chem Rev 98(2):549–561
11. Cloney LP, Bekkaoui DR, Hemmingsen SM (1993) Co-expression of plastid chaperonin genes and a synthetic plant Rubisco operon in *Escherichia coli*. Plant Mol Biol 23(6):1285–1290
12. Cobb RE, Sun N, Zhao H (2013) Directed evolution as a powerful synthetic biology tool. Methods 60(1):81–90
13. Currin A, Swainston N, Day PJ, Kell DB (2015) Synthetic biology for the directed evolution of protein biocatalysts: navigating sequence space intelligently. Chem Soc Rev 44(5):1172–1239
14. Durão P, Aigner H, Nagy P, Mueller-Cajar O, Hartl FU, Hayer-Hartl M (2015) Opposing effects of folding and assembly chaperones on evolvability of Rubisco. Nat Chem Biol 11(2):148–155
15. Eisenhut M, Ruth W, Haimovich M, Bauwe H, Kaplan A, Hagemann M (2008) The photorespiratory glycolate metabolism is essential for cyanobacteria and might have been conveyed endosymbiontically to plants. Proc Natl Acad Sci 105(44):17199–17204
16. Ellis RJ (1979) The most abundant protein in the world. Trends Biochem Sci 4(4):241–244
17. Evans JR (2013) Improving photosynthesis. Plant Physiol 162(4):1780–1793
18. Ezaki S, Maeda N, Kishimoto T, Atomi H, Imanaka T (1999) Presence of a structurally novel type Ribulose-bisphosphate carboxylase/oxygenase in the hyperthermophilic Archaeon,*Pyrococcus kodakaraensis* KOD1. J Biol Chem 274(8):5078–5082. doi:10.1074/jbc.274.8.5078
19. Falcone DL, Tabita FR (1991) Expression of endogenous and foreign ribulose 1,5-bisphosphate carboxylase-oxygenase (Rubisco) genes in a Rubisco deletion mutant of *Rhodobacter sphaeroides*. J Bacteriol 173(6):2099–2108
20. Farquhar G, von Caemmerer Sv, Berry J (1980) A biochemical model of photosynthetic CO_2 assimilation in leaves of C_3 species. Planta 149 (1):78–90
21. Feiz L, Williams-Carrier R, Wostrikoff K, Belcher S, Barkan A, Stern DB (2012) Ribulose-1,5-bis-phosphate carboxylase/oxygenase accumulation factor1 is required for holoenzyme assembly in maize. Plant Cell 24(8):3435–3446

22. Finn MW, Tabita FR (2004) Modified pathway to synthesize ribulose 1,5-bisphosphate in methanogenic archaea. J Bacteriol 186(19):6360–6366
23. Furbank RT, Quick WP, Sirault XRR (2015) Improving photosynthesis and yield potential in cereal crops by targeted genetic manipulation: prospects, progress and challenges. Field Crop Res 182:19–29
24. Goloubinoff P, Gatenby AA, Lorimer GH (1989) GroE heat-shock proteins promote assembly of foreign prokaryotic ribulose bisphosphate carboxylase oligomers in *Escherichia coli*. Nature 337(6202):44–47
25. Gready J, Kannappan B, Agrawal A, Street K, Stalker DM, Whitney S (2013) Status of options for improving photosynthetic capacity through promotion of Rubisco performance: Rubisco natural diversity and re-engineering, and other parts of C_3 pathways. Paper presented at the Proceedings of a workshop held at the Australian National University, Canberra, Australian Capital Territory, Australia, 2–4 Sept 2009
26. Greene DN, Whitney SM, Matsumura I (2007) Artificially evolved *Synechococcus* PCC6301 Rubisco variants exhibit improvements in folding and catalytic efficiency. Biochem J 404(3): 517–524
27. Groß M, Robinson CV, Mayhew M, Hartl FU, Radford SE (1996) Significant hydrogen exchange protection in GroEL-bound DHFR is maintained during iterative rounds of substrate cycling. Proc Sci 5(12):2506–2513
28. Hartman FC, Harpel MR (1994) Structure, function, regulation, and assembly of D-ribulose-1,5-bisphosphate carboxylase/oxygenase. Annu Rev Biochem 63:197–234
29. Hasse D, Larsson AM, Andersson I (2015) Structure of *Arabidopsis thaliana* Rubisco activase. Acta Crystallogr Sect D: Biol Crystallogr 71(Pt 4):800–808
30. Hauser T, Popilka L, Hartl FU, Hayer-Hartl M (2015) Role of auxiliary proteins in Rubisco biogenesis and function. Nat Plants 1
31. Hayer-Hartl M, Bracher A, Hartl FU (2016) The GroEL-GroES chaperonin machine: a nano-cage for protein folding. Trends Biochem Sci 41(1):62–76. doi:10.1016/j.tibs.2015.07.009
32. Hudson GS, Morell MK, Arvidsson YBC, Andrews TJ (1992) Synthesis of spinach phosphoribulokinase and ribulose 1, 5-bisphosphate in *Escherichia coli*. Aust J Plant Physiol 19:213–221
33. Hwang S-R, Tabita FR (1989) Cloning and expression of the chloroplast-encoded *rbc*L and *rbc*S genes from the marine diatom *Cylindrotheca* sp. strain N1. Plant Mol Biol 13: 69–79
34. Irani M, Maitra P (1974) Isolation and characterization of *Escherichia coli* mutants defective in enzymes of glycolysis. Biochem Biophys Res Commun 56(1):127–133
35. Jiao Z, Baba T, Mori H, Shimizu K (2003) Analysis of metabolic and physiological responses to *gnd* knockout in *Escherichia coli* by using C-13 tracer experiment and enzyme activity measurement. FEMS Microbiol Lett 220(2):295–301
36. Kanevski I, Maliga P, Rhoades DF, Gutteridge S (1999) Plastome engineering of ribulose-1,5-bisphosphate carboxylase/oxygenase in tobacco to form a sunflower large subunit and tobacco small subunit hybrid. Plant Physiol 119(1):133–141
37. Khlebnikov A, Datsenko KA, Skaug T, Wanner BL, Keasling JD (2001) Homogeneous expression of the PBAD promoter in *Escherichia coli* by constitutive expression of the low-affinity high-capacity *AraE* transporter. Microbiologica 147(12):3241–3247
38. Kitano K, Maeda N, Fukui T, Atomi H, Imanaka T, Miki K (2001) Crystal structure of a novel-type archaeal rubisco with pentagonal symmetry. Structure 9(6):473–481
39. Klenk C, Ehrenmann J, Schütz M, Plückthun A (2016) A generic selection system for improved expression and thermostability of G protein-coupled receptors by directed evolution. Sci Rep 6:28133
40. Kreel NE, Tabita FR (2015) Serine 363 of a hydrophobic region of Archaeal Ribulose 1,5-bisphosphate carboxylase/oxygenase from *Archaeoglobus fulgidus* and *Thermococcus kodakaraensis* affects CO_2/O_2 substrate specificity and oxygen sensitivity. PLoS ONE 10(9):e0138351

41. Kurek I, Chang TK, Bertain SM, Madrigal A, Liu L, Lassner MW, Zhu GH (2007) Enhanced thermostability of *Arabidopsis* Rubisco activase improves photosynthesis and growth rates under moderate heat stress. Plant Cell 19(10):3230–3241
42. Leggat W, Whitney S, Yellowlees D (2004) Is coral bleaching due to the instability of the zooxanthellae dark reactions? Symbiosis 37(1–3):137–153
43. Liu C, Young AL, Starling-Windhof A, Bracher A, Saschenbrecker S, Rao BV, Rao KV, Berninghausen O, Mielke T, Hartl FU (2010) Coupled chaperone action in folding and assembly of hexadecameric Rubisco. Nature 463(7278):197–202
44. Liu J-W, Boucher Y, Stokes HW, Ollis DL (2006) Improving protein solubility: the use of the *Escherichia coli* dihydrofolate reductase gene as a fusion reporter. Protein Expr Purif 47(1):258–263
45. Long Stephen P, Marshall-Colon A, Zhu X-G (2015) Meeting the global food demand of the future by engineering crop photosynthesis and yield potential. Cell 161(1):56–66
46. Morell MK, Paul K, Kane HJ, Andrews TJ (1992) Rubisco: maladapted or misunderstood? Aust J Bot 40:431–441
47. Mueller-Cajar O, Morell M, Whitney SM (2007) Directed evolution of Rubisco in *Escherichia coli* reveals a specificity-determining hydrogen bond in the form II enzyme. Biochemist 46(49):14067–14074
48. Mueller-Cajar O, Stotz M, Bracher A (2014) Maintaining photosynthetic CO_2 fixation via protein remodelling: the Rubisco activases. Photosynth Res 119(1–2):191–201
49. Mueller-Cajar O, Stotz M, Wendler P, Hartl FU, Bracher A, Hayer-Hartl M (2011) Structure and function of the AAA^+ protein CbbX, a red-type Rubisco activase. Nature 479(7372): 194–199
50. Mueller-cajar O, Whitney SM (2008) Directing the evolution of Rubisco and Rubisco activase: first impressions of a new tool for photosynthesis research. Photosynth Res 98(1–3):667–675
51. Mueller-Cajar O, Whitney SM (2008) Evolving improved *Synechococcus* Rubisco functional expression in *Escherichia coli*. Biochem J 414(2):205–214
52. Parikh MR, Greene DN, Woods KK, Matsumura I (2006) Directed evolution of Rubisco hypermorphs through genetic selection in engineered *E. coli*. Protein Eng Des Sel 19(3): 113–119
53. Parry MAJ, Andralojc PJ, Scales JC, Salvucci ME, Carmo-Silva AE, Alonso H, Whitney SM (2013) Rubisco activity and regulation as targets for crop improvement. J Exp Bot 64(3):717–730
54. Pellicer MT, Nunez MF, Aguilar J, Badia J, Baldoma L (2003) Role of 2-phosphoglycolate phosphatase of *Escherichia coli* in metabolism of the 2-phosphoglycolate formed in DNA repair. J Bacteriol 185(19):5815–5821
55. Portis AR, Li CS, Wang DF, Salvucci ME (2008) Regulation of Rubisco activase and its interaction with Rubisco. J Exp Bot 59(7):1597–1604
56. Price GD, Evans JR, von Caemmerer S, Yu J-W, Badger MR (1995) Specific reduction of chloroplast glyceraldehyde-3-phosphate dehydrogenase activity by antisense RNA reduces CO_2 assimilation via a reduction in ribulose bisphosphate regeneration in transgenic tobacco plants. Planta 195:369–378
57. Price GD, Howitt SM (2014) Plant science: towards turbocharged photosynthesis. Nature 513(7519):497–498
58. Sakikawa C, Taguchi H, Makino Y, Yoshida M (1999) On the maximum size of proteins to stay and fold in the cavity of GroEL underneath GroES. J Biol Chem 274(30):21251–21256
59. Satagopan S, Chan S, Perry LJ, Tabita FR (2014) Structure-function studies with the unique hexameric Form II ribulose-1, 5-bisphosphate carboxylase/oxygenase (Rubisco) from *Rhodopseudomonas palustris*. J Biol Chem 289(31):21433–21450
60. Satagopan S, Scott SS, Smith TG, Tabita FR (2009) A Rubisco mutant that confers growth under a normally "inhibitory" oxygen concentration. Biochemist 48(38):9076–9083
61. Satagopan S, Spreitzer RJ (2004) Substitutions at the Asp-473 Latch Residue of *Chlamydomonas* ribulosebisphosphate carboxylase/oxygenase cause decreases in carboxylation efficiency and CO_2/O_2 specificity. J Biol Chem 279(14):14240–14244

62. Satagopan S, Tabita FR (2016) RubisCO selection using the vigorously aerobic and metabolically versatile bacterium Ralstonia eutropha. FEBS J 283:2869–2880
63. Sato T, Atomi H, Imanaka T (2007) Archaeal type III RuBisCOs function in a pathway for AMP metabolism. Science 315(5814):1003–1006
64. Savir Y, Noor E, Milo R, Tlusty T (2010) Cross-species analysis traces adaptation of Rubisco toward optimality in a low-dimensional landscape. Proc Natl Acad Sci 107(8):3475–3480
65. Seta FD, Boschi-Muller S, Vignais M, Branlant G (1997) Characterization of *Escherichia coli* strains with *gapA* and *gapB* genes deleted. J Bacteriol 179(16):5218–5221
66. Sharwood R, von Caemmerer S, Maliga P, Whitney S (2008) The catalytic properties of hybrid Rubisco comprising tobacco small and sunflower large subunits mirror the kinetically equivalent source Rubiscos and can support tobacco growth. Plant Physiol 146:83–96
67. Sharwood RE, Ghannoum O, Whitney SM (2016) Prospects for improving CO_2 fixation in C_3-crops through understanding C_4-Rubisco biogenesis and catalytic diversity. Curr Opin Plant Biol 31:135–142
68. Sharwood RE, Sonawane BV, Ghannoum O, Whitney SM (2016) Improved analysis of C_4 and C_3 photosynthesis via refined in vitro assays of their carbon fixation biochemistry. J Exp Bot 67(10):3137–3148
69. Smith SA, Tabita FR (2003) Positive and negative selection of mutant forms of prokaryotic (cyanobacterial) ribulose-1, 5-bisphosphate carboxylase/oxygenase. J Mol Biol 331(3):557–569
70. Solaiman DK, Swingle BM, Ashby RD (2010) A new shuttle vector for gene expression in biopolymer-producing Ralstonia eutropha. J Microbiol Methods 82(2):120–123
71. Soo VW, Hanson-Manful P, Patrick WM (2011) Artificial gene amplification reveals an abundance of promiscuous resistance determinants in *Escherichia coli*. Proc Natl Acad Sci 108(4):1484–1489
72. Stemmer WPC (1994) DNA shuffling by random fragmentation and reassembly: in vitro recombination for molecular evolution. Proc Natl Acad Sci 91:10747–10751
73. Stotz M, Mueller-Cajar O, Ciniawsky S, Wendler P, Hartl FU, Bracher A, Hayer-Hartl M (2011) Structure of green-type Rubisco activase from tobacco. Nat Struct Mol Biol 18(12):1366–1370
74. Tabita FR, Hanson TE, Li H, Satagopan S, Singh J, Chan S (2007) Function, structure, and evolution of the Rubisco-like proteins and their Rubisco homologs. Microbiol Mol Biol Rev 71(4):576–599
75. Tamoi M, Miyazaki T, Fukamizo T, Shigeoka S (2005) The Calvin cycle in cyanobacteria is regulated by CP12 via the NAD(H)/NADP(H) ratio under light/dark conditions. Plant J 42(4):504–513
76. Tcherkez GGB, Farquhar GD, Andrews TJ (2006) Despite slow catalysis and confused substrate specificity, all ribulose bisphosphate carboxylases may be nearly perfectly optimized. Proc Natl Acad Sci 103:7246–7251
77. Tsai YC, Lapina MC, Bhushan S, Mueller-Cajar O (2015) Identification and characterization of multiple rubisco activases in chemoautotrophic bacteria. Nat Commun 6:8883. doi:10.1038/ncomms9883
78. Tsai YC, Mueller-Cajar O, Saschenbrecker S, Hartl FU, Hayer-Hartl M (2012) Chaperonin cofactors, Cpn10 and Cpn20, of green algae and plants function as hetero-oligomeric ring complexes. J Biol Chem 287(24):20471–20481
79. Vitlin Gruber A, Nisemblat S, Azem A, Weiss C (2013) The complexity of chloroplast chaperonins. Trends Plant Sci 18(12):688–694
80. Wang HH, Isaacs FJ, Carr PA, Sun ZZ, Xu G, Forest CR, Church GM (2009) Programming cells by multiplex genome engineering and accelerated evolution. Nature 460(7257):894–898
81. Whitney SM, Andrews TJ (1998) The CO_2/O_2 specificity of single-subunit ribulose-bisphosphate carboxylase from the dinoflagellate, *Amphidinium carterae*. Aust J Plant Physiol 25(2):131–138
82. Whitney SM, Andrews TJ (2001) Plastome-encoded bacterial ribulose-1, 5-bisphosphate carboxylase/oxygenase (RubisCO) supports photosynthesis and growth in tobacco. Proc Natl Acad Sci 98(25):14738–14743

83. Whitney SM, Baldet P, Hudson GS, Andrews TJ (2001) Form I Rubiscos from non-green algae are expressed abundantly but not assembled in tobacco chloroplasts. Plant J 26(5):535–547
84. Whitney SM, Birch R, Kelso C, Beck JL, Kapralov MV (2015) Improving recombinant Rubisco biogenesis, plant photosynthesis and growth by coexpressing its ancillary RAF1 chaperone. Proc Natl Acad Sci 112(11):3564–3569
85. Whitney SM, Houtz RL, Alonso H (2011) Advancing our understanding and capacity to engineer nature's CO_2-sequestering enzyme, Rubisco. Plant Physiol 155(1):27–35
86. Whitney SM, Sharwood RE (2008) Construction of a tobacco master line to improve Rubisco engineering in chloroplasts. J Exp Bot 59(7):1909–1921
87. Whitney SM, Sharwood RE (2014) Plastid transformation for Rubisco engineering and protocols for assessing expression. Methods Mol Biol 1132:245–262
88. Wilson R, Whitney S (2015) Photosynthesis: getting it together for CO_2 fixation. Nat Plants 1:15130
89. Wilson RH, Alonso H, Whitney SM (2016) Evolving *Methanococcoides burtonii* archaeal Rubisco for improved photosynthesis and plant growth. Sci Report 6:22284
90. Yamori W, von Caemmerer S (2009) Effect of Rubisco activase deficiency on the temperature response of CO_2 assimilation rate and rubisco activation state: insights from transgenic tobacco with reduced amounts of Rubisco activase. Plant Physiol 151(4):2073–2082
91. Yoshida S, Atomi H, Imanaka T (2007) Engineering of a type III rubisco from a hyperthermophilic archaeon in order to enhance catalytic performance in mesophilic host cells. Appl Environ Microbiol 73(19):6254–6261
92. Young JN, Heureux AMC, Sharwood RE, Rickaby REM, Morel FMM, Whitney SM (2016) Large variation in the Rubisco kinetics of diatoms reveals diversity among their carbon-concentrating mechanisms. J Exp Bot 67(11):3445–3456
93. Zhu X-G, Kurek I, Liu L (2010) Engineering photosynthetic enzymes involved in CO_2-assimilation by gene shuffling. In: Rebeiz C, Benning C, Bohnert H et al (eds) Advances in photosynthesis and respiration, The chloroplast, vol 31. Springer, Dordrecht, pp 307–322

Directed Evolution of Unspecific Peroxygenase

5

Patricia Molina-Espeja, Patricia Gómez de Santos, and Miguel Alcalde

Abstract

Unspecific peroxygenase (UPO) is a heme-thiolate peroxidase with mono(per)oxygenase activity for the selective oxyfunctionalization of C-H bonds. Fueled by catalytic concentrations of H_2O_2, which acts as both oxygen donor and as final electron acceptor, this stable, soluble, and extracellular enzyme is a potential biocatalyst for dozens of transformations that are of considerable interest in organic synthesis. In this chapter we describe the main attributes of this versatile enzyme while reflecting on the directed evolution campaigns recently followed in our laboratory that set out to enhance the functional expression of UPO in yeast and improve the activity, as well as approximating its properties to the required industrial standards.

5.1 Introduction

At the beginning of the twenty-first century, the first true natural peroxygenase was isolated from the basidiomycete *Agrocybe aegerita* (*Aae*UPO1) [54], an edible mushroom that produces white rot [24]. After its initial misclassification as an unusual alkaline lignin peroxidase, later as an haloperoxidase and often as an aromatic peroxygenase (APO), it was finally recognized as unspecific peroxygenase (UPO) and considered as the first member of a new sub-subclass of oxidoreductases (EC 1.11.2.1) (Table 5.1). What distinguished UPO in this sense is its ability to insert oxygen into unactivated carbon atoms (both in aliphatic and aromatic compounds) with high regio- and enantioselectivity, as well as its broad substrate

P. Molina-Espeja • P.G. de Santos • M. Alcalde (✉)
Department of Biocatalysis, Institute of Catalysis, Consejo Superior de Investigaciones Científicas (CSIC), 28049 Madrid, Spain
e-mail: malcalde@icp.csic.es

M. Alcalde (ed.), *Directed Enzyme Evolution: Advances and Applications*,
DOI 10.1007/978-3-319-50413-1_5

Table 5.1 EC-IUBMB 1.11 classification

Class	Subclass	Sub-subclass	Enzyme	Heme-thiolate peroxidases
EC 1 Oxidoreductases	EC 1.11 Peroxide as acceptor	EC 1.11.1 Peroxidases	EC 1.11.1.10 Chloroperoxidase (CPO)	←
			EC 1.11.1.13 Manganese peroxidase (MnP)	
			EC 1.11.1.14 Lignin peroxidase (LiP)	
			EC 1.11.1.16 Versatile peroxidase (VP)	
			EC 1.11.1.19 Dye decolorazing peroxidase (DyP)	
		EC 1.11.2 Peroxygenases	EC 1.11.2.1 Unspecific peroxygenase (UPO)	←
			EC 1.11.2.2 Myeloperoxidase (MPO)	
			EC 1.11.2.3 Plant seed peroxygenase	
			EC 1.11.2.4 Fatty-acid peroxygenase	
			EC 1.11.2.5 3-methyl-L-tyrosine peroxygenase	

specificity and the sole catalytic requirement for hydrogen peroxide, acting as both final electron acceptor and as oxygen donor (as demonstrated in studies using $H_2{}^{18}O_2$) [3, 27, 29, 30, 33, 55]. Previously, the only enzymes that could efficiently perform such oxyfunctionalizations were the cytochrome P450 monooxygenases (EC 1.14), albeit with an inconvenience that they rely on redox cofactors (NAD(P)H) and auxiliary flavoproteins (Fig. 5.1). Moreover, P450s are usually associated to membranes, while UPOs are soluble and extracellular. As such, from a catalytic point of view, UPO is considered as the "missing link" between P450s and the chloroperoxidase of *Caldariomyces fumago* (CPO, EC 1.11.1.10), being all of them heme-thiolate enzymes (i.e., with a cysteine residue as the axial ligand of the heme group). In this regard, strong efforts have been made to exploit the peroxide shunt pathway of P450s, through which, and like UPO, the enzyme works only supplied by H_2O_2 [13, 26]. However, the poor efficiency of this route, even for ad hoc evolved P450s, and the poor stability in the presence of H_2O_2 are two major hurdles that are not easily overcome [17, 51, 53]. CPO does naturally use H_2O_2 as a co-oxidant and oxygen donor, and hence, both CPO and UPO are placed within the group of heme-thiolate peroxidases. Nevertheless, when the traits of UPO and CPO are inspected

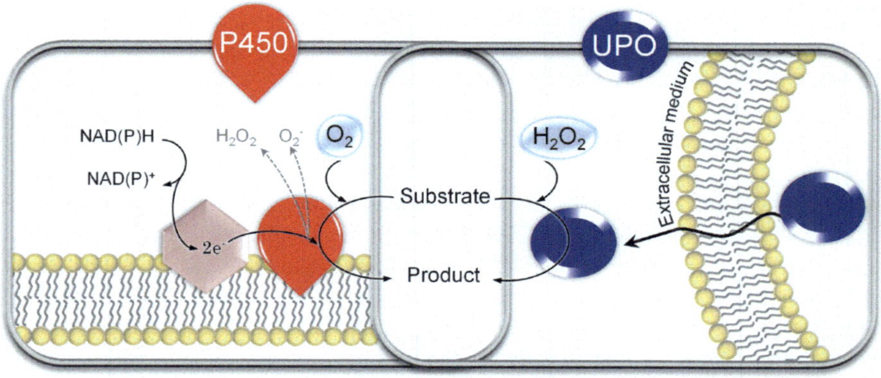

Fig. 5.1 Comparison between P450s and UPO. P450s (*left, red*) are intermembrane proteins with poor stability, which depend on auxiliary proteins (flavoproteins, *pink hexagon*) and redox cofactors as an electron source and that are partially deviated to the formation of unproductive oxygen species that reduces the reaction yield. By contrast, UPO (*right, blue*) is soluble and stable and only needs H_2O_2, overcoming the aforementioned disadvantages of P450s

in more detail, the substrate spectrum of the latter is much more restricted as it is incapable of oxygenating carbon atoms of alicyclic/aromatic rings or *n*-alkanes, which are very inert in chemical terms [20, 22].

*Aae*UPO1 has a relatively novel sequence, with little homology to P450s and sharing only 27% identity with CPO up to 35% in its N-terminal domain [44, 54]. In terms of other UPOs, more than 2500 putative UPO sequences from different fungi have been detected in genomic databases [21, 23]. Moreover, two additional UPOs have hitherto been identified and characterized, produced by *Coprinus* (*Coprinellus*) *radians* (*Cra*UPO) [2] and *Marasmius rotula* (*Mro*UPO) [18, 49]. Other fungi are also considered to be potential producers of UPOs, although they are still to be fully characterized [24]. In addition, Novozymes recently expressed a putative UPO gene from *Coprinopsis cinerea* (r*Cci*UPO) in *Aspergillus oryzae* [4], as well as an UPO from an undetermined mold (rNOVO) [46]. Given their widespread distribution in fungi, UPOs have been sorted into two structural groups: short and long UPOs. Short UPOs (like *Mro*UPO) are found throughout the world of fungi, with molecular weights of ~26 kDa and a histidine residue as a charge stabilizer at the active site. By contrast, long UPOs (like *Aae*UPO1 or *Cra*UPO) are found in basidiomycetes and ascomycetes, with molecular weights of ~44 kDa, an internal disulfide bridge and an arginine residue as a charge stabilizer [24, 44].

The role of UPO in its natural context remains uncertain, although several activities have been proposed. This enzyme might be involved in the synthesis of different metabolites, like antibiotics, as well as in detoxification processes or in the later stages of lignin and humus degradation [20, 24, 31]. Therefore, UPO can be included within the ligninolytic enzyme consortium involved in natural wood decay, along with high-redox potential peroxidases (versatile, lignin, and manganese peroxidases), laccases, and H_2O_2-supplying enzymes like aryl-alcohol oxidase.

5.2 Biochemical and Structural Features

To date, 16 UPO sequences have been detected in the *A. aegerita* genome, although as yet only *Aae*UPO1 has been characterized in depth [24, 44, 54]. With up to 22% glycosylation, this 46 kDa extracellular protein has different isoelectric points (pI) that range from 4.9 to 5.7. Its spectroscopic characteristics are determined by a peak at 420 nm (a Soret band representative of heme-containing proteins) and two maxima at 572 and 540 nm (CT1 and CT2 charge transfer bands, respectively) [54]. The crystallographic structure of *Aae*UPO1 was recently published at ~2 Å resolution [47], comprising ten α-helices and five ß-sheets, the latter formed by very few residues (Fig. 5.2). UPO possesses a heme group at the active site (protoporphyrin IX) whose iron is hexacoordinated showing the fifth position associated with a cysteine sulfur (Cys36) that acts as a proximal (axial) ligand, while the sixth position is associated to a water molecule (distal ligand) (Fig. 5.2a). The axial ligand in CPO and P450s is also a cysteine, unlike other heme-containing proteins such as classic peroxidases (His, His+Asp, Tyr, Tyr+Arg are the other possibilities) [50], and it is assumed that this axial Cys is responsible for the oxygen transfer reaction. The C-terminal is stabilized by a disulfide bridge formed between cysteines 278 and 319. UPO possesses a region for halide binding, close to the entrance to the catalytic pocket, and a structural Mg^{2+} ion (Fig. 5.2b). The entrance of the substrate to the catalytic pocket proceeds along a highly hydrophobic funnel-shaped channel that is made up of ten aromatic residues (nine Phe and one Tyr), these leading the substrate to the active center. Of these aromatic residues, five are of particular interest: three Phe residues that orient the substrate (Phe69, Phe121, and Phe199) and two more that act by delimiting its entrance to the channel (Phe76 and Phe191; Fig. 5.2c, d) [47]. Finally, the Glu196 and Arg189 of UPO act as an acid-base pair involved in catalysis.

UPO can catalyze peroxygenation reactions (i.e., insertion of an oxygen atom through two-electron oxidation, peroxygenase, or mono(per)oxygenase activity) or peroxidation reactions (i.e., abstraction of an electron, peroxidase, or peroxidative activity), and it also possesses certain catalase activity. As selective C-H oxyfunctionalizations are among the most solicited reactions in organic synthesis, the peroxygenase activity of UPO has been studied in depth, with 300 positive substrates already tested (a number that continues to grow). Accordingly, the peroxygenase reactions catalyzed by UPO include aromatic, alkylic, and aliphatic hydroxylations; aromatic and aliphatic olefin epoxidations; ether cleavage; N-dealkylations; sulfoxidations; N-oxidations; and brominations (Fig. 5.3). Indeed, the exploitation of such peroxygenase activity is generating great industrial interest. One of the areas in which UPO can be extremely useful is in the pharmaceutical sector, and indeed, the ability of this enzyme to produce drugs and human drug metabolites (HDMs) has been confirmed [5, 28, 35, 48]. While human P450s are known to metabolize drugs into complex HDMs in the liver, for which it is mandatory to perform pharmacokinetic and pharmacodynamic studies, its chemical synthesis is complicated. However, engineered UPOs could be used to produce HDMs with high selectivity in a more economic and efficient manner. Other interesting applications include the synthesis

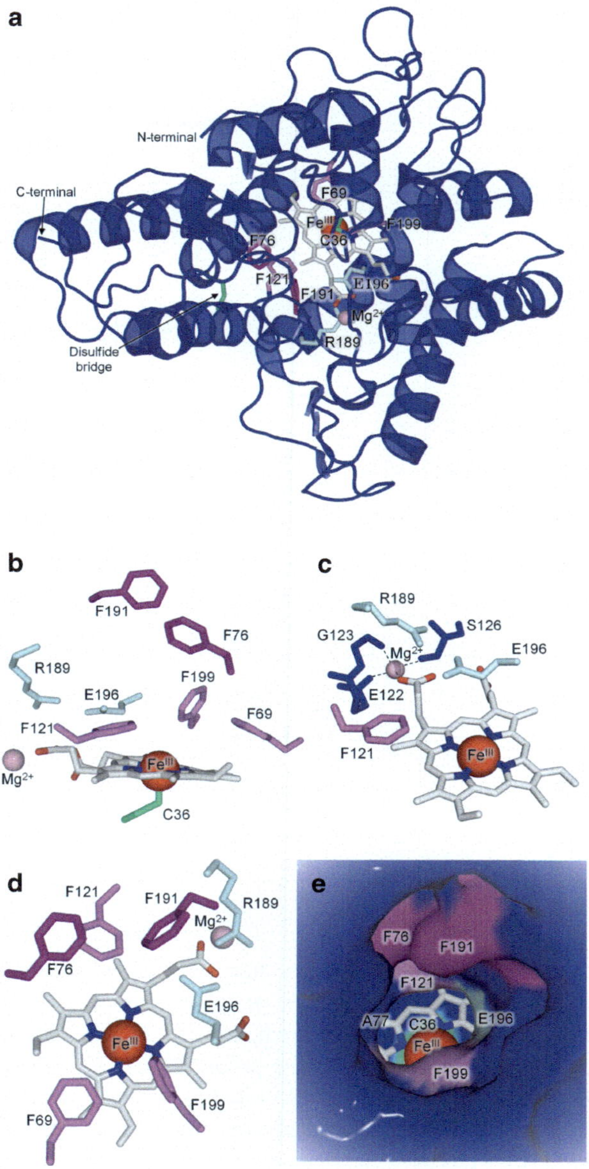

Fig. 5.2 *Aae*UPO1 crystal structure (PDB: 2YOR). (**a**) General view. The disulfide bridge and the axial Cys36 are marked in *green*, the catalytic Phe in *pink* (delimiting the entrance for substrates Phe76 and Phe191 (*dark pink*); orienting the substrate to the heme Phe69, Phe121, and Phe199 (*light pink*)), the structural Mg^{2+} in *light pink*, the acid-base pair (Glu196-Arg189) is represented in *light blue*, the Fe^{III} of the heme is in *red*, and the protoporphyrin IX of the heme is in CPK coloring. (**b**) Catalytic pocket of *Aae*UPO1 with detail of the heme environment. (**c**) Coordination of the Mg^{2+}. (**d**, **e**) Frontal view of the entrance to the heme channel with and without the surface, respectively

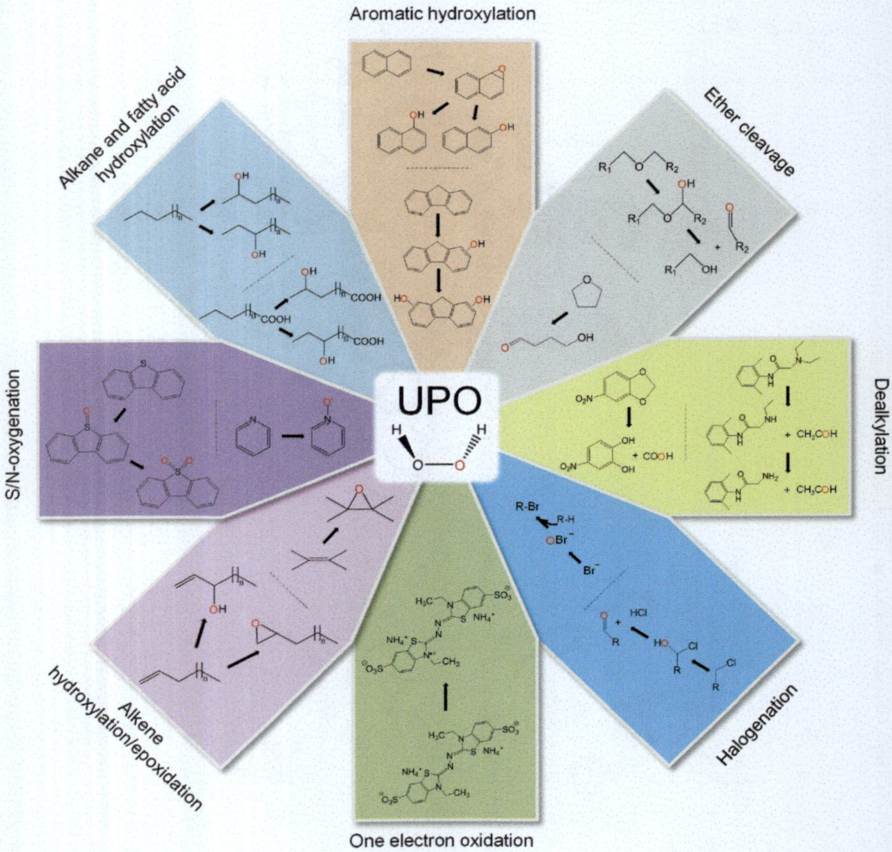

Fig. 5.3 Activities of UPO

of limonene, cyclohexanone, and hydroxylated fatty acids (fatty alcohols) to be used in cosmetics, food, solvents, polymers, or other materials (e.g., nylon) [19, 45, 46]. In this sense, the synthesis of 2,5-furandicarboxylic acid (FDCA), a renewable building block of particular interest for the production of biopolymers to replace traditional polyesters obtained from crude oil, has recently been achieved from 5-hydroxymethylfurfural through a cascade reaction that combined aryl-alcohol oxidase and UPO [12]. Moreover, the potential use of UPO for the functionalization of aliphatic alkanes into alcohols should not be underestimated, since this remains a clear challenge in contemporary chemistry. Besides, the functionalization of aromatic hydrocarbons is very relevant in obtaining added value products, for example, that of benzene to phenol [27] or of naphthalene to 1-naphthol [33]. Another practical example is the transformation of aromatic and aliphatic olefins into their corresponding epoxides, important intermediates in chemical synthesis due to their great versatility (e.g., to produce cosmetics, surfactants, or agrochemicals).

Although native UPO has been studied in the synthesis of the aforementioned compounds, engineering will be essential to improve its performance so that it may fully compete with chemical catalysts. Thus, one of the main goals for the design of this enzyme should include the modeling of its selectivity (e.g., for the terminal hydroxylation of alkanes or fatty acids), increasing its total turnover numbers in any given process, a reduction of its peroxidase activity as a side effect that can lead to unwanted by-products, and enhancement of its oxidative and operational stability or of its activity in the presence of organic cosolvents. To achieve such goals, it is first necessary to heterologously and functionally express the enzyme in a host suited to the tailoring of its properties by directed evolution.

5.3 Directed UPO Evolution

5.3.1 The Kickoff: Evolution Toward Functional Expression

One particular challenge in the directed evolution of ligninolytic enzymes (also known as ligninases) is to achieve their functional expression in the standard heterologous hosts used. All attempts to achieve functional expression of ligninases in *Escherichia coli* have ended up in the formation of inclusion bodies, and although they can be refolded in vitro for structure-function relationship studies, this solution is not practical given the high-throughput context in which directed evolution experiments take place. Differences in codon usage, missing chaperones, and the lack of suitable machinery to induce complex posttranslational modifications (e.g., glycosylation, processing of the N- and/or C-terminal ends) preclude the use of this bacterial host in the directed evolution of ligninases. Indeed, all attempts to express UPO in bacteria carried out to date have failed, not even achieving in vitro refolding from inclusion bodies (Prof. A.T. Martinez, CIB-CSIC, personal communication).

Conversely, the baker's yeast *Saccharomyces cerevisiae* is a more appropriate host to express eukaryotic ligninases as its physiology is much closer to that of its native fungi. As such, *S. cerevisiae* has been used for the directed evolution of the versatile peroxidase, medium- and high-redox potential laccases, aryl-alcohol oxidase, and now UPO ([1] and references herein). Among the main advantages offered by *S. cerevisiae* is its high transformation efficiency (yielding individual colonies with 10^7–10^8 transformants/μg DNA), as well as a broad portfolio of episomal uni- and bidirectional vectors, an efficient secretory system that directs the exocytosis of proteins into the culture medium (bypassing the lysis steps commonly used in *E. coli*), and a high frequency of homologous DNA recombination that incorporates proofreading to avoid the introduction of unwanted mutations. This latter feature allows us to employ a palette of library creation methods in directed evolution experiments.

Even considering all these advantages, the basal functional expression of ligninases in yeast is weak, and UPO is not an exception (in the mU/L range). This is mostly due to differences in the proteases and/or the signal peptidases along the secretory route, coupled to the longer residence time in the Golgi apparatus. This

Golgi retention is frequently associated to enhanced glycosylation that slows down its trafficking, often making the foreign protein toxic to the host. Therefore, we must first apply directed evolution as a "molecular purge" to foment the secretion of ligninases in yeast and, only subsequently, to drive them toward more specific challenges.

The main strategies and tools used for the directed evolution of UPO in *S. cerevisiae* include (i) the design of dual screening assays for both peroxygenase and peroxidase activities, (ii) the optimization of microculture conditions, (iii) signal peptide switching, and (iv) combined mutagenesis and DNA recombination for standard and focused evolution. Given the success achieved in this case study (see below), it is likely that a similar approach could now be adopted for CPO, the other heme-thiolate peroxidase discovered in the 1960s that has resisted the efforts of protein engineers to enhance its functional expression for decades.

When one is trying to drive evolution for secretion, the main objective is to maintain or even improve the different activities of the targeted enzyme while accumulating mutations that enhance secretion, without jeopardizing stability. To avoid the enzyme becoming too specific for a given substrate in the course of evolution for secretion, it is advisable to simultaneously screen with different compounds. In our efforts to direct the evolution of UPO, we used a dual screening assay for peroxygenase and peroxidase activities, while the kinetic thermostability was also measured during rescreening to rule out potential destabilizing mutations. Thanks to this approach, the final secretion mutant of the evolutionary pathway turned out to have equivalent kinetic parameters, stability, and spectroscopic features to the wild-type UPO (see below).

Since high-throughput culture conditions (i.e., the mutant libraries are grown and screened in 96 well plates) are far from optimal (in terms of stirring, oxygen availability, etc.), it is crucial to adjust some factors to ensure the best point of departure for directed UPO evolution, such as strain selection, incubation times, medium (i.e., heme supply and exogenous source of magnesium), and temperature [39].

It is also worth testing different signal peptides that might favor the trafficking of the foreign polypeptide in yeast. Indeed, in addition to the native signal peptide of UPO (n), a native prepro-leader of the α-factor from *S. cerevisiae* (α) [7] and a mutant of the latter that improves secretion (α*) [37] were attached to the native mature UPO in this study. The α-factor prepro-leader is commonly used for heterologous expression in yeast, and in particular, we have used it for the successful engineering of different laccases, versatile peroxidases, and aryl-alcohol oxidases [11, 14, 37, 56]. Surprisingly, the best result was obtained with the native UPO signal peptide (n-UPO), which produced 149 mU/L, followed by the α-UPO (74 mU/L) and the α*-UPO (negligible). It is possible that the evolved α* was not so effective when attached to mature UPO because the beneficial mutations were selected when it was linked to a fungal laccase [37], adjusting both polypeptides ad hoc to the requirements of the yeast's secretory pathway (i.e., the evolved prepro-leader and the laccase). Hence, this evolved α* leader cannot be considered as a universal peptide for the expression of other ligninases in yeast. Nevertheless, the α* leader does work when it is attached to laccases from several different sources and with distinct redox potentials (unpublished material), which suggests a strong connection must exist between the evolved signal peptide and the group of enzymes to which it is attached when pursuing improved secretion.

We very recently constructed chimeric prepro-leaders that combine regions of the α-factor and the K1 killer toxin prepro-leaders and that could be suitable starting points to enhance the secretion of other UPOs [56]. Introducing variability into the signal leader could be also an effective strategy to augment UPO expression, and we have approached this by using a homemade library creation method called MORPHING (mutagenic organized recombination process for homologous in vivo grouping) [16]. This simple tool, supported by the high frequency of homologous DNA recombination of *S. cerevisiae*, allows us to introduce random mutations and recombination events in stretches as small as ~20 amino acids while keeping the rest of the sequence intact. The mutational loads can be adjusted for each specific region according to its length, such that in vivo splicing of the different segments to the linearized plasmid in just a single transformation step occurs by using overhangs of ~40 bp that flank each PCR product. We have taken advantage of MORPHING to engineer versatile peroxidases [16], aryl-alcohol oxidase [56], rubisco (unpublished material), and UPO for different biotechnological needs, highlighting the usefulness of this method for focused evolution.

MORPHING at the UPO signal peptide yielded three almost consecutive and independent mutations (discovered in single mutants) that significantly enhanced secretion [16]. Those changes (F12Y, A14V, and R15G) lie in the hydrophobic core of this sequence, and they were combined with a previously identified A21D mutation [39]. In conjunction, these changes diminish the hydrophobicity of the region, and they may favor interplay between the signal peptide and the signal recognition particle (SRP) in the endoplasmic reticulum pathway. This evolved signal peptide was appropriate to obtain a reliable breakdown in the properties of the final variant. Thus, it was attached to the native UPO in order to achieve sufficient expression in yeast to compare its kinetics and secretion with the evolved UPO counterpart. In fact, we cannot exclude the use of this evolved leader for future efforts where new UPO genes from different sources must be functionally expressed.

Besides focusing evolution on the signal leader, in our experience it is important to combine random mutagenesis and DNA recombination to the whole DNA sequence so that beneficial mutations introduced into independent clones can be easily incorporated into the same gene scaffold in a drive toward total activity improvement (TAI, the product of specific activity and secretion). Accordingly, in five rounds of evolution, we mixed in vivo shuffling and error-prone PCR (with distinct mutational loads), as well as we used DNA polymerases with different mutational bias and thereafter performing in vivo assembly of mutant libraries. As a representative example of the significance of this approach, in the second round of evolution, we took advantage of the distances between the beneficial mutations found in the improved mutants from the first generation (1A11, L67F; 3C2, I248V-F311L) to favor homologous recombination in yeast (Fig. 5.4). In this way, clones

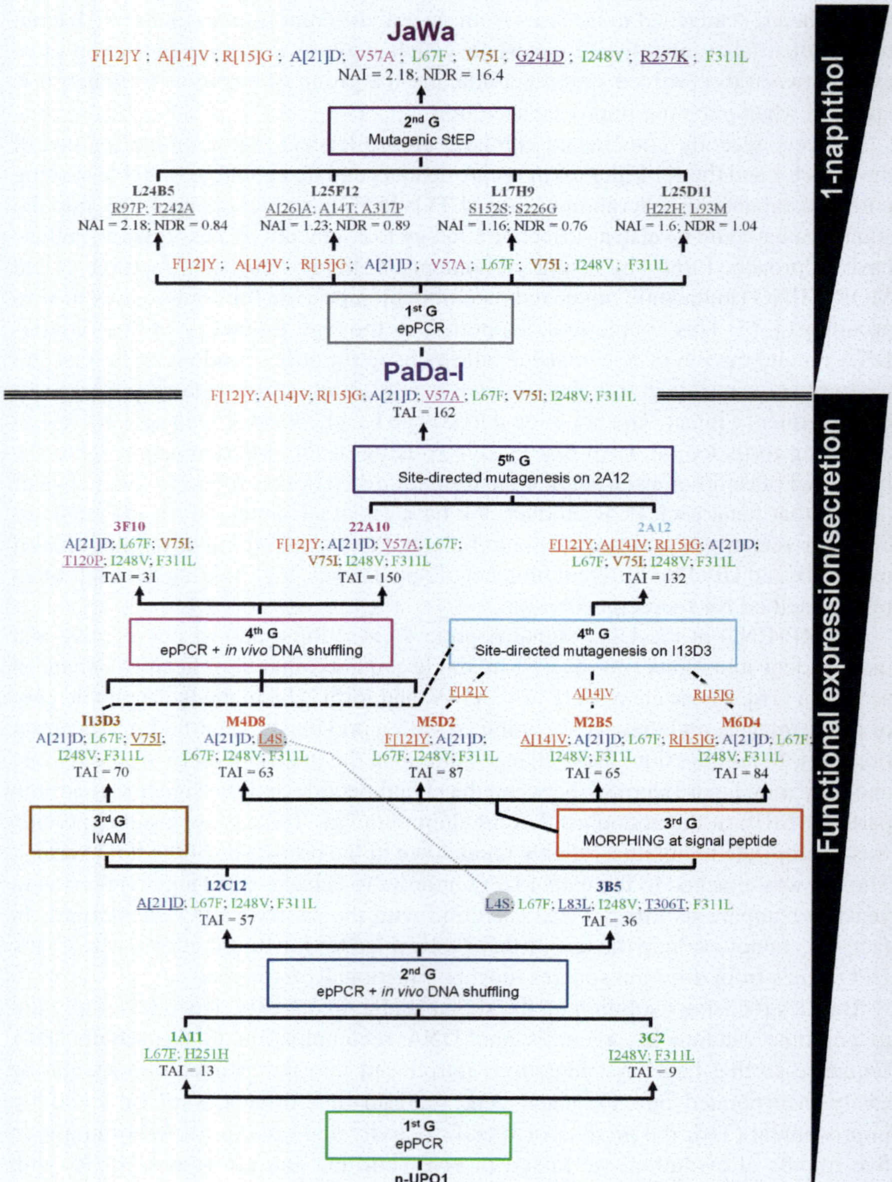

Fig. 5.4 Evolutionary tree of *Aae*UPO1 toward functional expression and 1-naphthol production. The new mutations that appeared in every generation (G) are *underlined*: *TAI* total activity improvement in fold for ABTS, *NAI* naphthalene activity improvement, *NDR* peroxygenase activity/peroxidase activity ratio

could be selected from this new round that incorporated the mutational backbone formed by L67F-I248V-F311, as well as new mutations introduced into either the leader sequence or the mature protein.

The final UPO variant in this evolutionary route toward secretion, PaDa-I, accumulated four mutations in the signal peptide and five more in the mature protein. PaDa-I had similar biochemical properties to the wild-type UPO in terms of N-terminal processing, isoelectric point, pH activity profiles, stability, and spectroscopic features. Its degree of glycosylation was slightly enhanced (from 22% in wild-type UPO to 30% in PaDa-I), possibly due to the tendency of *S. cerevisiae* to hyperglycosylate foreign proteins. In terms of the catalytic constants with the substrates tested (ABTS, NBD, veratryl alcohol, benzyl alcohol, and H_2O_2), both enzymes exhibited similar values, although the overall secretion of PaDa-I by *S. cerevisiae* increased 1114-fold to ~8 mg/L. When we transferred this mutant to the methylotrophic yeast *Pichia pastoris* for overproduction in a bioreactor, these values were enhanced to 217 mg/L, with the variant conserving all the evolved properties. It is worth noting that *P. pastoris* is not a suitable host for directed evolution, not least because of the lack of reliable episomal vectors and the poor transformation efficiencies. After some process engineering, we achieved ~1 g/L UPO from *P. pastoris* in a bioreactor (unpublished material), which comes closer to the application of this tandem-yeast expression system in practical industrial cases (*S. cerevisiae* for directed evolution and *P. pastoris* for large-scale fermentation) [40].

The use of an enzymatic cascade for the atom-efficient in situ generation of H_2O_2 was recently described with PaDa-I. In this approach, methanol works as a sacrificial electron acceptor in a cascade reaction with alcohol oxidase and formaldehyde dismutase to yield three equivalents of H_2O_2 for the complete oxidation of ethylbenzene by the UPO mutant. The total turnover numbers achieved in this system were as high as ~300,000 mol product mol biocatalyst^{-1}, emphasizing the potential application of this type of cascade to minimize UPO oxidative inactivation [42].

The PaDa-I crystal structure has just been resolved at a resolution of 1.5 Å (in collaboration with Prof. Julia Sanz, IQFR-CSIC, Madrid: unpublished data) highlighting some changes, including (i) a complete N-terminal in which the first three residues are lacking in the wild-type UPO crystal structure as a consequence of their potential proteolytic cleavage in *A. aegerita* EPG/LPP [47] and (ii) the F311L mutation that influences the position of Phe76, broadening the entrance to the catalytic pocket. More crystal structures have been resolved recently in association with an array of relevant compounds (naphthalene, veratryl alcohol, benzyl alcohol, 2,6-dimethoxyphenol, styrene, acetanilide, and diclofenac), with all the aromatic rings located at a van der Waals distance to the heme group. All of these are in the pipeline for future structure-guided evolution experiments.

Among the available directed evolution tools, the use of neutral genetic drift can help reveal promiscuous activities along with improved stabilities through the accumulation of a set of neutral mutations. However, unless ultrahigh-throughput screening assays are at hand, this method is very time-consuming as the generation of a neutral network (i.e., a polymorphic population of neutral variants with a substantial number of substitutions) requires considerable experimental effort.

We have coupled in vivo shuffling with genetic drift to harness the benefits of neutral evolution when engineering UPO ([43] and unpublished material). Using PaDa-I as departure point, we shuffled genetic drift, and after only eight rounds, the UPO sequence was modified by up to 11% (with an average of ~2 neutral mutations per clone analyzed). Among the most promising neutral variants, we identified clones with improved peroxygenase activity for different substrates, those with reduced peroxidase activity and those with enhanced stability (over 5 °C in the T_{50}).

5.3.2 Synthesis of 1-Naphthol: A Practical Case Study

A recent example of UPO evolution toward an industrial application was that of the efficient synthesis of 1-naphthol, a compound that is very relevant in the agrochemical, textile, and pharma sectors [6, 25]. The production of 1-naphthol traditionally requires chemical catalysts which, apart from their low efficiency and selectivity, consume considerable energy and generate toxic waste [8–10, 34, 52]. Wild-type UPO can oxygenate naphthalene yielding an intermediate epoxide that spontaneously hydrolyses at acid pH to generate 1-naphthol with only minor traces of 2-naphthol (Fig. 5.5) [32, 33]. Unfortunately, like all phenolic compounds, naphthols are substrates for the peroxidase activity of UPO, such that they are one electron oxidized by the enzyme to produce phenoxyl radicals and quinones. These products subsequently undergo nonenzymatic homopolymerization that affects the reaction yield. Hence, the question that arises is whether the peroxygenase activity of UPO can be conserved while quenching the unwanted peroxidase activity in this

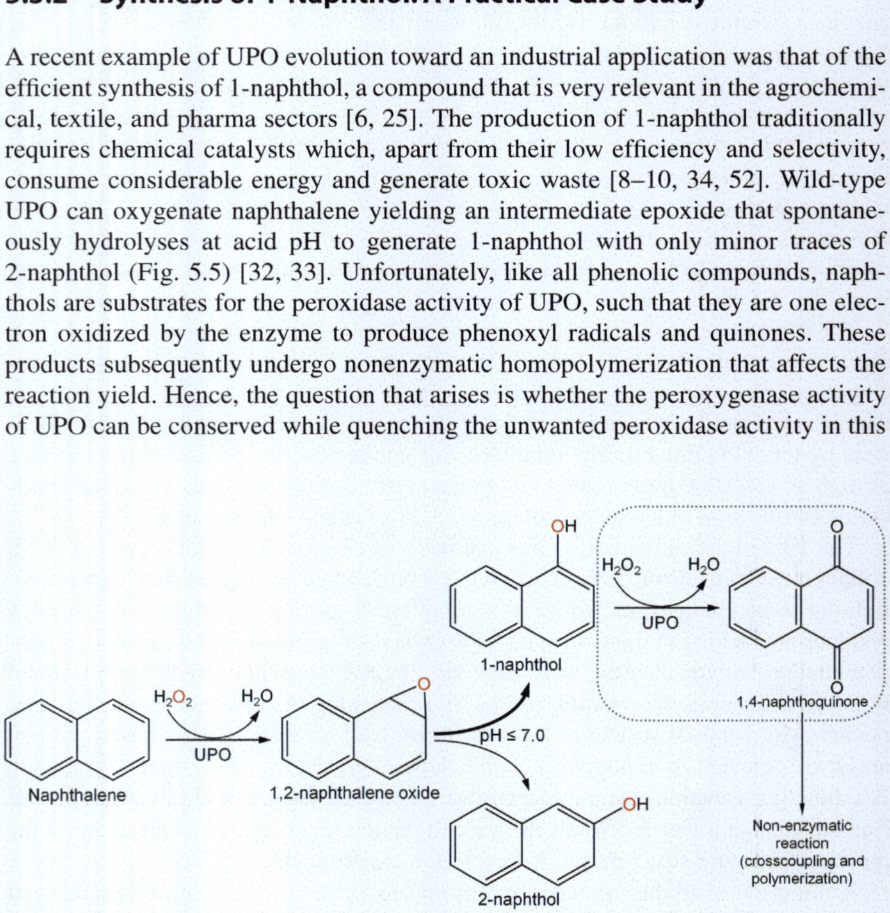

Fig. 5.5 Naphthalene oxygenation by UPO. First, naphthalene is transformed to 1,2-naphthalene oxide, an unstable epoxide that splits into 1-naphthol and minor traces of 2-naphthol. 1-naphthol is again used as substrate by the peroxidase activity of UPO, giving rise to 1,4-naphthoquinone. The resulting quinones undergo nonenzymatic polymerization reactions that produce visible precipitates. 2-naphthol can also follow a similar reaction mechanism (not shown)

process. To address this challenge, we prepared a dual screening assay based on the detection of 1-naphthol produced by UPO (positive selection criteria) and on the one-electron oxidation of phenolics (negative selection criteria). Accordingly, only those variants with improved/persistent peroxygenase activity to transform naphthalene into 1-naphthol and with reduced peroxidase activity to oxidize phenolics were selected as candidates for further engineering [41]. As in the previous evolutionary campaign, we also paid special attention to maintaining the robustness of the protein by measuring kinetic thermostability during rescreenings. Using this approach, we performed directed evolution using mutagenic PCR, in vivo shuffling, and StEP, identifying a new mutant named JaWa (G241D-R257K) with notably weaker peroxidase activity (3- to 12-fold lower in function of the substrate), slightly improved peroxygenase activity on naphthalene (~1.5-fold better catalytic efficiency), and enhanced thermostability (by 2 °C). As a result, JaWa had total turnover numbers of 50,000 mol product mol biocatalyst^{-1} and a high regioselectivity (97%) for 1-naphthol production. Computational analysis by QM/MM indicated that the mutation at position 241 rotates the backbone of the α-helix in which the acid-base pair (Glu196-Arg189) lies such that Arg189 is closer to the substrate in the catalytic site. In addition, the use of PELE (protein energy landscape exploration) simulations [36] indicated that the distance between the naphthalene and the heme was shorter for JaWa than for the parental PaDa-I, improving substrate affinity.

Due to its unique features, this JaWa variant has been benchmarked in our laboratory, along with other UPOs, for the synthesis of several phenolic HDMs like 5′-hydroxypropranolol 5′-OHP-, a process that requires the use of expensive radical scavengers like ascorbic acid to revert the peroxidase activity. Indeed, JaWa by far surpassed the performance of wild-type UPO, even in the absence of ascorbic acid, and it is currently being subjected to new rounds of structure-guided evolution for this synthetic application where peroxidase activity must be suppressed (unpublished material).

In this regard, it is noteworthy that the reduction in peroxidase activity is frequently associated to a drastic drop in stability, and as such, we are careful to select stable variants while decreasing undesired one-electron oxidation. For example, we identified the 3F10 mutant (T120P) in the course of evolution for secretion, with a ~fourfold improved peroxygenase to peroxidase ratio albeit at the cost of its stability (a decreased in T_{50} of ~7 °C). Similarly, we applied MORPHING to structural surface motives of UPO chosen by computational analysis in order to identify residues possibly involved in a long-range electron transfer (LRET) pathway to the heme for the peroxidase activity. Through this approach we again found changes at the same position but this time mutated to distinct residues (T120I; T120A) and with a drastic drop in stability. Further combinatorial saturation mutagenesis experiments unveiled the co-existence of several oxidations sites for peroxidative and peroxygenative activities in UPO [38]. We also explored several potential oxidizable residues involved in such LRET by QM/MM, including a mutated Trp24 (W24F). However, this change reduced both peroxidase and peroxygenase activities, emphasizing the complexity of this issue.

Our future endeavors to fully suppress peroxidase activity will include the introduction of several stabilizing mutations together with those that suppress peroxidase activity in a drive to obtain a unique mono(per)oxygenase with little peroxidase activity (ongoing work).

Conclusions

Since C-H oxyfunctionalization is one of the most desired reactions in organic synthesis, the use of UPO can potentially supplant chemical procedures to produce fine chemicals, drugs, or building blocks. Whether or not UPO becomes the next-generation replacement of P450s remains to be seen. Nevertheless, since its discovery in 2004, it has become clear that UPO may help simplify some complex reactions that are particularly significant in organic chemistry and that until now have been the exclusive terrain of P450s. UPO is extracellular and stable, and it harnesses the limited peroxide shunt pathway of P450s in a highly proficient manner, achieving thousands of total turnover numbers within a range of selective oxygen transfer reactions. Even though UPO reunites some very attractive features, its practical use in real biotechnological processes is still to be established. However, it is easy to envisage how the directed evolution of UPO could represent a vehicle to transfer this biocatalyst from the laboratory bench to industry. Indeed, UPO has now been adapted to the secretory machinery of *S. cerevisiae* and *P. pastoris* by means of directed evolution, and it has been further engineered for the synthesis of agrochemicals and HDMs, which we are confident is only the beginning.

We foresee that the robust tandem-yeast expression system designed for UPO evolution will allow us to sculpt new enzyme attributes, involving both strong peroxygenase activity with minimal oxidative inactivation and the full suppression of peroxidase activity. We truly visualize a future in which the catalog of UPO applications will be widened toward nonnatural chemistry or activities in nonnatural environments and whereby it may also be integrated into a fully consolidated bioprocessing microbe (white-rot yeast) in a drive toward more eco-friendly solutions for the production of chemicals and fuels [15].

Acknowledgments We acknowledge the funding and financial support obtained from the European Commission projects FP7-KBBE-2013-7-613549-INDOX, H2020-BBI-PPP-2015-2-720297-ENZOX2, the COST-Action CM1303, and the Spanish Government projects BIO2013-43407-R-DEWRY and BIO2016-79106-R-LIGNOLUTION.

References

1. Alcalde M (2015) Engineering the ligninolytic enzyme consortium. Trends Biotechnol 33:155–162
2. Anh DH, Ullrich R, Benndorf D, Svatoś A, Muck A, Hofrichter M (2007) The coprophilous mushroom *Coprinus radians* secretes a haloperoxidase that catalyzes aromatic peroxygenation. Appl Environ Microbiol 73:5477–5485
3. Aranda E, Kinne M, Kluge M, Ullrich R, Hofrichter M (2009) Conversion of dibenzothiophene by the mushrooms *Agrocybe aegerita* and *Coprinellus radians* and their extracellular peroxygenases. Appl Microbiol Biotechnol 82:1057–1066

4. Babot ED, del Río JC, Kalum L, Martínez AT, Gutiérrez A (2013) Oxyfunctionalization of aliphatic compounds by a recombinant peroxygenase from *Coprinopsis cinerea*. Biotechnol Bioeng 110:2323–2332
5. Barková K, Kinne M, Ullrich R, Hennig L, Fuchs A, Hofrichter M (2011) Regioselective hydroxylation of diverse flavonoids by an aromatic peroxygenase. Tetrahedron 67:4874–4878
6. Booth G (2012) Naphthalene derivatives. In: Ullmann's encyclopedia of industrial chemistry. Wiley-VCH Verlag GmbH & Co. KGaA, Weinheim.
7. Brake AJ, Merryweather JP, Coit DG, Heberlein UA, Masiarz FR, Mullenbach GT, Urdea MS, Valenzuela P, Barr PJ (1984) α-factor-directed synthesis and secretion of mature foreign proteins in *Saccharomyces cerevisiae*. Biochemistry 81:4642–4646
8. Calinescu I, Avram R (1994a) 1-Naphthol production from tetraline. 1. Tetraline catalytic-oxidation to 1-tetralene. Rev Chem 45:97–103
9. Calinescu I, Avram R (1994b) 1-naphthol production from tetraline. 3. Tetralone dehydrogenation to 1-naphthol. Rev Chem 45:865–867
10. Calinescu I, Avram R, Lovu H (1994) 1-naphthol production from tetraline. 2. Tetraline catalytic-oxidation to 1-tetralone – kinetic constants determination. Rev Chem 45:299–305
11. Camarero S, Pardo I, Cañas AI, Molina P, Record E, Martínez AT, Martínez MJ, Alcalde M (2012) Engineering platforms for directed evolution of laccase from *Pycnoporus cinnabarinus*. Appl Environ Microbiol 78:1370–1384
12. Carro J, Ferreira P, Rodríguez L, Prieto A, Serrano A, Balcells B, Ardá A, Jiménez-Barbero J, Gutiérrez A, Ullrich R, Hofrichter M, Martínez AT (2015) 5-hydroxymethylfurfural conversion by fungal aryl-alcohol oxidase and unspecific peroxygenase. FEBS J 282: 3218–3229
13. Cirino PC, Arnold FH (2002) Protein engineering of oxygenases for biocatalysis. Curr Opin Chem Biol 6:130–135
14. Garcia-Ruiz E, Gonzalez-Perez D, Ruiz-Dueñas FJ, Martínez AT, Alcalde M (2012) Directed evolution of a temperature-, peroxide- and alkaline pH-tolerant versatile peroxidase. Biochem J 441:487–498
15. Gonzalez-Perez D, Alcalde M (2014) Assembly of evolved ligninolytic genes in *Saccharomyces cerevisiae*. Bioengineered 5:254–263
16. Gonzalez-Perez D, Molina-Espeja P, Garcia-Ruiz E, Alcalde M (2014) Mutagenic organized recombination process by homologous in vivo grouping (MORPHING) for directed enzyme evolution. PLoS ONE 9:e90919
17. Grinkova YV, Denisov IG, McLean MA, Sligar SG (2013) Oxidase uncoupling in heme monooxygenases: human cytochrome P450 CYP3A4 in nanodiscs. Biochem Biophys Res Commun 430:1223–1227
18. Gröbe G, Ullrich R, Pecyna MJ, Kapturska D, Friedrich S, Hofrichter M, Scheibner K (2011) High-yield production of aromatic peroxygenase by the agaric fungus *Marasmius rotula*. AMB Express 2011:1–31
19. Gutiérrez A, Babot ED, Ullrich R, Hofrichter M, Martínez AT, del Río JC (2011) Regioselective oxygenation of fatty acids, fatty alcohols and other aliphatic compounds by a basidiomycete heme-thiolate peroxidase. Arch Biochem Biophys 514:33–43
20. Hofrichter M, Ullrich R (2006) Heme-thiolate haloperoxidases: versatile biocatalysts with biotechnological and environmental significance. Appl Microbiol Biotechnol 71:276–288
21. Hofrichter M, Ullrich R, Pecyna MJ, Liers C, Lundell T (2010) New and classic families of secreted fungal heme peroxidases. Appl Microbiol Biotechnol 87:871–897
22. Hofrichter M, Ullrich R (2013) Oxidations catalyzed by fungal peroxygenases. Curr Opin Chem 19:116–125
23. Hofrichter M, Ullrich R, Kellner H, Upadhyay RC and Scheibner K (2014) Fungal unspecific peroxygenases: a new generation of oxygen-transferring biocatalysts. In: Proceedings of the 8th international conference on Mushroom Biology and Mushroom Products (ICMBMP8), pp 172–181
24. Hofrichter M, Kellner H, Pecyna MJ and Ullrich R (2015) Fungal unspecific peroxygenases: heme-thiolate proteins that combine peroxidase and cytochrome P450 properties. In: Hrycay

EG, Bandiera SM (eds) Monooxygenase, peroxidase and peroxygenase properties and mechanisms of cytochrome P450. Adv Exp Med Biol 851:341–368.
25. Jegannathan KR, Nielsen PH (2013) Environmental assessment of enzyme use in industrial production – a literature review. J Clean Prod 42:228–240
26. Joo H, Lin Z, Arnold FH (1999) Laboratory evolution of peroxide-mediated cytochrome P450 hydroxylation. Nature 399:670–673
27. Karich A, Kluge M, Ullrich R, Hofrichter M (2013) Benzene oxygenation and oxidation by the peroxygenase of *Agrocybe aegerita*. AMB Express 3:5
28. Kinne M, Poraj-Kobielska M, Aranda E, Ullrich R, Hammel KE, Scheibner K, Hofrichter M (2009) Regioselective preparation of 5-hydroxypropranolol and 4′-hydroxydiclofenac with a fungal peroxygenase. Bioorg Med Chem Lett 19:3085–3087
29. Kinne M, Poraj-Kobielska M, Ralph SA, Ullrich R, Hofrichter M, Hammel KE (2009) Oxidative cleavage of diverse ethers by an extracellular fungal peroxygenase. J Biol Chem 284:29343–29349
30. Kinne M, Zeisig C, Ullrich R, Kayser G, Hammel KE, Hofrichter M (2010) Stepwise oxygenations of toluene and 4-nitrotoluene by a fungal peroxygenase. Biochem Biophys Res Commun 397:18–21
31. Kinne M, Poraj-Kobielska M, Ullrich R, Nousiainen P, Sipilä J, Scheibner K, Hammel KE, Hofrichter M (2011) Oxidative cleavage of non-phenolic ß-0-4 lignin model dimers by an extracellular aromatic peroxygenase. Holzforschung 65:673–679
32. Kluge M, Ullrich R, Scheibner K, Hofrichter M (2007) Spectrophotometric assay for detection of aromatic hydroxylation catalyzed by fungal haloperoxidase-peroxygenase. Appl Microbiol Biotechnol 75:1473–1478
33. Kluge M, Ullrich R, Dolge C, Scheibner K, Hofrichter M (2009) Hydroxylation of naphthalene by aromatic peroxygenase from *Agrocybe aegerita* proceeds via oxygen transfer from H_2O_2 and intermediary epoxidation. Appl Microbiol Biotechnol 81:1071–1076
34. Kudo K, Ohmae T, Uno A (1976) U S Patent 3, 935, 282
35. Lucas F, Babot ED, Cañellas M, del Río JC, Kalum L, Ullrich R, Hofrichter M, Guallar V, Martínez AT, Gutiérrez A (2016) Molecular determinants for selective C25-hydroxylation of vitamins D2 and D3 by fungal peroxygenases. Catal Sci Technol 6:288–295
36. Madadkar-Sobhani A, Guallar V (2013) PELE web server: atomistic study of biomolecular systems at your fingertips. Nucleic Acids Res 41:322–328
37. Mate DM, García-Burgos C, García-Ruiz E, Ballesteros AO, Camarero S, Alcalde M (2010) Laboratory evolution of high-redox potential laccases. Chem Biol 17:1030–1041
38. Mate DM, Palomino MA, Mollina-Espeja P, Martin-Diaz J, Alcalde M (2017) Modification of the peroxygenative: peroxidative activity ratio in the unspecific peroxygenase from *Agrocybe aegerita* by structure-guided evolution. Protein Eng Des Sel. In press. https://doi.org/10.1093/protein/gzw073
39. Molina-Espeja P, Garcia-Ruiz E, Gonzalez-Perez D, Ullrich R, Hofrichter M, Alcalde M (2014) Directed evolution of unspecific peroxygenase from *Agrocybe aegerita*. Appl Environ Microbiol 80:3496–3507
40. Molina-Espeja P, Ma S, Mate DM, Ludwig R, Alcalde M (2015) Tandem-yeast expression system for engineering and producing unspecific peroxygenase. Enzym Microb Technol 73-74:29–33
41. Molina-Espeja P, Cañellas M, Plou FJ, Hofrichter M, Lucas F, Guallar V, Alcalde M (2016) Synthesis of 1-naphthol by a natural peroxygenase engineered by directed evolution. ChemBioChem 17:341–349
42. Ni Y, Fernández-Fueyo E, Gomez Baraibar A, Ullrich R, Hofrichter M, Yanase H, Alcalde M, van Berkel WJH, Hollmann F (2016) Peroxygenase-catalyzed oxyfunctionalization reactions promoted by the complete oxidation of methanol. Angew Chem Int Ed Eng 54:1–5
43. Paret C, Martin-Diaz J, Alcalde M (2015) Shuffling the neutral drift of unspecific peroxygenase expressed in yeast. Master degree dissertation in Molecular and Cellular Biology, Universidad Autónoma de Madrid

44. Pecyna MJ, Ullrich R, Bittner B, Clemens A, Scheibner K, Schubert R, Hofrichter M (2009) Molecular characterization of aromatic peroxygenase from *Agrocybe aegerita*. Appl Microbiol Biotechnol 84:885–897
45. Peter S, Kinne M, Ullrich R, Kayser G, Hofrichter M (2013) Epoxidation of linear, branched and cyclic alkenes catalyzed by unspecific peroxygenase. Enzym Microb Technol 52:370–376
46. Peter S, Karich A, Ullrich R, Gröbe G, Scheibner K, Hofrichter M (2014) Enzymatic *one-pot* conversion of cyclohexane into cyclohexanone: comparison of four fungal peroxygenases. J Mol Catal B Enzym 103:47–51
47. Piontek K, Strittmatter E, Ullrich R, Gröbe G, Pecyna MJ, Kluge M, Scheibner K, Hofrichter M, Plattner DA (2013) Structural basis of substrate conversion in a new aromatic peroxygenase cytochrome P450 functionality with benefits. J Biol Chem 288:34767–34776
48. Poraj-Kobielska M, Kinne M, Ullrich R, Scheibner K, Kayser G, Hammel KE, Hofrichter M (2011) Preparation of human drug metabolites using fungal peroxygenases. Biochem Pharmacol 82:789–796
49. Poraj-Kobielska M (2013) Conversion of pharmaceuticals and other drugs by fungal peroxygenases. PhD Dissertation. Technical University of Dresden, Germany
50. Rydberg P, Sigfridsson E, Ryde U (2004) On the role of the axial ligand in heme proteins: a theoretical study. J Biol Inorg Chem 9:203–223
51. Sakaki T (2012) Practical application of cytochrome P450. Biol Pharm Bull 35:844–849
52. Schuster L, Seid B (1979) U S Patent 4, 171, 459
53. Shoji O, Watanabe Y (2014) Peroxygenase reactions catalyzed by cytochromes P450. J Biol Inorg Chem 19:529–539
54. Ullrich R, Nüske J, Scheibner K, Spantzel J, Hofrichter M (2004) Novel haloperoxidase from the agaric basidiomycete *Agrocybe aegerita* oxidizes aryl alcohols and aldehydes. Appl Environ Microbiol 70:4575–4581
55. Ullrich R, Dolge C, Kluge M, Hofrichter M (2008) Pyridine as novel substrate for regioselective oxygenation with aromatic peroxygenase from *Agrocybe aegerita*. FEBS Lett 582: 4100–4106
56. Viña-Gonzalez J, Gonzalez-Perez D, Ferreira P, Martinez AT, Alcalde M (2015) Focused directed evolution of aryl-alcohol oxidase in *Saccharomyces cerevisiae* by using chimeric signal peptides. Appl Environ Microbiol 81:6451–6462

Recent Advances in Directed Phytase Evolution and Rational Phytase Engineering

6

Amol V. Shivange and Ulrich Schwaneberg

Abstract

Phytases are hydrolytic enzymes that initiate stepwise removal of phosphate from phytate. Phytate is the major phosphorous storage compound in cereal gains, oilseeds, and legumes and is indigestible by monogastric animals such as poultry and swine. Supplementation of phytase in animal feed proved to improve animal nutrition and decrease phosphorous pollution. Several phytases were discovered in the last century, and today a highly competitive market situation emerged the demands for phytases that are redesigned to excellently match industrial demands. Phytase engineering by directed evolution and rational design has offered a robust approach to tailor-made phytases with high specific activity, broad thermal and pH profile, and protease resistance. In this chapter, we summarized challenges and successful approaches employed in phytase engineering. Factors influencing phytase thermostability, pH stability, pH optima, and protease resistance have been discussed with respect to structural perspective and potential molecular mechanism for improvement. Importance of cooperative substitutions and a way to identify these interactions are discussed. Recent development in screening technology and molecular insights in combining key beneficial substitutions are detailed. In addition, strategies and approaches for rapid and efficient evolution of phytases and to understand structure function relationships on a molecular level have been proposed.

A.V. Shivange, PhD (✉)
Division of Biology and Biological Engineering, California Institute of Technology, Mail Code 156-29, 1200 E. California Blvd, Pasadena, CA 91125, USA
e-mail: shivange@caltech.edu

U. Schwaneberg
Lehrstuhl für Biotechnologie, RWTH Aachen University, Worringerweg 1, 2074 Aachen, Germany

M. Alcalde (ed.), *Directed Enzyme Evolution: Advances and Applications*, DOI 10.1007/978-3-319-50413-1_6

6.1 Introduction

Phytase catalyzes a partial or complete removal of inorganic phosphate from phytic acid (*myo*-inositol hexakisphosphates) or its salt phytate by stepwise hydrolysis of phosphomonoester bonds (Fig. 6.1). Phytate is a primary storage form of phosphate in plants, stored in cereal gains, oilseeds, and legumes which accounts for 50–80% of total phosphate in the plant [1]. Unfortunately, phytate is indigestible by monogastric animals such as swine, poultry, fish, and human beings due to lack of adequate phytase activity in the gastrointestinal tract. In addition, a strong chelating property of phytate makes complexes with divalent cations, proteins, and amino acids [2, 3]. Consequently, supplementation of digestible phosphate is required resulting in a large amount of inorganic phosphate excretion (2% of the dry matter content) causing environmental pollution (60 million pigs are produced in the United States annually, and each pig excretes 1.23 kg of phosphorous in the full life cycle) [4]. Supplementation of phytase as a feed additive has proven to improve nutrition uptake and reduce phosphate excretion by 50% [5]. An estimated supplementation with 300–600 phytase activity units/kg of the diet may replace phosphorous (1.0 g/kg of diet) supplied as monocalcium phosphate [4]. Additionally, supplemental phytase improves calcium, zinc, and iron utilization by animals [6, 7]. The first phytase was reported in 1907 [4] and the phytase research persisted slowly over a century because the use of phytase provided no cost benefits over the use of inorganic phosphorous as a feed supplement. In the last decade, a potential sixfold increase in inorganic phosphorous cost (phosphate rocks forecasted to be eventually mined out [8]), awareness, and recent worldwide regulations on controlling agricultural pollution has made phytase supplementation much more cost-effective and attractive. Currently, phytases represent more than 60% of the total feed enzyme market [9] with a market value of US ~$350 million per year and an impressive growth rate of 10% per annum [10]. In addition to the feed industry, phytases have a great potential in food processing [11], biofuel production [12], alcohol production [13], human nutrition [1], and synthesis of lower *myo*-inositol phosphate [14].

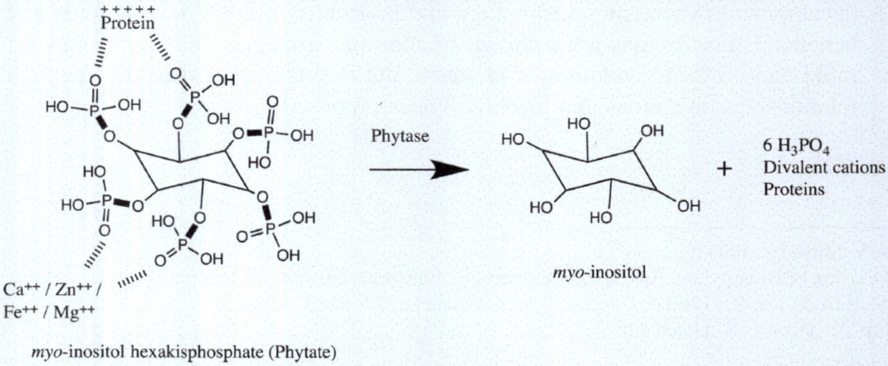

Fig. 6.1 Schematic illustration of the phytase hydrolytic reaction. A stepwise removal of phosphate is catalyzed by hydrolytic cleavage of a phosphomonoester bond (*bold lines*) releasing inorganic phosphate, less phosphorylated *myo*-inositol, divalent metal ions, and bound proteins

Phytases are widely distributed and found in animals, plants, and microorganisms. Based on the crystal structure and catalytic mechanism, a commonly accepted phytase nomenclature divides them into four classes: histidine acid phosphatase (HAP), β-propeller phytase (BPP), purple acid phosphatase (PAP), and protein tyrosine phosphatase (PTP) also referred as cysteine phytase. Several extensively studied and commercialized fungal and bacterial phytases belong to HAP class that shares conserved active site sequence motifs (RHGXRXP and HD) and two structural domains (α-domain and α/β-domain). Another extensively studied phytase from BPP class, also known as alkaline phytases, has β-sheet propeller structure, exhibits strict activity for calcium-phytate complex, and shows high thermostability [15]. Industrial application of phytases requires high specific activity, higher thermal tolerance for feed pelleting process (60–80 °C), and higher pH stability and protease resistance to withstand gastric environments. Phytases from HAP class possess high specific activity (100–3000 U/mg) but have moderate thermostability, whereas phytases from BPP class are more thermostable but exhibit lower catalytic efficiency (specific activity <50 U/mg). Therefore, an ideal phytase may rarely, if at all, be found in nature, and there is a constant need to evolve phytases for industrial demands. In this chapter, we summarize recent development in phytase engineering by directed evolution and rational design strategies. Novel strategies and approaches for rapid and efficient evolution of phytases are additionally discussed based on success stories in phytase reengineering.

6.2 Challenges in Directed Phytase Evolution

6.2.1 Limitations of Random Mutagenesis Methods Employed for Directed Phytase Evolution

Despite the advances in "smart" mutant library generation and screening methods, directed phytase evolution campaigns have been a challenging and laborious process. Thermostability improvement for phytases has been a primary goal for industrial demands of phytases. The main challenge in designing directed phytase evolution experiment for improving thermal resistance is selection of a "right" method for mutagenesis and screening. Over the past decade, several phytases have been evolved for higher thermostability. A key observation in the success stories of directed phytase evolution has been that the majority of amino acid substitutions are "chemically" different than the wild-type amino acids, for example, glycine to aspartic acid, aspartic acid to asparagine [16], lysine to glutamate/methionine [17], threonine to lysine, glutamine to histidine, and lysine to glutamine [18]. This chemical diversity in the substitution pattern indicates a prerequisite to employ a mutagenesis method that generates diverse mutational spectra enriched with functional traits. Organization of the genetic code and the naturally occurring mutational bias favor strongly chemically similar substitution patterns [19] agreeing with the ambiguity reduction theory [20], which proposes that the genetic code was organized to reduce incorporation of chemically diverse substitutions. Widely used polymerase-based mutagenesis methods have a bias toward transitions (mutation from purine ➔ purine or pyrimidine ➔ pyrimidine), resulting in hampered chemical diversity in the generated mutant library. For example, a transition at the third position of a codon encodes in almost all the codons the wild-type amino acid. In contrast, transversion (A/G ➔ T/C or T/C ➔ A/G)-enriched mutagenesis methods have been shown to generate chemical diversity enriched with diverse mutational spectra [21, 22]. Screening several epPCR libraries (6800 clones) failed to yield a thermostable variant from *Yersinia mollaretii* phytase (Ymphytase); however, generating a chemically diverse library with high mutational load resulted in an identification of a thermostable Ymphytase variant M1 with moderate screening efforts (2350 clones) [22]. Two (T77K and V298M) out of five (D52N, T77K, K139E, G187S, V298M) substitutions found in M1 were unobtainable by the employed epPCR method. These results provide an insight that employing a mutagenesis method which incorporates chemically diverse substitutions will enrich the library population with functional phytase variants and reduce screening efforts in the directed phytase evolution.

6.2.2 Phytase Fitness Landscape: Improving More than One Property in Directed Phytase Evolution

Enzyme evolution for industrial demands frequently requires an optimization of more than one property. An ideal phytase would be thermostable, highly active, protease resistant and low pH resistant. Directed phytase evolution campaigns have usually been aimed to improve one property at a time. An essential property for industrial feed pelleting process is the high thermal tolerance. A known fact is that the phytases from thermophilic organism either have undesirable high optimal temperature (>50 °C) or lower specific activity [23], whereas mesophilic phytases have higher specific activity but moderate thermostability [17]. A "fitness" for phytases can be defined as a sequence that satisfies all the criteria for a phytase to function as desired. Criteria in the screening process may include high activity at low pH and a high thermal and protease resistance. Phytase evolution can then be envisioned as an uphill walk from one functional phytase to the desired fitness (from Wt to M6 in Fig. 6.2a). The fitness landscape for a phytase with more than one criterion in which two properties are being improved is shown in Fig. 6.2. It is obvious that a fitness landscape for one property (fitness versus combination of substitutions) cannot overlap with other property of interest indicating a separate fitness landscape exists for each property. Thermostability and activity improvement of a phytase are inversely proportional to each other and can be correlated well with the flexibility of the enzyme. Reducing the flexibility of a phytase by improving intramolecular noncovalent interactions has been shown to enhance thermal resistance [18, 24], while high flexibility in the active site has been proposed to be a prerequisite for higher catalytic efficiency [25]. On a path of uphill walk of the phytase thermostability fitness (Fig. 6.2a), several variants that have low or even undesirable fitness for activity exist (Fig. 6.2b). During iterative rounds of directed evolution, it is crucial to

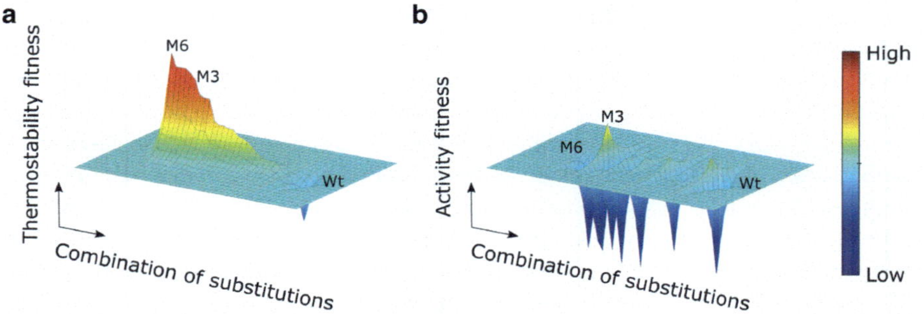

Fig. 6.2 A fitness landscape of combining key substitutions in directed Ymphytase evolution. The mapping of genotype (combination of substitutions) to phenotype (experimentally determined fitness for desired property) for thermostability (**a**) and phytase activity (**b**) revealed a nonoverlapping fitness landscape, indicating a separate fitness landscape exists for each property. A walk from *blue*, to *cyan*, to *green*, to *yellow*, and to *red* represents an increased phytase fitness. On the uphill walk for thermostable variant (from Wt to M6), several variants with undesirable activity exist. Epistatic effects play a major role in optimizing enzymes for more than one fitness; a combination of substitution in M3 improved two properties as a result of epistatic interactions

select phytase variants with more than one selection pressure and combine the key substitutions to identify variants with higher fitness for more than one property (M3 in Fig. 6.2).

6.2.3 Importance of Cooperative Substitutions (Epistatic Effects) in Directed Phytase Evolution

Epistatic effects can be defined as nonadditive interactions between substitutions that affect enzyme properties. These substitutions act cooperatively and can only be discovered if more than one amino acid is being mutated simultaneously. Over the past decades, it is becoming clear that there are two pathways in directed evolution, (i) a widely known pathway of accumulating beneficial substitutions in each round of mutagenesis and screening and (ii) a pathway in which cooperative/epistatic effects between substitutions result in additive or synergistic effects. Epistatic effects in directed evolution have been underestimated despite their high importance [18, 26–28]. The latter might be attributed to (i) a low mutational load of mutant libraries (classical directed evolution experiment), (ii) a low robustness toward mutations depending on the protein structure and catalyzed reaction, and (iii) methodological limitations to incorporate more than one substitution per protein. Classical directed evolution achieves accumulation of a beneficial substitution in iterative rounds of mutagenesis and screening, in which enzymes are often improved by exchanging one amino acid per round of evolution [29]. Several improved phytases possess more than one residue substitution per round of directed evolution [30]. A study elucidating individual role of each substitution as well as the role of each combination of substitutions revealed cooperative effects between substitutions. A detrimental substitution G187S (affecting activity and thermostability) in Ymphytase improved thermostability in combination with distantly located T77K substitution, supporting the hypothesis that non-beneficial substitutions may also be vital in directed phytase evolution. Another substitution Q154H had no effect on activity; however, the specific activity was increased by ~200 U/mg after being combined with G187S-T77K variant [18]. A surface substitution K265E had a negligible effect on *Bacillus* phytase activity; however, the specific activity was improved when K265E was combined with either D24G or S51G (active site substitutions) [28]. Interestingly, a multisite saturation of five distantly located sites that were identified in a directed phytase evolution improved thermostability as well as pH stability of Ymphytase that might be due to possible epistatic effects [26]. The sites for saturation mutagenesis were identified from a previous directed evolution experiment employing random mutagenesis method. A threefold higher pH stability (at pH 2.8/3 h) and a small improvement in thermal resistance were obtained when compared to previously identified combination of substitutions (M1). Discovery of a novel combination by screening only 1100 clones from an OmniChange library proved to be a promising strategy to re-interrogate the substitutions for possible epistatic effects. These results demonstrate an interesting and vital role of the second pathway in directed evolution and encourage to incorporate strategies to study and improve enzymes by cooperative substitutions.

6.3 Exploring Phytase Engineering by Directed Evolution and Rational Design Strategies

Directed evolution and rational design strategies have emerged concomitantly as an ideal method to tailor-make a phytase for industrial demands. Over the past years, several microbial phytases with high catalytic efficiency and moderate thermostability have been discovered [31]. Especially, the phytases (HAP) from the *Enterobacteriaceae* family have attracted industrial interest due to their high specific activity. The major goal in phytase design has been engineering phytases with higher activity and thermostability. Phytases engineered by directed evolution and rational design strategies have been summarized in Tables 6.1 and 6.2, respectively. Structural flexibility plays an important role in enzyme activity and thermostability and usually depends on the nature of intra-protein interactions. Improving thermostability typically affects activity of phytases and vice versa; therefore, implementing a "smart" strategy is a prerequisite in engineering phytases for industrial applications. A flowchart on the design and discovery of novel phytases is shown in Fig. 6.3 illustrating several successful approaches used to engineer phytases.

6.3.1 Engineering Thermal Resistance in Phytases

6.3.1.1 Hydrogen Bonding Network

Hydrogen bonds in a protein structure are considered a dominant factor for thermal stabilization [47, 48]. Several directed phytase evolution experiments yielded

Table 6.1 Directed phytase evolution campaigns employed in improving phytases for industrial applications

Organism name	Phytase name (family)	Specific activity (U/mg)	Mutagenesis method (clones screened)	*Tm* (°C)	Substitutions	Improvements	Potential mechanism	Reference
Aspergillus niger N25	phyA (HAP)	Parent: 204,514	epPCR (5500)	NA	E156G, T236A, Q396R	61% higher specific activity	E156G, T236A: decreased number of hydrogen bonds ➔ improved phytase flexibility	[32]
		Variant: 330,064				81% higher catalytic efficiency[a]	Q396R: improved affinity to negatively charged phytate	
Escherichia coli	AppA2 (HAP)	wt: 1003	epPCR (5000)	wt: 62	Variant-1: K46E	20% increased residual activity at 80 °C/10 min	K46E: increased hydrogen bonds	[17]
		Variant-1: 742		Variant-1: 68.5	Variant-2: K65E, K97M, S209G	6–7 °C increased *Tm*	K65E: increased hydrogen bonds ➔ stabilizing interaction between α- and α/β-domains	
		Variant-2: 905		Variant-2: 69.8		56% and 152% improved catalytic efficiency	K97M/S209G: reduced structural hindrance to neighboring residues	

(continued)

Table 6.1 (continued)

Organism name	Phytase name (family)	Specific activity (U/mg)	Mutagenesis method (clones screened)	*Tm* (°C)	Substitutions	Improvements	Potential mechanism	Reference
Escherichia coli	AppA (HAP)	wt and Phy9X 1700	Gene site saturation mutagenesis (GSSM) (158,608)	wt: 63.7	Q84W, Y277D, W68E, K97C, R181Y, N226C, A95P, S168E	12 °C increased *Tm*	Reduced flexibility in loops and random coiled structures	[33, 34]
				Phy9X: 75.7		3.5-fold higher gastric stability	Increased surface polarity	
							Elimination of N-glycosylation site close to the active site ➔ reduced steric hindrance for substrate entry	
Penicillium sp.*Q7*	phyA (HAP)	wt: 32	epPCR with $MnCl_2$ and then epPCR with dITP (5300)	NA	2–28: T11A, G56E, L65F, Q144H, L151S	~Fourfold increased specific activity	Increased number of intra-protein hydrogen bonds	[35]
		Variant 2–28: 133.3			2–249: T11A, H37Y, G56E, L65F, Q144H, L151S, N354D	55% (2–28) and 74% (2–249) increased residual activity at 80 °C/5 min	L151S: improved catalytic residue (H352) exposure to the substrate	
		Variant 2–249: 136.6				Retained pepsin resistance (96.7% activity after 2 h at 0.01 pepsin/phytase (w/w) ratio)	N354D: decreased hydrogen bonds to catalytic residue D353 ➔ improved activity	
							Reduced pKa of the side chain (N354D) affected pKa of the catalytic center ➔ shift on the optimal temperature	

Escherichia coli	AppA (HAP)	wt: 1610	epPCR (NA)	NA	I408L	23.3% increased residual activity at 80 °C/5 min	Stabilizing terminal part of the phytase	[36]
		I408L: 1653						
Yersinia mollaretii	Ymphytase (HAP)	wt: 1073	SeSaM (2350)	wt: 63	D52N, T77K, K139E, G187S, V298M	~20% increased residual activity at 58 °C/20 min	T77K: incorporation of a salt bridge	[18, 22]
		M1: 993		M1: 64.5		1.5 °C increased *Tm*	G187S: improved hydrogen bonding network to an adjust loop	
Yersinia mollaretii	Ymphytase (HAP)	wt: 1043	OmniChange (1100)	wt: 59	D52E, K139T, G187S, V298F	32% increased residual activity at 58 °C/20 min	D52E, G187S, K139E: increased number of hydrogen bonds	[26]
		Variant Omni1: 1034		Omni1: 61		2 °C increased *Tm*	V298F: incorporation of aromatic-aromatic interactions	
						Twofold higher pH stability (pH 2.8; 3 h)	Possible epistatic interactions	

(continued)

Table 6.1 (continued)

Organism name	Phytase name (family)	Specific activity (U/mg)	Mutagenesis method (clones screened)	*Tm* (°C)	Substitutions	Improvements	Potential mechanism	Reference
Yersinia mollaretii	Ymphytase (HAP)	wt: 1073	SeSaM (1600)	wt: 58.5	M3: T77K, G187S, Q154H	M6: 54% increased residual activity at 58 °C/20 min	Increased intra-protein hydrogen bonding network	[18]
		M3: 1274	SSM (200)	M3: 61	M6: T77K, Q154H, G187S, K289Q	M6: 3 °C increased *Tm*	T77K: incorporated salt bridge ➔ stabilizing two adjacent helices	
		M6: 1017	SDM	M6: 61.5		Specific activity of M3 increased by ~200 U/mg	T77K and K289Q: increased interactions between adjacent helices (holding tightly) ➔ reduced flexibility in the loop situated next to the helix ➔ improved thermostability	
			KeySIDE				Q154H: improved solvent exposure ➔ improved thermostability	
							Increased flexibility of the active site loop	
							epistatic effects between T77K and G187S (thermostability) and T77K, G187S, Q154H (activity)	

Bacillus subtilis 168	phy168 (BPP)	wt: 13.8	epPCR (NA)	NA	D24G, K70R, K111E, N121S	42.8% improvement in specific activity	D24G: widening of the binding pocket and elimination of steric hindrance for phytate binding	[28]
		Variant: 19.7				2.3-fold improved catalytic efficiency	Epistatic effects between D24G and K265E/K265N and S51G and K265E	
						Retention time to reach 40% residual activity at 80 °C was improved by 17 min	Increased number of hydrogen bonds ➔ improved thermostability	
Synthetic	Ymphytase	wt: 1043	ProCoS (1050)	NA	Synthetic: 34 substitutions	Broadened pH optima	20% increased negative polar surface	[52]
		ProCoS-2: 401				3.8-fold improved pH stability at pH 2.8 (37 °C/3 h)	8% increased positive polar surface	
							pH stability increased by increased fraction of polar surface and decreasing neutral surface (20%)	

[a]*kcat/Km*, *Tm* melting temperature, *HAP* histidine acid phosphatase, *BPP* β-propeller phytase, *NA* not available

Table 6.2 Approaches employed to improve phytases by rational and semi-rational design

Organism name	Phytase name (family)	Specific activity (U/mg)	*Tm* (°C)	Substitutions	Rationale	Strategy	Results	Reference
Synthetic	Consensus phytase (HAP)	Parent: NA	Parent: 56–63	Synthetic	Consensus sequence	Design of a consensus sequence from 13 fungal phytase sequences	Optimum temperature increased from 45–55 to 71 °C	[24]
		Consensus: ~30	Consensus 78				15–22 °C increased *Tm*	
							Stabilized a surface loop and N-terminus by increased number of hydrogen bonds	
Escherichia coli	EcAppA (HAP)	wt: 2163	NA	N204A	Sequence alignment with the consensus of highly active phytases	Altering binding pocket and glycosylation pattern based on *Citrobacter* phytases	N204A or S206A: elimination of N-glycosylation site close to the active site → reduced steric hindrance for substrate entry	[34]
		N204A or S206A: ~2370		S206A				
				KKG → NQT				
				HPP → NGT				
				P173S				
				G311S			33% increased residual activity at 66 °C/5 min for the triple mutant	

Aspergillus niger	PhyA (HAP)	wt: 80	wt: 63.3	A58E, P65S, Q191R, T271R	Structure alignment with a thermostable phytase	Adopting hydrogen bonding network and ionic interactions form a thermostable homolog	20% increased residual activity at 100 °C/10 min	[37]
		Variant: 83	Variant: ~73				7 °C increased *Tm*	
							Improved affinity (decreased *Km*)	
							Improved inorganic phosphate release from phytate in soybean meal	
Synthetic	Beta-propeller phytase (BPP)	FTE: ~28	NA	Synthetic	Structure-guided consensus sequence	A homology model of the consensus sequence from 15 bacillus phytases was used to design sequences	Optimum temperature between 45 and 70 °C	[38]
		FTEII: ~28					Variants possessed higher thermostability at pH 5.5 than pH 7.5	
		FBA: ~37					Higher thermostability at 5 mM $CaCl_2$ than 1 mM $CaCl_2$	
							Retained ~80% residual activity (80 °C/10 min) at pH 5.5, 5 mM $CaCl_2$	

(continued)

Table 6.2 (continued)

Organism name	Phytase name (family)	Specific activity (U/mg)	*Tm* (°C)	Substitutions	Rationale	Strategy	Results	Reference
Escherichia coli	AppA (HAP)	NA	NA	NA	Deletion of flexible region	Deletion of seven C-terminal residues	39% increased residual activity at 80 °C/10 min	[39]
Escherichia coli	AppA (HAP)	wt: 1541	NA	Y311K, I427L	Structure-based design	Selection of flexible surface residues based on an MD simulation (RMSF)	30% increased residual activity at 80 °C/10 min	[40]
		Variant: 1274					Increased affinity (decreased *Km*)	
						Incorporation of salt bridges	Epistatic effects (synergy) between Y311K and I427L	
Yersinia frederiksenii	APPA (HAP)	wt: 500	NA	S51I	Sequence alignment	Binding site analysis based on a homology model and sequence alignment	S51I or S51T: shift in the optimum pH from 2.5 to 4.5	[41]
		S51I: 3333					Side chain structure (not the charge) affected shift in the optimum pH	

Bacillus sp. MD2	phytase (BPP)	wt: 32	NA	E229V	Binding site charge modification	Decreasing negatively charged residues near binding site, widening binding pocket	E229V: 19% increased specific activity	[42]
		E229V: 38		K77R, K179R			K77R-K179R: ~20% increased residual activity (pH stability) at pH 2.6	
		K77R-K179R: 7.5						
Escherichia coli	AppA (HAP)	wt: 37	NA	C2	Altering glycosylation pattern	Incorporation of N-glycosylation sites	25% increased residual activity at 80 °C/15 min (no alteration in glycosylation)	[43]
		C200N/D207N/S211N:65					Increased specific activity may be due to eliminating a disulfide bond (C200N) ➔ increased flexibility of α-domain	

(continued)

Table 6.2 (continued)

Organism name	Phytase name (family)	Specific activity (U/mg)	*Tm* (°C)	Substitutions	Rationale	Strategy	Results	Reference
Bacillus licheniformis	PhyL (BPP)	wt:		PhyL$^{G\text{-}A}$: G117A, G266A	Restricting conformational flexibility	Incorporation of rigidifying substitutions (Xaa ➔ Pro, Gly ➔ Ala) at consensus position	PhyL$^{G\text{-}A}$: 2.5 fold improved free energy of unfolding (ΔG_u)	[44]
		PhyL$^{G\text{-}A}$:		PhyL$^{X\text{-}P}$: H32P, S256P, K304P, K324P, S353P			Gly ➔ Ala substitution at consensus positions was more promising	
		PhyL$^{X\text{-}P}$:						
Synthetic (*Aspergillus niger* and *A. fumigatus*)	Afp, Anp (HAP)	Anp: 110	Anp: 62.1	NA	Structure-based fragment shuffling	Swapping of fragments based on structural elements in two homologous phytases (Anp: ABCDEFG and Afp: 1234567)	Variant AB345F7 showed highest specific activity	[45]
		Afp: 40	Afp: 59				Variant A234567: 7.8 °C increased *Tm* compared to Afp	
		A234567: 40	A234567: 66.8					
		AB345F7: 90						
Bacillus amyloliquefaciens DSM 1061	Phytase (BPP)	wt: 15		D148E	Sequence alignment and structure visualization	Increasing glutamic acid content ➔ improves thermostability	D148E: 27% increased residual activity at 70 °C/10 min	[46]
		D148E: 20		S197E			D148E: increased specific activity	
				N156E				
				D52E				

Tm melting temperature, *HAP* histidine acid phosphatase, *BPP* β-propeller phytase, *NA* not available/applicable, *MD* molecular dynamics

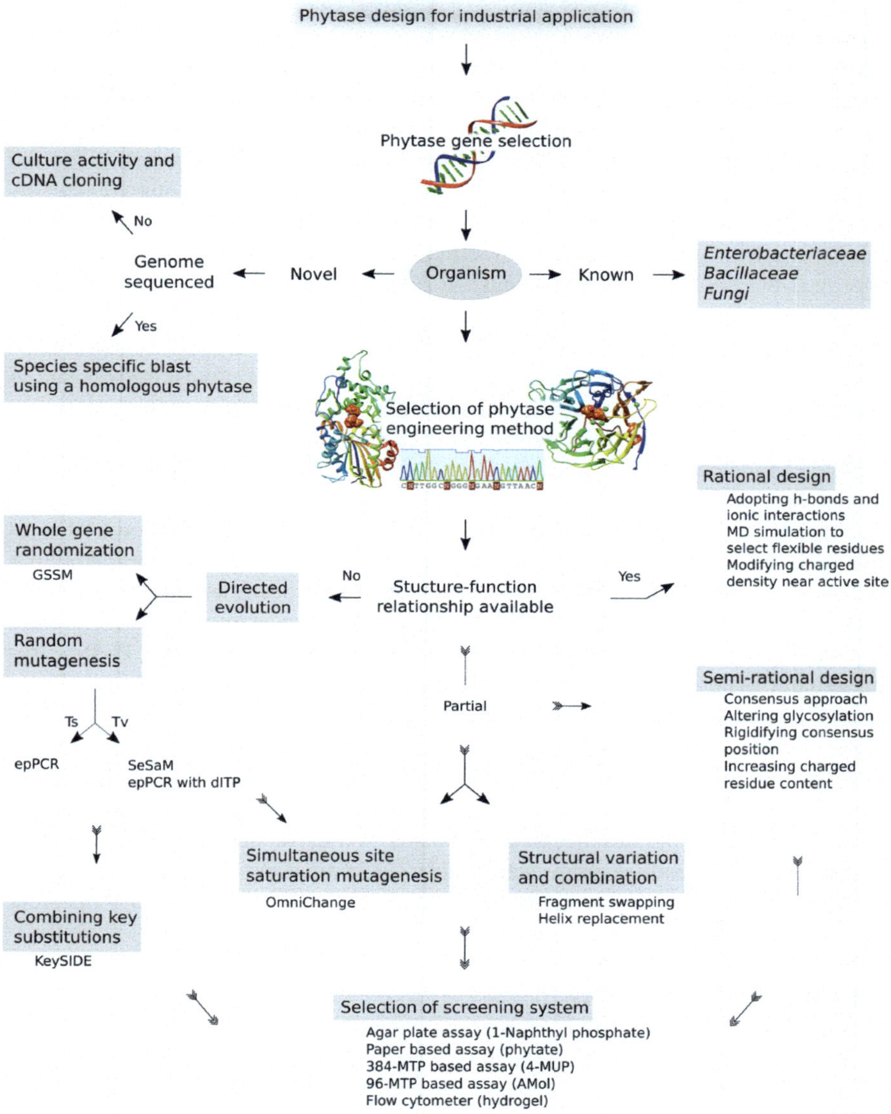

Fig. 6.3 A flowchart representing a design and discovery path of novel phytase by selection of a "smart" phytase engineering strategy. Successful approaches employed in directed phytase evolution are illustrated. (*Ts* transition, *Tv* transversion-biased mutagenesis methods, *4-MUP* 4-methylumbelliferyl phosphate assay, *AMol* ammonium molybdate assay)

thermally improved variants with a "stronger" hydrogen bonding network. Thermostability of a phytase (AppA2) from *E. coli* was improved by 20% (80 °C/10 min) via a strengthened hydrogen bonding network to adjacent secondary structures [17]. A charged substitution (K46E) fostered intra-protein hydrogen bonds resulting in 6 °C improved melting temperature (*Tm*). Additionally, in another variant, a substitution K65E introduced hydrogen bonds that might stabilize the interactions between α- and αβ-domain resulting in a 7 °C *Tm* improvement. A protease-resistant phytase (phyA) from *Penicillium* was evolved using epPCR employing Mn^{2+}-dITP for higher thermal resistance. Thermal resistance was improved by 74% (80 °C/5 min), and the substitutions G56E, L65F, Q144H, and L151S introduced new hydrogen bonds to neighboring residues [35]. Recently, a phytase from *Yersinia mollaretii* was evolved using a transversion-enriched mutational pattern generated by the SeSaM (sequence saturation mutagenesis) random mutagenesis method and a subsequent KeySIDE (iterative key-residues interrogation of the wild type with substitutions identified in directed evolution) approach to combine key beneficial substitutions [18]. A combination of substitutions yielded a variant M6 with 54% improved thermostability (58 °C/20 min) and 3 °C higher melting temperature. Molecular dynamics simulations on M6 variants revealed an increased hydrogen bonding network in the phytase, stabilizing two adjacent helices (K289Q) and a surface loop (G187S). Adopting a hydrogen bonding network and ionic interactions from a thermostable *A. fumigatus* phytase into the *A. niger*, phytase improved melting temperature by 7 °C. Substitutions A58E and P65S introduced new hydrogen bonds that contributed to improved thermostability. These results suggest employing an evolution strategy directed toward introducing hydrogen bonds in a phytase as a highly beneficial strategy for improving thermal resistance. Almost all substitutions in the improved phytase variants were chemically different ones (opposite charge or polar), suggesting to employ random mutagenesis method which incorporates diverse amino acids. A transversion-biased mutagenesis method able to introduce subsequent nucleotide mutations or a codon exchange mutagenesis method offers a higher chance of incorporating chemically different amino acids than routinely used epPCR method with a transition bias.

6.3.1.2 Phytase Flexibility

Phytase flexibility especially in loop regions (quantified via MD simulations) is an important criterion to be considered during selection and evolution of a phytase for higher thermal tolerance or activity [18, 40]. Simulations on the improved Ymphytase variant revealed a reduced flexibility in loop regions compare to the wild-type Ymphytase. Interestingly, these loops were located next to helices that possess substitutions (e.g., reduced flexibility in a loop located after helix-K, substitution K289Q belongs to helix-K stabilizing helix-K and L). Similarly, T77K (on helix-B) introduced a salt bridge that holds two helices (helix-B and C) together resulting in a reduced flexibility of a loop located next to helix-B [18]. The active site loop studies illustrated a vital role of the loop flexibility in highly active phytases (specific activity 1000–3000 U/mg). This active site loop region was replaced by a helix in phytases with lower specific activity (~100 U/mg; fungal and

Klebsiella phytase) [25], suggesting that the local flexibility in the active site is a prerequisite for higher catalytic activity, and therefore the loops situated away from the active site can be targeted for mutagenesis. Randomizing helices that are located distantly from the active site loop and next to flexible loops may enrich the library with thermostable phytase variants. A classical study of whole-gene randomization, saturating each position in the *E. coli* phytase (AppA) and screening (158,608 clones) with 384-well-based system, resulted in a phytase variant with 12 °C improved melting temperature and 3.5-fold increased gastric resistance; the improved resistance toward gastric proteases might be due to reduced flexibility in the loops and random coils [33]. In a semi-rational phytase design, a consensus phytase designed using a sequence alignment of 13 fungal phytases showed 15–22 °C improved melting temperature. A loop that was unresolved (non-visible electron density map) in the *Aspergillus niger* phytase was clearly seen in the consensus phytase suggesting the substitution (V251D) in the improved variant stabilized this loop by an increased number of hydrogen bonds with the neighboring residues [24]. These studies suggest employing an evolution approach targeting flexible loops may also enrich the library population and reduce the screening efforts in the directed phytase evolution. Restricting the sequence space by sampling only a part of the protein sequence might be a good strategy for rapid phytase evolution.

6.3.1.3 Ionic and Hydrophobic Interactions

Ionic interactions also known as salt bridge interactions are associated with thermostability; the correlation between the number of salt bridge and thermostability was considered not to be as strong as hydrogen bonds [48]. However, recently several studies have established a good correlation, showing that the increase in the number of salt bridges improves protein thermostability (7 °C increased T_m) [49]. Incorporation of new salt bridge interactions (Q191R & T271R) in *A. niger* phytase based on a homolog *A. fumigatus* structure as a template improved thermostability (13% increased residual activity at 100 °C/10 min, for a triple mutant A58E, Q191R, and T271R) [37]. Using a rational approach based on MD simulations, a phytase from *E. coli* (AppA) was designed to introduce new salt bridge interactions. Variant Q206E and Y311K showed 8–9% improved thermostability [40]. Screening of a transversion-biased mutant library (SeSaM) of Ymphytase yielded a variant M6 with 3 °C increased melting temperature; substitution T77K introduced a salt bridge interaction with adjacent helix and turned out to be unobtainable by epPCR [18, 22]. Therefore, employing a mutagenesis method that introduces chemically diverse amino acid substitutions is crucial for phytase thermostabilization.

Incorporation of hydrophobic interactions has been a barely seen phenomenon in phytase engineering for improving thermostability. Despite the low occurrences of aromatic amino acid in protein sequences (may be due to fewer codons), aromatic interactions showed a thermostabilizing tendency in 166 studied proteins [50]. A phytase from *Y. mollaretii* evolved using OmniChange yielded incorporation of aromatic-aromatic interactions (V298F) that might contribute to improved thermal and pH resistance [26].

6.3.1.4 Surface Engineering

Role of surface charge-charge interactions of protein to solvent molecules has received less attention; however, over the past decade, it became evident that the surface modification plays a vital role in protein stabilization [51]. Recently, a phytase from *Y. mollaretii* has been engineered with a computer-assisted method entitled ProCoS (protein consensus-based surface engineering). Screening of only 1050 clones from an in vitro recombined library of synthetic genes (designed to incorporate 30–40 surface substitutions) yielded an Ymphytase variant (harboring 34 substitutions) with a 3.8-fold increased pH stability (pH 2.8/3 h). The improvement was achieved by an increased fraction of polar surface, a significant increase in charged surface (28%), and especially an increase in negative surface (20%) of the Ymphytase variant. The neutral surface was decreased by 20% when compared to wild-type Ymphytase [52]. In another study, a surface substitution (Q154H) improved thermostability in the Ymphytase which might be attributed to increased surface exposure to the solvent (decreased intra-protein hydrogen bonds in a MD simulation) [18]. These studies suggest that phytase surface modifications and reengineering may provide an alternate route for improving phytase stability.

6.3.2 Improving pH-Activity Profile and pH Stability

Application of phytase in animal feeding requires a decent gastric tolerance. Stomach pH for many poultry animal ranges between pH 1.9 and pH 4.5 and a gastric emptying time of 1–2 h depending on the diet components [53]. Optimum pH of the first commercial phytase (phyA) from *A. niger* was shifted from pH 5.5 to the more acidic side. Several variants targeting areas near the binding pocket were generated by swapping residue charges; the pH optima for these variants ranged from 2.5 to 5.5. The best variant E228K showed the pH optimum at 3.8 [54]. Another phytase from *A. niger*, phyB, has a pH optimum at 2.5; swapping the charge near binding pocket (E272K, from negative to positive) shifted the pH optimum to 3.2 that might be due to improved affinity for the negatively charged phytate [55]. These studies indicate a convincing role of charged residues and ionization near binding pocket of phytase that affects substrate binding at various pH values. However, a bacterial phytase from *Yersinia frederiksenii* (pH optimum 2.5) was evolved by substituting a non-charged residue (Ser51) near the binding pocket. The residue Ser51 was found to be dissimilar to the close homologs (pH optimum 4.5) which were identified using a sequence alignment. Substitutions S51T and S51I shifted the pH optimum from 2.5 to 4.5 [41], suggesting the facts that a side chain structure can also contribute to the phytase's pH optimum. In a directed phytase evolution experiment, substitutions far away from the active site yielded an Ymphytase variant with twofold improved pH stability (pH 2.8/3 h). The latter novel combination of substitutions was discovered by a multisite saturation mutagenesis (OmniChange) of previously identified sites from a thermally improved Ymphytase variant [26]. As aforementioned, a surface engineering of Ymphytase using the ProCoS method yielded a variant with 3.8-fold increased pH stability by increasing (20%) the number of negatively charged amino acids on the phytase surface [52]. Therefore, altering the overall charge by surface engineering and modifying the local environment near the binding pocket are promising strategies to consider in directed phytase evolution, to alter pH optima, and to improve pH stability.

6.3.3 Enhancing Protease Resistance

Protease resistance in phytases has still not been a completely addressed challenge in the field of directed phytase evolution. Surface exposed loops in *A. fumigatus* phytase were targeted to remove protease cleavage sites; for instance, the substitution (S126N) yielded an ~eightfold (50 °C/90 min) improved protease resistance against proteases present in the NW205 culture media used for phytase expression [56]. A whole-gene randomization (GSSM) on *E. coli* phytase yielded a thermostable variant that showed a 3.5-fold improved stability in the simulated gastric fluid (pepsin 3.2 mg/ml, pH 1.2) [33]. Few patented phytases have been reported to possess a higher protease resistance. A phytase from *Hafnia* species was evolved for higher tolerance against trypsin and pepsin [57]. A bacterial phytase (rAppA) showed 30% increased residual activity after pepsin treatment [58]. Synthetic phytase variants (from *Yersinia mollaretii* and *Hafnia* species) designed for higher stability retained 80–90% residual activity after pepsin treatment (pepsin 30 U/ml, 37 °C/30 min) [59]. Several wild-type phytases have been reported to possess desirable resistance against pepsin and trypsin [60]. Phytase from *Penicillium* species [35] was found to be resistant to protease (pepsin) treatment; however, the specific activity is low (32–136 U/mg) when compared to phytases from *Enterobacteriaceae* family (1000–3000 U/mg) [60]. A bacterial phytase from Malaysian wastewater was reported to retain ~95% of phytase activity after being treated with pepsin (3000 U/37 °C/30 min, pH 2.0). This pepsin resistance was found to be equivalent to the pepsin resistance of *E. coli* phytase [61]. Recently, phytases from *Yersinia* species have been engineered for higher pepsin resistance by incorporating N-glycosylation sites that obstructed pepsin accessibility to peptic cleavage sites [62]. On the contrary, a phytase (Nov9X) was rationally designed to incorporate protease cleavage sites (in the surface exposed loops) that generated protease susceptible phytase variants [63], suggesting a possibility to design phytases for improved protease resistance by removing protease cleavage sites on the phytase surface especially in loop regions.

6.3.4 Combining Key Substitutions Identified in Directed Phytase Evolution

It has always been a major challenge to identify substitutions that contribute to the desired property improvement in a directed evolution campaign. In the pool of improved variants, the number of substitutions that will lead to the desired property improvement is unknown. The latter unpredictability intensifies and renders it even more challenging to identify cooperative effects (epistatic effects) between distantly located substitutions. Therefore, implementing a strategy to combine substitutions identified in iterative rounds of directed phytase evolution is a prerequisite for efficient and rapid evolution of phytases. Presently, the additive or synergistic effects are impossible to predict. Several substitutions identified in directed phytase evolution demonstrated a reliable proof on the second pathway of directed evolution in which the epistatic effects play an important role (see Sect. 6.2.3). KeySIDE (iterative key-residues interrogation of the wild type with substitutions identified in directed evolution) was recently reported to iteratively combine substitutions that were identified in directed Ymphytase evolution [18]. G187S mutation was identified as one of the substitution (out of five, D52N, T77K, K139E, G187S, V298M) in the first round of directed evolution that improved thermostability of Ymphytase. Incorporation of G187S in the wild-type phytase decreased activity (by ~50%) as well as thermostability (20% less than wild type). However, when the G187S substitution was combined with T77K, the thermostability of the M2 variant (G187S-T77K) was better than T77K alone, and the specific activity was restored to wild-type phytase. Another substitution Q154H improved thermostability when combined with T77K (20% improved residual activity) or incorporated into the wild type (16% improved residual activity), without altering specific activity, whereas combining Q154H to the M2 variant (G187S-T77K) improved specific activity by ~200 U/mg. These results suggest that cooperative effects play an important role in modulating phytase properties. This is in evidence with the study employing site saturation mutagenesis of all individual position in *E. coli* phytase and combining substitutions that improved stability. Incorporation of two stabilizing substitutions led to lower stability for the double mutant than that of single substitution. Therefore, substitutions were combined by adding each substitution to the most stable variant identified in the previous round of site-directed mutagenesis [33]. However, this approach is limited by the possibility of missing combinations that act cooperatively by ignoring detrimental substitution (e.g., G187S was detrimental substitution for Ymphytase but improved thermostability in combination with T77K). Therefore, combining substitutions iteratively by the KeySIDE approach to decipher the role of each substitution and the role of possible combinations is a preferable approach to rapidly and efficiently improve phytase properties.

6.3.5 Screening Methods for Directed Phytase Evolution

Successful directed evolution experiments rely on a robust high-throughput screening systems. Prescreening of a mutant library using bacterial colonies has been a desirable option to reduce screening efforts in directed evolution experiments. High-throughput screening methods developed for bacterial colonies were successful to select only active or thermostable phytase variants for subsequent screening in 96-well microtiter plates. A widely employed synthetic substrate in directed phytase evolution, 4-methylumbelliferyl phosphate (4-MUP), has been one of the most reliable substrates used in prescreening. A phytase from *Y. mollaretii* was evolved by developing a 384-well microtiter plate-based prescreening method to select thermostable variants (increased *Tm* by 3 °C). In this method, mutant libraries were expressed in an auto-induction media [64] which omit steps involved in IPTG induction. Further, bacterial colonies expressing Ymphytase variants were directly subjected to heat treatment, omitting protein expression and lysis steps [22]. A simple agar plate assay employing 1-naphthyl phosphate and Fast Garnet salt was used to identify the *Klebsiella* phytase from soil samples [65]. A filter paper-based assay for "colony screening" was also developed for screening mutant libraries employing the natural/application substrate (phytate) [22, 66]. The use of natural/application substrate is always desirable to avoid evolving enzymes toward nonnatural substrate [67]. Recently, an ultrahigh-throughput screening method (fur-shell technology) employing fluorescence-activated cell sorting (FACS) was developed as a prescreening system for phytases. A coupled reaction of Ymphytase and glucose oxidase produces a hydroxyl radicle by utilizing glucose-6-phosphate as a substrate. The hydroxyl radicle initiates a poly(ethylene-glycol)-acrylate-based polymerization incorporating a fluorescent molecule during polymerization and ultimately forming a fluorescent hydrogel shell surrounding the *E. coli* that expresses an active Ymphytase. Further, screening of ~10 million Ymphytase variants validated the fur-shell technology by identification of an Ymphytase variant with improved specific activity (increased by 97 U/mg) [68]. Screening conditions optimized for 96-well microtiter plate-based screening platform employing either phytate or 4-MUP as a substrate showed reliable true standard deviations (below 15%) [22]. These advances in the prescreening and screening methods facilitate a rapid, efficient, and non-laborious identification of novel phytase variants with desirable properties.

6.3.6 Expression Host for Phytase Engineering and Production

Host organism used for phytase expression exhibits an immense influence on phytase activity and thermostability mainly due to posttranslational modifications. Glycosylation has been considered an important feature for altering activity or thermostability. Elimination of N-glycosylation sites (N204A or S206A) close to the active site of *E. coli* phytase improved activity by reducing the steric hindrance for phytate entry into the binding pocket [34]. A detailed study to alter the glycosylation pattern in *E. coli* phytase (AppA) resulted in an identification of a phytase variant (C200N, D207N, S211N) with improved specific activity (65 U/mg) when compared to the wild type (37 U/mg) [43]. Glycosylation has been shown to improve stability of numerous proteins [69, 70]. However, the variant (C200N, D207N, S211N) showed 25% higher thermostability (increased residual activity at 80 °C/15 min), despite having the same level of glycosylation. Whereas in another study, an incorporation of two glycosylation sites in *E. coli* phytase showed >40% enhanced thermostability (80 °C/10 min) and 4–5 °C higher melting temperature [71]. These results suggest the posttranslational modifications to play a vital role in phytase properties. The *E. coli* phytase expressed in yeast (*Pichia pastoris*) [60, 72] showed ~threefold higher specific activity (~3100 U/mg) when compared to the expression of the same phytase in *E. coli* (975 U/mg) [73]. Despite an advantageous role of yeast in phytase expression, *E. coli* is a preferred host for directed evolution due to its faster growth, nonchromosomal integration (as happens when using *P. pastoris*), and high plasmid transformation efficiency, which shortens time required for each round of directed phytase evolution. *Saccharomyces cerevisiae* as a host offers an advantage of nonchromosomal integration when using episomal vectors. However, *P. pastoris* is a widely adopted host for industrial production of phytases due to correct posttranslational modifications, very high density cultures, and high protein yields.

6.4 Structural Perspectives in Phytase Engineering

Rational phytase engineering has become an attractive strategy due to increased knowledge on structural information and structural dynamics studies. Several studies have demonstrated a very good correlation between phytase flexibility and activity or thermostability (Tables 6.1 and 6.2). Rigidifying a protein in part or overall by improving the intra-protein hydrogen bonding network has proven to be crucial in improving phytase performance. A direct evidence on structural features or overall fold of a phytase determining its activity or thermostability does not exist to date. However, recent improvements in phytases by directed evolution and rational design strategies have advanced our knowledge on factors affecting phytase properties and could be employed in designing novel phytases for industrial demands. Several approaches led to improved phytases ranging from simple sequence alignment to MD simulation studies (Table 6.2). A fungal consensus phytase designed by sequence alignment yielded a variant with 15–22 °C increased melting temperature [24]. A structure-guided consensus sequence was designed using a homology model for beta-propeller phytase resulted in a variant with ~80% residual activity (at 80 °C/10 min) [38]. Flexibility within the phytase enzyme is a critical parameter; in detail a local flexibility in the binding pocket determines catalytic efficiency/specific activity; overall protein flexibility or surface loop flexibility governs thermostability. Fungal and bacterial (*Enterobacteriaceae*) HAP phytases have a similar overall fold (α-domain, α/β-domain, and binding pocket at the interface of these two domains), but the *Enterobacteriaceae* phytases are ~30-fold more active than fungal phytases that might be due to local flexibility in the binding pocket. The flexible active site loop in the binding pocket of *Enterobacteriaceae* phytases has been replaced by a less flexible helical structure in the fungal phytase, suggesting that a local flexibility in the binding pocket is prerequisite for higher specific activity [25]. In addition, conformational flexibility in a *Bacillus* phytase was restricted by incorporating rigidifying substitutions and resulted in 2.5-fold improved free energy that indicates a significantly more stable variant [44]. Deletion of a flexible C-terminal part of *E. coli* phytase increased its residual activity (80 °C/10 min) by 39% [39]. MD simulation studies on a thermostable *Y. mollaretii* phytase variant discovered by directed evolution revealed an overall decreased flexibility in the loop regions [18]. A structure-based design of *E. coli* phytase using MD simulation to target flexible residues resulted in 30% increased residual activity (80 °C/10 min) [40]. These studies suggest a strong correlation of phytase structural flexibility with its activity and thermostability. Additionally, surface charge modifications of phytases are emerging as a novel strategy for phytase engineering. For instance, surface engineering of the Ymphytase resulted in a variant with an increased negative polar surface (20%) that showed 3.8-fold improved pH resistance (pH 2.8/3 h) [52]. Whereas a local surface modification near to the binding site affects phytase activity and pH optima. A substitution Q396R (positively charged) in *A. niger* phytase improved affinity to phytate (negatively charged) [32], and substitution N354D in *Penicillium* phytase affected the pKa of the catalytic center that shifted the pH optima [35].

6.5 Conclusions and Future Prospective for Phytases Engineering

Although many phytases have been discovered from nature in the last century, the potential to improve phytases for industrial applications is barely explored and especially potentials of epistatic effects are barely realized. Nevertheless, directed evolution and rational design methodologies provide an excellent tool to tailor-made phytases with high specific activity, thermal and pH profile, and protease resistance. Epistatic effects played an important role in the directed phytase evolution; interrogation of substitutions identified in directed evolution and rationally combining these substitutions may provide an alternate and rapid route for the uphill walk on the phytase fitness landscape. Phytase engineering for improved thermostability revealed the following parameters for efficient improvements: (i) the higher the chemical diversity in the mutant libraries the better, (ii) target flexible loops away from the binding pocket, (iii) perform surface engineering, and (iv) alter glycosylation pattern. Furthermore, surface charge modifications have a great influence on the pH optima and pH stability of phytases, suggesting an attractive route for phytase stabilization. Optimal pH could be altered by charge modification near the binding pocket, whereas overall charge modification may improve pH stability. Elimination of potential protease-cleaving sites and stabilization of surface loops represent additional strategies to improve gastric tolerance.

References

1. Bohn L, Meyer AS, Rasmussen SK (2008) Phytate: impact on environment and human nutrition. A challenge for molecular breeding. J Zhejiang Univ Sci B 9(3):165–191
2. Harland BF, Oberleas D (1999) Phytase in animal nutrition and waste management. BASF Ref Man 237–240.
3. Sebastian S, Touchburn SP, Chavez ER (1998) Implications of phytic acid and supplemental microbial phytase in poultry nutrition: a review. World Peoult Sci J 54(01):27–47
4. Lei XG, Weaver JD, Mullaney E et al (2013) Phytase, a new life for an "old" enzyme. Ann Rev Anim Biosci 1:283–309

5. Lei XG, Ku PK, Miller ER et al (1993) Supplementing corn-soybean meal diets with microbial phytase maximizes phytate phosphorus utilization by weanling pigs. J Anim Sci 71(12):3368–3375
6. Lei XG, Ku PK, Miller ER et al (1994) Calcium level affects the efficacy of supplemental microbial phytase in corn-soybean meal diets of weanling pigs. J Anim Sci 72(1):139–143
7. Lei X, Ku PK, Miller ER et al (1993) Supplemental microbial phytase improves bioavailability of dietary zinc to weanling pigs. J Nutr 123(6):1117–1123
8. Gilbert N (2009) Environment: the disappearing nutrient. Nature 461(7265):716–718
9. Adeola O, Cowieson AJ (2011) BOARD-INVITED REVIEW: opportunities and challenges in using exogenous enzymes to improve nonruminant animal production. J Anim Sci 89(10):3189–3218
10. Cowieson A, Cooper R (2010) Introduction to the event and overview of the phytase market. In: International phytase summit. International Phytase Summit, Washington, DC
11. Meyer AS (2010) Enzyme technology for precision functional food ingredient processes. Ann N Y Acad Sci 1190:126–132
12. Hubenova Y, Georgiev D, Mitov M (2014) Stable current outputs and phytate degradation by yeast-based biofuel cell. Yeast 31(9):343–348
13. Fujita J, Fukuda H, Yamane Y-I et al (2001) Critical importance of phytase for yeast growth and alcohol fermentation in Japanese sake brewing. Biotechnol Lett 23(11):867–871
14. Billington DC (1993) In: Billington DC (ed) The Inositols phosphates: chemical synthesis and biological significance. VCH Verlagsgesellschaft, Weinheim
15. Fu S, Sun J, Qian L et al (2008) Bacillus phytases: present scenario and future perspectives. Appl Biochem Biotechnol 151(1):1–8
16. Kim MS, Weaver JD, Lei XG (2008) Assembly of mutations for improving thermostability of *Escherichia coli* AppA2 phytase. Appl Microbiol Biotechnol 79(5):751–758
17. Kim MS, Lei XG (2008) Enhancing thermostability of *Escherichia coli* phytase AppA2 by error-prone PCR. Appl Microbiol Biotechnol 79(1):69–75
18. Shivange AV, Roccatano D, Schwaneberg U (2016) Iterative key-residues interrogation of a phytase with thermostability increasing substitutions identified in directed evolution. Appl Microbiol Biotechnol 100(1):227–242
19. Wong TS, Roccatano D, Zacharias M et al (2006) A statistical analysis of random mutagenesis methods used for directed protein evolution. J Mol Biol 355(4):858–871
20. Di Giulio M (2005) The origin of the genetic code: theories and their relationships, a review. Biosystems 80(2):175–184
21. Zhao J, Kardashliev T, Joelle Ruff A et al (2014) Lessons from diversity of directed evolution experiments by an analysis of 3,000 mutations. Biotechnol Bioeng 111(12):2380–2389
22. Shivange AV, Serwe A, Dennig A et al (2012) Directed evolution of a highly active Yersinia mollaretii phytase. Appl Microbiol Biotechnol 95(2):405–418
23. Singh B, Satyanarayana T (2009) Characterization of a HAP-phytase from a thermophilic mould Sporotrichum thermophile. Bioresour Technol 100(6):2046–2051
24. Lehmann M, Kostrewa D, Wyss M et al (2000) From DNA sequence to improved functionality: using protein sequence comparisons to rapidly design a thermostable consensus phytase. Protein Eng 13(1):49–57
25. Shivange AV, Schwaneberg U, Roccatano D (2010) Conformational dynamics of active site loop in *Escherichia coli* phytase. Biopolymers 93(11):994–1002
26. Shivange AV, Dennig A, Schwaneberg U (2014) Multi-site saturation by OmniChange yields a pH- and thermally improved phytase. J Biotechnol 170:68–72
27. Salverda ML, Dellus E, Gorter FA et al (2011) Initial mutations direct alternative pathways of protein evolution. PLoS Genet 7(3):e1001321
28. Chen W, Ye L, Guo F et al (2015) Enhanced activity of an alkaline phytase from Bacillus subtilis 168 in acidic and neutral environments by directed evolution. Biochem Eng J 98: 137–143
29. Tracewell CA, Arnold FH (2009) Directed enzyme evolution: climbing fitness peaks one amino acid at a time. Curr Opin Chem Biol 13(1):3–9

30. Chen C-C, Cheng K-J, Ko T-P et al (2015) Current progresses in phytase research: three-dimensional structure and protein engineering. Chem Biol Eng Rev 2(2):76–86
31. Yao MZ, Zhang YH, Lu WL et al (2012) Phytases: crystal structures, protein engineering and potential biotechnological applications. J Appl Microbiol 112(1):1–14
32. Liao Y, Zeng M, Wu ZF et al (2012) Improving phytase enzyme activity in a recombinant phyA mutant phytase from Aspergillus niger N25 by error-prone PCR. Appl Biochem Biotechnol 166(3):549–562
33. Garrett JB, Kretz KA, O'Donoghue E et al (2004) Enhancing the thermal tolerance and gastric performance of a microbial phytase for use as a phosphate-mobilizing monogastric-feed supplement. Appl Environ Microbiol 70(5):3041–3046
34. Wu TH, Chen CC, Cheng YS et al (2014) Improving specific activity and thermostability of *Escherichia coli* phytase by structure-based rational design. J Biotechnol 175:1–6
35. Zhao Q, Liu H, Zhang Y (2010) Engineering of protease-resistant phytase from Penicillium sp.: high thermal stability, low optimal temperature and pH. J Biosci Bioeng 110(6): 638–645
36. Zhu W, Qiao D, Huang M et al (2010) Modifying thermostability of appA from *Escherichia coli*. Curr Microbiol 61(4):267–273
37. Zhang W, Mullaney EJ, Lei XG (2007) Adopting selected hydrogen bonding and ionic interactions from Aspergillus fumigatus phytase structure improves the thermostability of Aspergillus niger PhyA phytase. Appl Environ Microbiol 73(9):3069–3076
38. Viader-Salvado JM, Gallegos-Lopez JA, Carreon-Trevino JG et al (2010) Design of thermostable beta-propeller phytases with activity over a broad range of pHs and their overproduction by Pichia pastoris. Appl Environ Microbiol 76(19):6423–6430
39. Fei B, Cao Y, Xu H et al (2013) AppA C-terminal plays an important role in its thermostability in *Escherichia coli*. Curr Microbiol 66(4):374–378
40. Fei B, Xu H, Cao Y et al (2013) A multi-factors rational design strategy for enhancing the thermostability of *Escherichia coli* AppA phytase. J Ind Microbiol Biotechnol 40(5):457–464
41. Fu D, Huang H, Meng K et al (2009) Improvement of Yersinia frederiksenii phytase performance by a single amino acid substitution. Biotechnol Bioeng 103(5):857–864
42. Tran TT, Mamo G, Buxo L et al (2011) Site-directed mutagenesis of an alkaline phytase: influencing specificity, activity and stability in acidic milieu. Enzym Microb Technol 49(2): 177–182
43. Rodriguez E, Wood ZA, Karplus PA et al (2000) Site-directed mutagenesis improves catalytic efficiency and thermostability of *Escherichia coli* pH 2.5 acid phosphatase/phytase expressed in Pichia pastoris. Arch Biochem Biophys 382(1):105–112
44. Tung ET, Ma HW, Cheng C et al (2008) Stabilization of beta-propeller phytase by introducing Xaa-->Pro and Gly-->Ala substitutions at consensus positions. Protein Pept Lett 15(3):297–299
45. Bei J, Chen Z, Fu J et al (2009) Structure-based fragment shuffling of two fungal phytases for combination of desirable properties. J Biotechnol 139(2):186–193
46. Xu W, Shao R, Wang Z et al (2015) Improving the neutral phytase activity from Bacillus amyloliquefaciens DSM 1061 by site-directed mutagenesis. Appl Biochem Biotechnol 175(6):3184–3194
47. Vogt G, Argos P (1997) Protein thermal stability: hydrogen bonds or internal packing? Fold Des 2(4):S40–S46
48. Vogt G, Woell S, Argos P (1997) Protein thermal stability, hydrogen bonds, and ion pairs. J Mol Biol 269(4):631–643
49. Karshikoff A, Nilsson L, Ladenstein R (2015) Rigidity versus flexibility: the dilemma of understanding protein thermal stability. FEBS J 282(20):3899–3917
50. Folch B, Dehouck Y, Rooman M (2010) Thermo- and mesostabilizing protein interactions identified by temperature-dependent statistical potentials. Biophys J 98(4):667–677
51. Strickler SS, Gribenko AV, Gribenko AV et al (2006) Protein stability and surface electrostatics: a charged relationship. Biochemistry 45(9):2761–2766

52. Shivange AV, Hoeffken W, Haefner S, Schwaneberg U (2016). Protein consensus based surface engineering (ProCoS): a computer-assisted method for directed protein evolution. Biotechniques 61(6):305–314
53. Svihus B (2014) Function of the digestive system. J Appl Poult Res 23(2):306–314
54. Kim T, Mullaney EJ, Porres JM et al (2006) Shifting the pH profile of Aspergillus niger PhyA phytase to match the stomach pH enhances its effectiveness as an animal feed additive. Appl Environ Microbiol 72(6):4397–4403
55. Weaver JD, Mullaney EJ, Lei XG (2007) Altering the substrate specificity site of Aspergillus niger PhyB shifts the pH optimum to pH 3.2. Appl Microbiol Biotechnol 76(1):117–122
56. Wyss M, Pasamontes L, Friedlein A et al (1999) Biophysical characterization of fungal phytases (myo-inositol hexakisphosphate phosphohydrolases): molecular size, glycosylation pattern, and engineering of proteolytic resistance. Appl Environ Microbiol 65(2):359–366
57. Lassen SF, De Maria L, Friis EP et al. (2012) Hafnia phytase variants. (US20120225468) Novozymes A/S
58. Lei X (2003) Enzymes with improved phytase activity. (US6511699 B1) Cornell Research Foundation, Inc
59. Haefner S, Welzel A, and Thummer R (2014) Synthetic phytase variants. (US20140044835) BASF SE
60. Huang H, Luo H, Wang Y et al (2008) A novel phytase from Yersinia rohdei with high phytate hydrolysis activity under low pH and strong pepsin conditions. Appl Microbiol Biotechnol 80(3):417–426
61. Greiner R, Farouk AE (2007) Purification and characterization of a bacterial phytase whose properties make it exceptionally useful as a feed supplement. Protein J 26(7):467–474
62. Niu C, Luo H, Shi P et al (2015) N-glycosylation improves the pepsin resistance of HAP phytases by enhancing the stability at acidic pH and reducing the pepsin accessibility to peptic cleavage sites. Appl Environ Microbiol 82:1004–1014
63. Basu SS and Zhang S (2010) Engineering enzymatically susceptible proteins. (US20100273198 A1) Syngenta Participations Ag
64. Studier FW (2005) Protein production by auto-induction in high density shaking cultures. Protein Expr Purif 41(1):207–234
65. Sajidan A, Farouk A, Greiner R et al (2004) Molecular and physiological characterisation of a 3-phytase from soil bacterium Klebsiella sp. ASR1. Appl Microbiol Biotechnol 65(1): 110–118
66. Senn AM, Wolosiuk RA (2005) A high-throughput screening for phosphatases using specific substrates. Anal Biochem 339(1):150–156
67. Aharoni A, Thieme K, Chiu CP et al (2006) High-throughput screening methodology for the directed evolution of glycosyltransferases. Nat Methods 3(8):609–614
68. Pitzler C, Wirtz G, Vojcic L et al (2014) A fluorescent hydrogel-based flow cytometry high-throughput screening platform for hydrolytic enzymes. Chem Biol 21(12):1733–1742
69. Mitra N, Sinha S, Ramya TN et al (2006) N-linked oligosaccharides as outfitters for glycoprotein folding, form and function. Trends Biochem Sci 31(3):156–163
70. Sola RJ, Griebenow K (2009) Effects of glycosylation on the stability of protein pharmaceuticals. J Pharm Sci 98(4):1223–1245
71. Yao MZ, Wang X, Wang W et al (2013) Improving the thermostability of *Escherichia coli* phytase, appA, by enhancement of glycosylation. Biotechnol Lett 35(10):1669–1676
72. Huang H, Luo H, Yang P et al (2006) A novel phytase with preferable characteristics from Yersinia intermedia. Biochem Biophys Res Commun 350(4):884–889
73. Miksch G, Kleist S, Friehs K et al (2002) Overexpression of the phytase from *Escherichia coli* and its extracellular production in bioreactors. Appl Microbiol Biotechnol 59(6):685–694
74. Suzuki U, Yoshimura K, Takaishi M (1907) Über ein enzym 'Phytase' das anhydro-oxymethylen diphosphorsaure' spalter. Tokyo Imp Univ Coll Agric Bull 7:503–512

Strain Development by Whole-Cell Directed Evolution

7

Tong Si, Jiazhang Lian, and Huimin Zhao

Abstract

Due to limited knowledge of complicated cellular networks, directed evolution has played critical roles in strain improvement, especially for complex traits with hundreds of genetic determinants and for organisms with few genetic tools. Directed evolution mimics natural evolution in the laboratory via iterative cycles of diversity generation and functional selection or screening to isolate evolved mutants with desirable phenotypes. In this chapter, we summarize recent technological advances and applications of directed evolution in strain development, focusing on the efforts for accelerating evolution workflows, expanding the range of target phenotypes, and facilitating mechanistic understanding of evolved mutations.

[$]Tong Si and Jiazhang Lian contributed equally to this work

T. Si
Carl R Woese Institute for Genomic Biology, University of Illinois at Urbana-Champaign, 1206 W Gregory Dr, Urbana, IL 61801, USA

J. Lian
Carl R Woese Institute for Genomic Biology, University of Illinois at Urbana-Champaign, 1206 W Gregory Dr, Urbana, IL 61801, USA

Department of Chemical and Biomolecular Engineering, University of Illinois at Urbana-Champaign, 600 South Mathews Ave, Urbana, IL 61801, USA

H. Zhao (✉)
Carl R Woese Institute for Genomic Biology, University of Illinois at Urbana-Champaign, 1206 W Gregory Dr, Urbana, IL 61801, USA

Department of Chemical and Biomolecular Engineering, University of Illinois at Urbana-Champaign, 600 South Mathews Ave, Urbana, IL 61801, USA

Departments of Chemistry, Biochemistry, and Bioengineering, University of Illinois at Urbana-Champaign, 600 South Mathews Ave, Urbana, IL 61801, USA
e-mail: zhao5@illinois.edu

M. Alcalde (ed.), *Directed Enzyme Evolution: Advances and Applications*,
DOI 10.1007/978-3-319-50413-1_7

7.1 Introduction

Metabolic engineering aims at "the improvement of cellular activities by manipulation of enzymatic, transport, and regulatory functions of the cell" [8] and has been successfully applied to create efficient microbial cell factories for producing fuels, drugs, and value-added chemicals [59, 80, 83, 102, 117]. There are two fundamental approaches for metabolic engineering: rational design and directed evolution. For rational design, metabolic bottlenecks are identified and addressed via directed genetic manipulations for improved production of a target molecule [8, 9]. However, this approach is associated with four major limitations. First, effective rational designs often require comprehensive understanding of metabolic pathways and cellular metabolism, which is incomplete most of time. In particular, there is limited knowledge of some complex phenotypes, such as stress tolerance and chemical inhibitor resistance [122, 161]. Second, designed mutations may result in unexpected metabolic outcomes, due to the complicated nature of biological networks. Third, genetic manipulation is generally difficult in many industrial strains. Finally, there are regulatory issues that limit the use of genetically modified organisms (GMOs) with rationally designed mutations in certain applications, such as the food industry [18]. On the contrary, directed evolution mimics natural evolution in laboratory settings, where iterative rounds of genetic diversity generation and phenotypic selection or screening are performed to isolate strains with improved traits (Fig. 7.1) [18, 32, 68, 120, 150, 159]. Whole-cell directed evolution, also known as evolutionary engineering, can overcome the aforementioned limitations of rational design. It requires no prior knowledge on genotype-phenotype relationships and achieves relatively defined phenotypic outcomes dependent on selection/screening schemes. Mutant strains can be generated using spontaneous mutations without genetic manipulations and hence considered as non-GMOs for higher public acceptance [18].

Adaptive evolution (AE) is the most widely used and well-established approach for whole-cell directed evolution [32, 159]. In a typical AE experiment, a growing asexual culture is maintained for a prolonged period of time via serial transfer in a batch format or continuous dilution in a chemostat (see Sect. 7.2.2 for details). Strain variants arise due to spontaneous mutations. Under a certain selection pressure, variants with improved competitive fitness outgrow the parental and less-fit sibling populations (Fig. 7.1). Genotypes of evolved strains are analyzed at individual or population levels for understanding the molecular mechanisms conferring improved phenotypes.

Although successfully applied in improving a wide range of industrially important traits in microbial strains, AE suffers from various limitations, including low rates and relatively restricted spectrums of mutations, prolonged cultivation, limited ranges of target phenotypes, and challenges to differentiate beneficial mutants from neutral to deleterious hitchhikers. In this chapter, we aim to provide an update on the technical advances in the application of directed evolution for strain development, and different methods are organized according to how they overcome various limitations of the classical AE (Fig. 7.1). We focus on practical considerations on

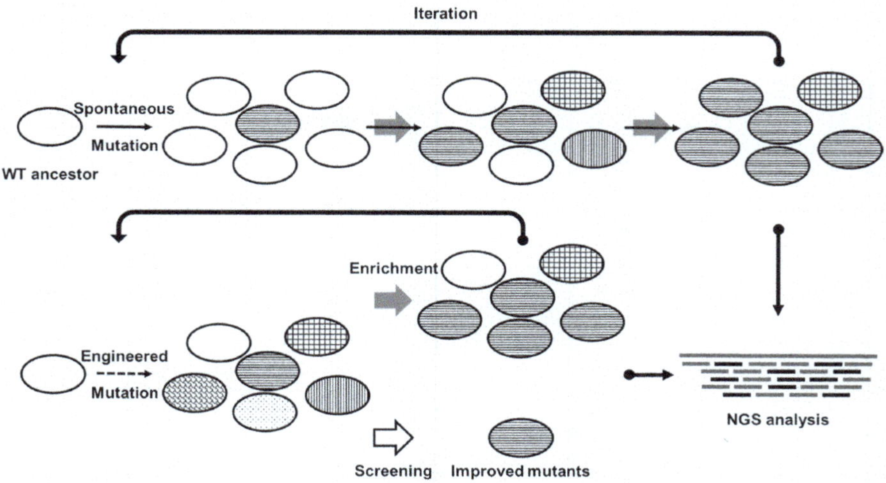

Fig. 7.1 Overview of directed evolution processes consisting of iterative cycles of diversity generation and functional screening or selection. For the classical AE, diversity is generated by spontaneous mutations (*thin, solid arrow*). During prolonged cultivation under selection pressures, mutants with improved competitive fitness are enriched (*thick, closed arrow*). New methods are developed to introduce systematic and multiplex mutations (*thin, dashed arrow*) for accelerating evolution and to perform high-throughput screening (*thick, open arrow*) for engineering traits that are not linked to growth. Genetic mutations are then analyzed using next-generation sequencing (*NGS*)

the choice of appropriate techniques (Table 7.1). We then discuss recent examples of directed evolution in improving various cellular properties and provide prospective on the future directions.

7.2 Recent Technology Advances in the Application of Directed Evolution for Strain Development

7.2.1 Diversity Generation

Classical AE experiments depend on spontaneous mutations, primarily in the form of single-base mutations and at a rate of 10^{-10}–10^{-9} substitutions per base pair per replication [76, 78, 88, 143]. Other types of mutations, such as insertion and deletion of one or a few bases, gene duplication, and chromosomal rearrangement, occur at even lower frequencies [93]. The rare and uncontrollable nature of naturally occurring mutations renders AE as a very time-consuming process, often requiring tens of thousands of replicative generations. Furthermore, the limited types of spontaneous mutations (mostly single-base substitutions) [93] exhibit modest impacts on competitive fitness [45]. These relatively small steps of evolutionary paths on a fitness landscape are prone to be trapped at local optimum [150]. To overcome these obstacles, new methods are developed to increase random mutation rates, introduce

Table 7.1 Comparison of mutagenesis methods for directed evolution in strain development

Mutagenesis methods	Host range	Need genetic tool	Need genome sequence	Mutational spectrum	Systematic mapping	Multiplex mutation
Adaptive evolution	Prokaryote/eukaryote	No	No	Mostly single-base substitution	No	Yes
Chemical/radiation mutagenesis	Prokaryote/eukaryote	No	No	Diverse	No	Yes
Mutator strain	Prokaryote/eukaryote	Yes	No	Diverse	No	Yes
Transposon insertion	Prokaryote/eukaryote	Yes	No	Mostly knockout	Yes	No
RNA interference[a]	Eukaryote	Yes	No	Knockdown	Yes	Yes
Recombineering	Prokaryote[b]	Yes	Yes	Diverse	Yes	Yes
CRISPR-Cas	Prokaryote/eukaryote	Yes	Yes	Diverse[c]	Yes	Yes
Transcription/translation factor engineering	Prokaryote/eukaryote	Yes	No	Diverse	No	Yes
Genome shuffling	Prokaryote/eukaryote	Yes	No	Diverse	No	Yes

[a]Other regulatory RNA mechanisms, such as small RNAs and antisense RNAs, can also be used for directed evolution (see [133] for a summary)
[b]The efficiency of oligonucleotide-mediated recombineering is not high enough for practical applications in yeast [28]
[c]In addition to gene knockout, which was primarily discussed in this chapter, genetic activation and repression can also be enabled using CRISPR-Cas (see [133] for a summary)

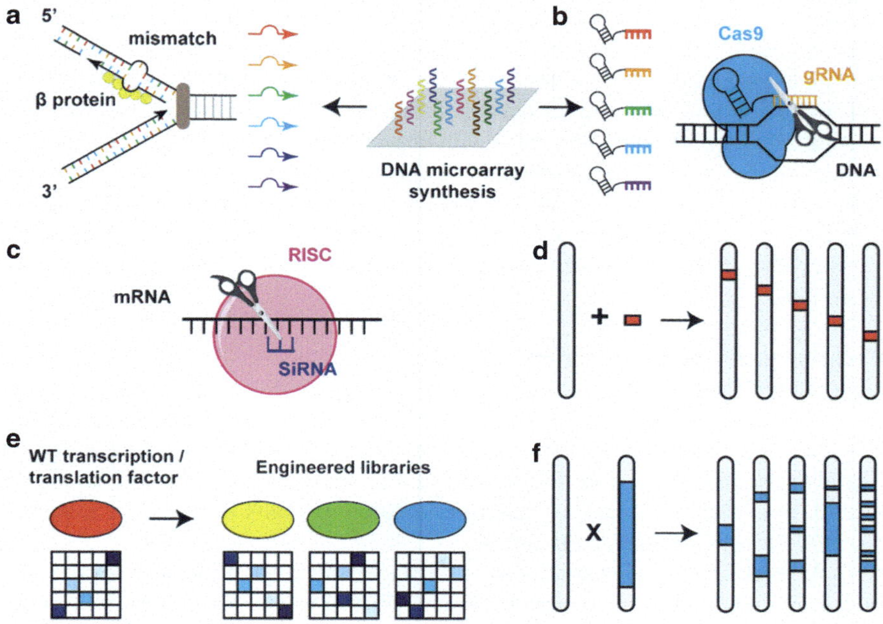

Fig. 7.2 Schemes of selected methods for diversity generation. (**a**) Recombineering. (**b**) CRISPR-Cas. (**c**) RNA interference. (**d**) Transposon insertional mutagenesis. (**e**) Global transcription and translation machinery engineering. (**f**) Genome shuffling

systematic perturbations, and create large-scale and multiplex genetic diversities (Figs. 7.1 and 7.2). A comparison of various diversity generation methods is summarized in Table 7.1.

Increasing random mutagenesis rates

Increasing random mutagenesis rates is proposed to improve the probability of generating beneficial mutations and accelerate the evolutionary processes [159]. Chemical mutagens and radiation are widely used to induce DNA damages that lead to genetic mutagenesis [123]. In addition, mutator strains with impaired DNA replication or repair systems are employed to enhance genetic diversity [44, 125]. However, increased mutagenesis rates may also result in more frequent occurrence of deleterious mutations in already adapted strains, rendering the fixation of beneficial mutations difficult in an evolving population. To solve this dilemma, dynamic control of mutagenesis rates is proposed, whereby mutagenesis rates are raised transiently and then decreased to normal levels after adaptation is achieved. In one example, GREACE (Genome Replication Engineering Assisted Continuous Evolution) introduced defective components of a DNA polymerase complex on a plasmid, converting the target strain into a mutator [92]. After evolution, the plasmid was cured to return to normal mutation rates, allowing for isolation of stable strains for analysis. In another example, FREP (feedback-regulated evolution of phenotype) inversely coupled the mutagenesis rates with the concentrations of a target metabolite [23]. In particular, a genetic sensor activated expression of a proofreading exonuclease mutant at low target metabolite concentrations, and the activation effect diminished with increasing production. Application of this strategy successfully identified mutant strains with improved production of tyrosine and isopentenyl diphosphate [23]. Notably, the constant or dynamic use of mutator phenotypes is limited to organisms with known DNA replication and repair mechanisms. Controlling mutagenesis rates in less-characterized strains may be achieved by varying the dosage of chemical and physical mutagens prior to or during evolution.

Creating genome-scale diversity

Genome-wide strain libraries prove to be instrumental for studying genotype-phenotype relationships [19, 134]. Compared with AE that generates random mutations in limited numbers of genes, strain libraries allow systematic investigation on nearly every gain-/reduction-/loss-of-function mutation within a target genome [7, 42, 50, 94]. Also, it is easier to interpret screening results using single-mutation strain libraries, compared with hundreds of changes observed in evolved genomes from AE experiments. However, conventional methods require each mutant strain to be created individually in an arrayed manner [7, 42], which is prohibitively tedious and costly for iterative screening in directed evolution. The main technical hurdle is the low efficiency of homologous recombination (HR) for allelic replacement on a genome.

To stimulate HR and accelerate library construction, two cellular mechanisms have been utilized. Using bacteriophage-based recombination systems, recombineering (recombination-based genetic engineering) enables efficient genome editing with single-stranded DNA oligonucleotides as homologous donors (Fig. 7.2a) [34, 127]. For example, combined with microarray-based DNA synthesis, promoter replacement of every gene in *Escherichia coli* was achieved using a pool of DNA oligonucleotides in a single round of transformation [157]. Furthermore, HR efficiency can be enhanced via introduction of double-stranded breaks at targeted genomic loci, using programmable DNA nucleases derived from ZFPs (zinc finger proteins), TALEs (transcription activator-like effectors), and CRISPR-Cas (clustered regularly interspaced short palindromic repeats and CRISPR-associated proteins) [36, 134]. In particular, recognition of a target DNA sequence by the CRISPR-Cas nuclease Cas9 is mediated by a *trans-acting* guide RNA (gRNA) via Watson-Crick base pairing (Fig. 7.2b) [25, 96]. Using gRNA expression libraries synthesized by DNA microarray (Fig. 7.2b), genome-scale screening has been demonstrated in human cells [154], and similar approaches should be readily applied in engineering microbial systems.

HR-independent methods are also developed for constructing strain libraries. First, *trans-acting* RNAs have been employed for facile introduction of genome-wide perturbation (Fig. 7.2c) [133]. For example, the RNA interference (RNAi) machinery has been reconstituted in *Saccharomyces cerevisiae* for genome-scale knockdown screening. This RNAi-assisted genome evolution (RAGE) strategy identified mutations conferring resistance towards fermentation inhibitors [133, 163]. Second, genome-scale insertion mutagenesis can be conducted using transposons, which are a class of mobile genetic elements (Fig. 7.2d) [2, 55]. For positive mutants exhibiting improved fitness, the inserted transposon sequence can act as a tag to identify insertion positions and affected genes. Most transposition events lead to gene disruption, but may also enable gene activation if equipped with outward-oriented promoters [123].

Notably, CRISPR-Cas and RNAi may be particularly useful for creating libraries of industrial strains, many of which are polyploidy. CRISPR-Cas has been used for simultaneous disruption of two alleles of a gene in several industrial yeast strains [142], which is very challenging using traditional methods. In addition, RNAi-based gene silencing targets mRNA transcripts and hence can regulate gene expression without modifying multiple copies of the same gene.

Introducing large-scale and multiplex mutations

Given the positive correlation between the probability of finding improved mutants and the degree of genetic diversity [66], it can be beneficial to introduce large-scale and combinatorial mutations to a genome, especially when targeting complex phenotypes (e.g., tolerance) that have many genetic determinants [122, 161]. Although AE is capable of accumulating hundreds of mutations, it requires prolonged cultivation and generates limited types of mutations. Therefore, it is desirable to develop new methods for generating many mutations of diverse types in a short period of time.

One strategy is to introduce mutations in cellular components that are involved in transcription or translation processes (Fig. 7.2e). For example, gTME (global transcriptional machinery engineering) introduced random mutations to master transcription factors (TFs) via error-prone PCR [4, 5], altering expression levels of hundreds of genes. Large perturbation in transcriptomic profiles can also be enabled upon incorporation of artificial TFs containing tandem repeats of ZFPs [109, 110]. In addition, screening with lethal concentrations of ribosome-targeting antibiotics can isolate genetic mutations in ribosomal components. Ribosomal mutations result in perturbed proteomes for emergence of improved cellular phenotypes [53, 106, 130, 136].

Another strategy is to create multiplex mutations for combinatorial diversity. Three aforementioned mechanisms—recombineering, CRISPR-Cas, and *trans-acting* regulatory RNAs— are also widely applied for this purpose. Mediated by the ssDNA-binding protein (β) from the λ-red bacteriophage, oligonucleotide pools targeting the RBSs of 24 genes were transformed recursively into *E. coli* for allelic replacement, creating over 4.3 billion combinatorial genomic variants per day [152]. Using CRISPR-Cas, multiplex gene disruption or integration can be achieved with high efficiency, whereby multiple targeting gRNAs and HR donors are introduced in a single cell [10, 57, 58, 129]. Moreover, iterative rounds of RNAi screening result in the accumulation of beneficial knockdown mutations that act synergistically to improve acetic acid tolerance in *S. cerevisiae* [133].

Large-scale chromosomal rearrangement is another useful source for increasing diversity. First, genome shuffling promotes HR between genomes following protoplast fusion, sexual mating, and transformation of whole-genome fragments (Fig. 7.2f), generating combinations of mutations from a genetically diverse collection of parental genomes [12]. Genetic diversity among parents can be derived from directed evolution or natural sources such as different strains of the same species or even different species [12]. Second, targeted genome deletion can be achieved following a generation of DNA breaks (DSBs or nicks) at two genomic loci by CRISPR-Cas [24, 140]. Furthermore, during construction of a synthetic chromosome arm in yeast, a number of LoxPsym sequences were inserted after the stop codon of every nonessential gene or near major genomic landmarks such as repetitive sequences or telomeres [33]. With this SCRaMbLE (synthetic chromosome rearrangement and modification by LoxP-mediated evolution) design, recombination occurred randomly between two LoxPsym sequences to produce inversions or deletions, resulting in formation of structurally distinct genomes [33, 128].

7.2.2 Selection and Screening

High-throughput selection and screening are critical in isolating mutants with improved phenotypes following diversity generation (Fig. 7.1). For classical AE experiments, evolving populations are maintained via serial or continuous dilution [45, 123]. For serial dilution in batch cultures, transfer is typically conducted during the exponential phase prior to the onset of the stationary phase [148], and cells experience fluctuating selection pressures due to ever-changing concentrations of nutrients and cells. On the contrary, a steady-state cell culture is kept in a chemostat, whereby fresh medium addition and culture removal are performed at defined rates [105]. The cell density is determined by a limiting nutrient of a defined medium [171], which acts as a constant selection pressure in a chemostat.

While readers are directed to a recent review for comprehensive comparison between serial transfer and chemostat [45], we would like to discuss some practical considerations based on the different types of selection pressures. It is arguably easier to interpret the genetic basis of adaptation in a chemostat, thanks to the consistency of selection pressures, whereas heterogeneous selection resulting from dynamic environments in batch cultures renders explanation of functional causality difficult. For example, competitive fitness in batch cultures may result from reduced duration in the lag phase, increased growth rates during the exponential phase, or enhanced capability to divide in the stationary phase [84]. As a result, entangled mechanisms make interpretation of adaptive mutations very challenging for batch selection [144]. However, constant selection pressures in a chemostat may be problematic, as isolated mutants may perform poorly in a different condition, a phenomenon known as overfitting. For example, robust stress responses allow microbes to enter a quiescent state for long-term survival upon starvation (e.g., a limiting nutrient in a chemostat). While this capacity is beneficial in natural or industrial environments that fluctuate in nutrient availability, activation of quiescence stops proliferation and confers strong disadvantages in a chemostat. Therefore, loss of stress response pathways is repeatedly observed in chemostat selection [95, 104], rendering isolated mutants unsuitable in an industrial setting due to reduced robustness. Taken together, serial transfer in batch cultures may be preferred to isolate mutant strains for practical applications, whereas continuous cultivation in a chemostat is more suitable for studying genetic determinants of a target phenotype.

Both batch and continuous selections are based on competitive fitness and therefore useful to engineer traits that are related to growth or survival, such as utilization of unnatural feedstock substrates, tolerance towards harsh industrial conditions, and resistance to high concentrations of substrates, products, or inhibitors. However, target molecule production is generally not linked to fitness. To screen for enhanced productivity, three major strategies have been devised. First, product formation can be coupled with growth advantages by improving redox balancing [39, 139], enhancing resistance to toxic metabolite analogs [137], increasing tolerance to oxidative stress [118], or rescuing engineered auxotrophy [6, 15]. Second, colorimetric or fluorometric assays can be developed based on either the optical properties of target compounds or chemical and enzymatic conversions linking concentrations to spectrum signals [29]. Third, riboswitch or TF-based metabolite-sensing modules can be used to link molecular concentrations to the abundance of a fluorescence protein or an essential metabolite [29, 134], and strain libraries harboring these genetic biosensors can then be subjected to screening or selection [30, 86, 98, 153].

In addition to expand the range of phenotypes for screening, it is also desirable to accelerate directed evolution by automation. Currently, manual efforts are required for culture maintenance, archival storage, contamination tests, selective pressure adjustment, and phenotype analysis [159]. Inevitable human interventions limit throughput and reproducibility and introduce subjective bias. To circumvent these limitations, liquid handling robots were applied for automated batch evolution in microtiter plates [52, 77]. A microfluidic platform was devised to maintain thousands of microdroplet chemostat systems in parallel [56]. Feedback control systems were also equipped to monitor cell growth and then dynamically regulate selection pressures [147]. For these approaches, however, cautions need to be taken on how transferable the improved phenotypes are during scale-up processes from a microwell or a droplet to a real fermenter.

7.2.3 Mutation Analysis

Thanks to the advances in next-generation sequencing (NGS) (Fig. 7.1), acquired genetic mutations during AE can be readily analyzed via whole-genome sequencing (WGS) [11]. However, the main challenge is how to distinguish beneficial mutations from neutral to deleterious hitchhikers [26]. Due to the lack of "golden standard" workflows, it is advisable to integrate information from multiple analytical schemes.

First, both endpoint and time-course WGS analysis should be performed for individual clones and the whole populations [11, 75]. Sequencing of an individual clone can reveal all mutations in the specific evolved genome [11], but these mutations represent only a random subset of all genetic variants in a population [65, 74]. Therefore, sequencing of several clones is recommended, but it requires proper multiplexing techniques for reducing the analysis cost [72]. On the other hand, population sequencing provides a more comprehensive survey on mutation frequency across different subpopulations, as well as information on evolutionary trajectory if performed at different stages during evolution [11]. However, sequencing/alignment errors may occur during experimental and computational analysis of WGS. In particular, it is more technically challenging to differentiate low-frequency mutations from sequencing/alignment ambiguity for whole-population sequencing [75]. In general, WGS accuracy can be improved via higher sequencing coverage and special sequencing techniques including paired-end sequencing [40], circular sequencing [91], and long-read sequencing [71]. Moreover, time-course whole-population sequencing can enhance detection of low-frequency alleles in evolving populations [77]. Given the trade-offs in clonal and population sequencing, it may be beneficial to combine both to assist mechanistic studies [75].

Second, statistical analysis can help distinguish between adaptive drivers and neutral or deleterious hitchhikers. Multiple evolution experiments can be performed in parallel, and mutations appearing in replicate populations are more likely adaptive [77]. In addition, mutations can be grouped by their functions (e.g., gene ontology (GO) enrichment analysis), and cellular processes that are key to a specific trait can be revealed [119, 144]. Furthermore, expected ratios of synonymous/non-synonymous mutations can be calculated under the assumption of neutral selection, and underrepresentation of synonymous mutations is indicative of adaption [11, 13].

Third, other genome-wide analyses should be performed in addition to WGS [108, 160]. Transcriptomic analysis is necessary due to the difficulty of predicting the impact of some genetic mutations, especially for noncoding sequences and regulatory proteins [47, 73]. For example, a global TF variant led to differential expression of hundreds of genes, which can only be revealed using transcriptional profiling [4]. Moreover, proteomics [132] and metabolomics [145, 160] have been applied to reveal molecular mechanisms of evolved traits, as proteins and metabolites are the direct actuators of many cellular phenotypes. Given the challenges in interpreting large-scale omics datasets, it is desirable to develop advanced computational tools for data integration [38, 121].

Finally, reconstruction of mutations from evolved strains in isolation or in combination in an unevolved ancestor background should provide the most direct observations on genotype-phenotype relationship. However, this approach is almost impossible in strains lacking genetic tools. For genetically tractable organisms, it is also very challenging to reconstitute the multitude of mutations obtained from

AE. To accelerate mutant creation, recombineering and CRISPR coupled with microarray-based DNA synthesis may be helpful for large-scale and multiplex introduction of mutations as discussed previously (Sect. 7.2.1). In addition, back-crossing of isolated mutants with the ancestor via genome shuffling can separate evolved mutations into a strain collection [11, 116], and progenies exhibiting improved phenotypes are more likely to harbor causative mutations. To increase throughput of phenotyping, the use of the molecular bar code technology allows rapid profiling of competitive fitness of every mutant in a mixed population, whereby population dynamics can be monitored via frequency quantification of mutation-associated bar codes using NGS [53, 157]. Aforementioned automation technology can also be applied to accelerate phenotyping of reconstructed mutants.

7.3 Examples of Directed Evolution for Construction and Optimization of Cell Factories

Directed evolution approaches have been proven to be effective in creating industrial microorganisms with extended substrate ranges, improved cellular properties, and enhanced production [18, 83, 141, 159]. In this part of the chapter, examples in using directed evolution approaches for strain improvement in recent years will be discussed.

7.3.1 Extension of Substrate Utilization

Mostly driven by environmental and energy security considerations, there is a growing interest in engineering microorganisms for production of fuels and chemicals from renewable feedstocks, such as lignocellulose and macroalgae. Notably, besides glucose, these renewable feedstocks contain different sugar components, such as xylose and arabinose from cellulosic biomass, galactose from red algae, and mannitol and 4-deoxy-L-erythro-5-hexoseulose urinate (DEHU) from brown algae. As substrate utilization can be readily coupled to cellular growth, directed evolution has been extensively applied to construct efficient fermentation strains with expanded substrate scopes, especially of *S. cerevisiae*, which is a preferred cell factory for many industrial applications.

As the most abundant raw material, lignocellulose has attracted increasing attention for its conversion to fuels and chemicals. Although hexoses (such as glucose) can be efficiently fermented by most microorganisms, the utilization of pentoses (mainly xylose and arabinose), which constitute more than 30% of the total carbohydrate, occurs with a much lower efficiency, even after extensive pathway and strain engineering. For example, the fungal oxidoreductase pathway containing xylose reductase (XR), xylitol dehydrogenase (XDH), and xylulose kinase (XKS) has been introduced into *S. cerevisiae* to enable xylose fermentation. Unfortunately, several bottlenecks including low xylose uptake, cofactor imbalance, and limited metabolic fluxes of the pentose phosphate pathway result in non-optimal xylose utilization and biofuel production (Fig. 7.3). Thus, directed evolution has been applied to construct efficient xylose-fermenting yeast strains [63, 67, 124, 167]. One of the most successful examples was demonstrated by Kim et al. in which serial transfer in xylose-containing media was performed to construct a yeast strain with the highest xylose-fermenting capability reported to date [63]. Whole-genome sequencing of the evolved strains indicated that the loss of function of *PHO13* played a dominant role in efficient xylose utilization. Follow-up studies confirmed that *PHO13* deletion induced the activation of pentose phosphate pathway especially the *TAL1* gene encoding the sedoheptulose-7-phosphate:D-glyceraldehyde-3-phosphate transaldolase [64] and prevented the accumulation of sedoheptulose [164] (Fig. 7.3). To bypass the cofactor imbalance issue of the fungal xylose utilization pathway, researchers have switched their focus to the bacterial xylose isomerase (XI) pathway [27, 82, 115, 151, 169]. While initial attempts were not very successful, probably due to the low activity and/or poor expression of XI, directed evolution strategies were adopted to construct yeast strains that could convert xylose to ethanol at high yields. Besides *S. cerevisiae*, directed evolution has been applied

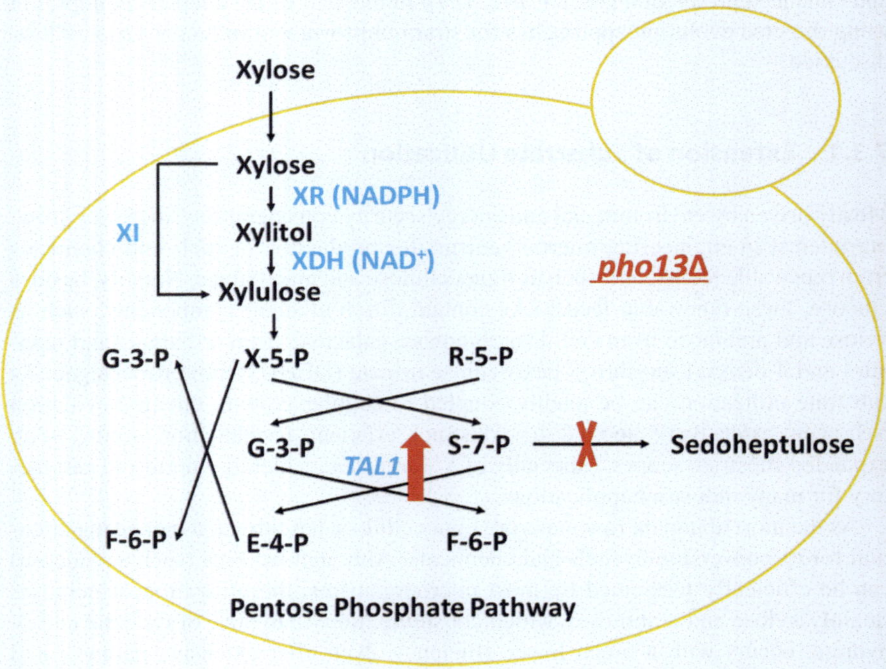

Fig. 7.3 Construction of an efficient xylose-fermenting *S. cerevisiae* strain using directed evolution. *XR* xylose reductase, *XDH* xylitol dehydrogenase, *TAL1* sedoheptulose-7-phosphate:D-glyceraldehyde-3-phosphate transaldolase, *X-5-P* xylulose-5-phosphate, *R-5-P* ribose-5-phosphate, *G-3-P* glyceraldehyde-3-phosphate, *S-7-P* sedoheptulose-7-phosphate, *E-4-P* erythrose-4-phosphate, *F-6-P* fructose-6-phosphate

to enable efficient xylose fermentation in other hosts, such as *Pichia pastoris* [85], an industrially important cell factory for recombinant protein production.

Recently, the use of marine macroalgae as a renewable feedstock has attracted increasing attention mainly because it does not require the arable lands and fresh water, and the absence of lignin makes the depolymerization of seaweed rather simple and straightforward [156, 158]. Among several types of macroalgae, red algae and brown algae are considered as ideal sustainable feedstocks for the production of biofuels and chemicals. The major sugar components are glucose and galactose for red algae and glucose, mannitol, and DEHU for brown algae. Although many microorganisms including *S. cerevisiae* can ferment galactose, its utilization and the corresponding biofuel production are still not efficient enough. Therefore, several directed evolution efforts have been attempted to improve galactose fermentation [79, 81]. Lee et al. introduced a genome-wide perturbation library into *S. cerevisiae* and isolated fast galactose-fermenting strains. It was found that the overexpression of the truncated *TUP1* gene encoding a global transcriptional repressor resulted in the most remarkable improvement of galactose fermentation [81]. Another directed evolution study found that a mutation in the global carbon-sensing Ras/PKA pathway led to significantly improved galactose fermentation [81]. Both studies highlighted the significance of the alteration of global regulatory networks for efficient galactose fermentation in *S. cerevisiae* [81]. Recently, Lee et al. also applied directed evolution to construct *S. cerevisiae* mutants with enhanced ability to produce bioethanol from both galactose and red algae hydrolysate [79]. On the contrary, the utilization of brown algae-derived sugars especially DEHU is not well explored. A synthetic yeast platform for converting brown algae sugars into bioethanol was constructed by combining several strategies. The endogenous mannitol transporter and mannitol-2-dehydrogenase were activated for mannitol utilization. A DEHU transporter and a DEHU reductase were introduced to reduce DEHU to 2-keto-3-deoxy-D-gluconate (KDG). Finally, a KDG kinase and a KDG-6-phosphate aldolase were included to enable DEHU fermentation [35]. However, the utilization of DEHU was poor and only possible under aerobic condition. Then this yeast platform was further adapted to grow on mannitol and DEHU under anaerobic condition, which yielded a yeast strain capable of producing ethanol from mannitol and DEHU with a titer of 36.2 g/L and 83% of the theoretical yield [35].

Although directed evolution strategies have been extensively used to increase the fermentation of a single sugar, it is desirable to engineer a platform strain capable of consuming a mixture of sugars simultaneously to increase fermentation productivity. Directed evolution of a xylose-fermenting *S. cerevisiae* strain lacking the major hexose transporter genes yielded a mutant showing improved growth on xylose, which was due to the expression of a normally silent *HXT11* gene. Further selection for growth on xylose based on a hexokinase deletion strain at high glucose concentrations resulted in a mutation at N366 of Hxt11p, which reversed the transporter specificity from glucose into xylose. The Hxt11p mutant was found to enable efficient co-fermentation of xylose and glucose at industrially relevant sugar concentrations [131]. Similarly, directed evolution was also performed in a hexokinase-deficient xylose-fermenting *S. cerevisiae* strain for growth on xylose in the presence of high glucose concentrations, which resulted in a mutation at N367 in the endogenous chimeric Hxt36p transporter. Using the Hxt36p^{N367} variant, efficient co-consumption of glucose and xylose was achieved [103]. Notably, co-consumption of glucose and xylose was only possible when the endogenous hexose transporter genes were disrupted. Therefore, more engineering efforts, including directed evolution, are needed to construct yeast strains capable of co-fermenting glucose and xylose.

7.3.2 Improvement of Cellular Properties

Construction of robust cell factories with resistance to multiple stresses is highly desirable due to the harsh conditions in industrial biotechnological processes. The molecular basis of stress resistance is complicated, making it difficult to construct multiple stress-resistant strains by rational approaches. On the contrary, directed evolution has been proven successful in engineering the tolerance to inhibitors in raw material hydrolysates, final products at high concentrations, and other industrial harsh conditions.

After pretreatment and depolymerization of the sustainable raw materials, many undesirable compounds arise in the hydrolysates, such as acetic acid and furfural, whose concentrations are sufficient to dramatically inhibit the host growth. Tolerance to these toxic compounds is generally engineered using the classical AE by serial transfer or continuous culture [48, 99, 162]. The recently developed RAGE method has been used to increase the tolerance to both acetic acid [133] and furfural [163] (Fig. 7.4). By introducing a genome-wide RNAi library into a *S. cerevisiae* strain followed by iterative rounds of screening under gradually increased stress conditions, Si et al. identified three gene knockdown targets (*PTC6*, YPR086W, and tRNA$^{Val(AAC)}$) that acted synergistically to confer an engineered yeast strain with substantially improved acetic acid tolerance [133] (Fig. 7.4). Similarly, the same RNAi-based directed evolution was applied to engineer furfural tolerance, and *SIZ1*, a gene encoding the E3 SUMO-protein ligase, was identified as a novel determinant of furfural tolerance [163] (Fig. 7.4). Besides the resistance to a single inhibitor, directed evolution has been successfully used to construct cell factories with significantly improved growth in the presence of a mixture of inhibitors [22] or even biomass hydrolysates [3, 49, 70, 112–114].

For economically feasible industrial processes, final products are produced at high concentrations, especially for biofuels and bulk chemicals, which may result in slower or arrested fermentation. The tolerance of the producing host to the desired product is one of the determinants in developing a successful biotechnological process. Although *S. cerevisiae* has a long history as the host for ethanol fermentation and shows the highest ethanol tolerance in nature, its ethanol tolerance can still be further improved by directed evolution [138, 155]. Snoek et al. developed a large-scale, robot-assisted genome shuffling strategy to increase the ethanol tolerance of the industrial *S. cerevisiae* strains. In their work, a large collection of *Saccharomyces* yeasts was characterized in detail, and eight parental strains were chosen for genome

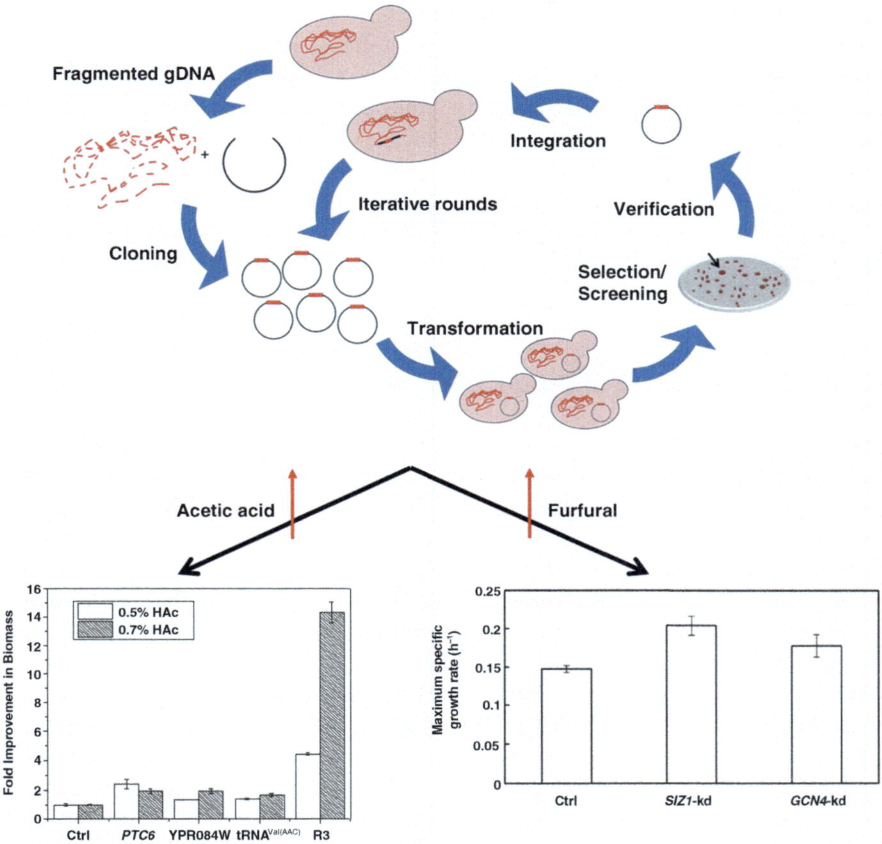

Fig. 7.4 Engineering acetic acid and furfural tolerance using RNAi-assisted genome evolution (*RAGE*) (Reprinted from *ACS Synthetic Biology* 2015, 4, 283–291, Copyright 2014, with permission from American Chemical Society and *Biotechnology for Biofuel* 2014, 7:78)

shuffling, which yielded several novel hybrids outperforming the currently used industrial yeast strains. To increase the *n*-butanol tolerance in *E. coli*, GREACE and stress-induced mutagenesis-based AEs were developed, both of which yielded *n*-butanol-tolerant strains in a short period of time [92, 170]. Ling et al. engineered the transcription factors related to the pleiotropic drug resistance in *S. cerevisiae* to improve the resistance to alkanes [89]. The construction of replacement jet fuel-tolerant yeast strains using directed evolution has also been reported [14].

It is highly desirable for the industrial fermentation processes to be operated at high temperatures, so as to reduce cooling costs and prevent contamination. Through directed evolution, Caspeta et al. obtained several *S. cerevisiae* strains that demonstrated much improved growth at a high temperature (>40 °C) [20]. Systematic characterization of the evolved strains using system biology tools revealed that a change in sterol composition, from ergosterol to fecosterol, caused by mutations in the sterol desaturase gene and increased expression of genes involved in sterol biosynthesis, contributed the most to the thermotolerant phenotype. Shui et al. performed a proteomic analysis of the evolved thermotolerant yeast strains, which led to a comprehensive understanding of the molecular basis of thermotolerance, and identified novel targets for further improvement [132].

Besides the application in industrial biotechnological processes, directed evolution approaches are also applied in food biotechnology. For example, in wine making, directed evolution has been applied to reduce alcohol levels [146], to increase the synthesis of aromas [16, 17], and to enhance the fermentation capability at low temperatures [90].

7.3.3 Enhancement of Product Formation

The formation of a desired product at high titer and yield is the ultimate goal of most biotechnological processes. Unfortunately, unlike substrate utilization and cellular tolerance, product formation cannot be easily coupled to cellular growth and even impairs cellular growth in many cases. In other words, a generally applicable high-throughput screening strategy of improved production is not readily available.

Metabolic engineering has been proven effective in enhancing the yield of the desired product, but often at the cost of cellular growth and fitness. In this case, directed evolution and metabolic engineering can be combined to enhance both the yield and productivity. The construction of a *S. cerevisiae* strain with abolished ethanol production serves as one of such examples (Fig. 7.5). To eliminate ethanol formation, pyruvate decarboxylases (PDCs) have to be inactivated. Unfortunately, the Pdc$^-$ strain (*pdc1*Δ *pdc5*Δ *pdc6*Δ) is notorious for its inability to grow on glucose as the sole carbon source, requiring the supplementation of a C_2 compound (acetate or ethanol) to synthesize cytosolic acetyl CoA [37] (Fig. 7.5). Several studies have reported to evolve C_2-independent Pdc$^-$ yeast strains growing in glucose as the sole carbon source [61, 149, 168]. Whole-genome sequencing of the evolved strains revealed that an internal deletion [107] or point mutations (Ala81Pro [61] or Ala81Asp [168]) in the *MTH1* coding sequence enabled the growth of Pdc$^-$ strain on glucose (Fig. 7.5). Alternatively, overexpression of *MTH1* or the truncated *MTH1* on a multi-copy plasmid resulted in a Pdc$^-$ strain with similar properties [87]. The evolved Pdc$^-$ strain was able to accumulate pyruvate to a level as high as 135 g/L [149], which can be further developed into an important platform cell factory to produce a wide range of biofuels and value-added chemicals other than ethanol (Fig. 7.5). Lactate with a titer up to 110 g/L could be obtained in a Pdc$^-$ strain expressing an *LDH* gene from *Lactobacillus casei* in 1 L of fermenter under aerobic conditions [1] (Fig. 7.5). High titer and yield production of 2,3-butanediol was also reported using an evolved Pdc$^-$ strain overexpressing an acetolactate synthase, an acetolactate decarboxylase, and a butanediol dehydrogenase from various carbon sources, such as glucose [61], galactose [87], cellobiose [101], and xylose [62]. The highest production was achieved using glucose- and galactose-fed-batch fermentation, with a titer around 100 g/L [87] (Fig. 7.5). Fermentative production of malate

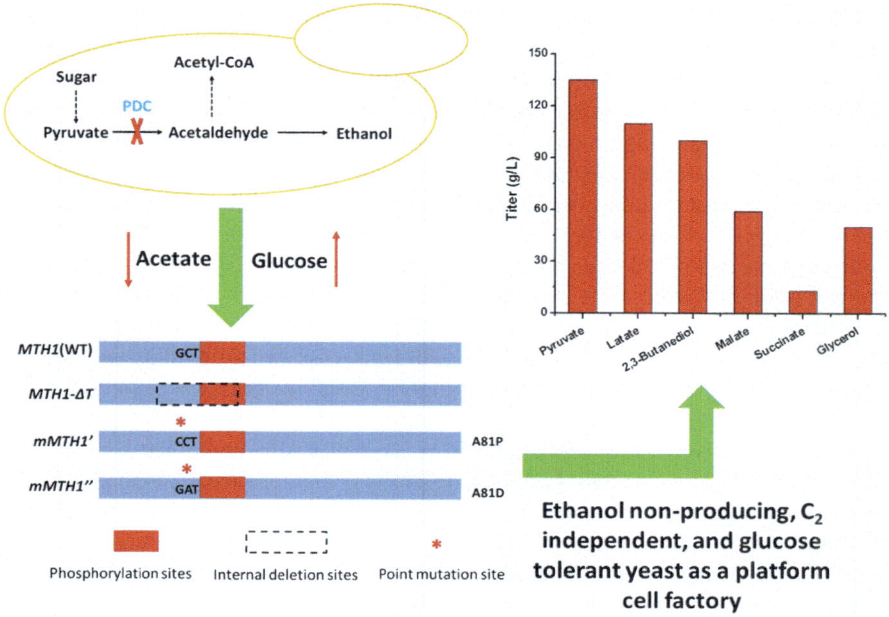

Fig. 7.5 Construction of an ethanol nonproducing, C_2-independent, and glucose-tolerant yeast platform strain by combining metabolic engineering with directed evolution

using the Pdc⁻ strain was also attempted by combined overexpression of a pyruvate carboxylase (*PYC2*), a cytosolic malate dehydrogenase (*MDH3ΔSKL*), and a malate transporter from *Schizosaccharomyces pombe* (Sp*MAE1*). Malate titer of up to 59 g/L was reached with a yield of 0.42 mol/mol glucose in shake flask fermentation [166] (Fig. 7.5). Recently, the malate producer was further engineered to produce succinate. Under optimal conditions in a bioreactor, the engineered strain produced around 13 g/L of succinate with a yield of 0.21 mol/mol glucose at low pH [165] (Fig. 7.5). Although glycerol is not directly derived from pyruvate, the elimination of ethanol formation in the Pdc⁻ strain may redirect the metabolic fluxes to glycerol formation, especially under low-oxygen or anaerobic conditions. By further engineering cytosolic NADH availability and overexpressing *GPD2*, a titer of higher than 50 g/L and a yield as high as 1.08 mol/mol glucose for glycerol production were achieved in aerobic and glucose-limited chemostat cultures with formate co-feeding [41] (Fig. 7.5).

Another example of combining directed evolution with metabolic engineering is the effort to increase the yield of ethanol in *S. cerevisiae* by eliminating glycerol formation [46]. Glycerol production is required for redox-cofactor balancing in anaerobic cultures. Acetate reduction was found to replace glycerol formation under anaerobic condition for NADH re-oxidation. However, the acetate-reducing (*mhpF* from *E. coli* overexpression) and glycerol-nonproducing (*GPD1* and *GPD2* deletion) yeast strain is sensitive to high sugar concentrations. Directed evolution enabled the isolation of an evolved strain that grew anaerobically at 1 M of glucose, and the ethanol yield on sugar increased from 79% of the theoretical maximum in the reference strain to 92% in the evolved strain.

In some cases, it is possible to take advantage of the unique properties of the final products to develop growth-based directed evolution strategies. For example, glutathione is an antioxidant, and directed evolution can be performed by coupling the enhanced glutathione accumulation phenotype with the acrolein resistance phenotype [111]. The evolved strain accumulated glutathione in 3.3-fold higher concentration compared to its parental strain and reached a particularly high glutathione content of almost 6%. Similarly, by taking advantage of the antioxidative properties of carotenoids, directed evolution was designed based on periodic hydrogen peroxide shocking, and a threefold increase in carotenoids production (from 6 mg/g dry cell weight to up to 18 mg/g dry cell weight) was achieved in the evolved strain [118].

Increased production of isobutanol in *E. coli* was also attempted by directed evolution [137]. The isobutanol production ability is closely related to the metabolic flux through the valine biosynthetic pathway, which can be coupled to the cellular resistance to the valine analog norvaline. Using this strategy, a final isobutanol titer of 21.2 g/L was achieved in 99 h with a yield of 0.31 g isobutanol/g of glucose or 76% of theoretical maximum, in comparison with a production of 5.3 g/L obtained with the wild-type strain.

7.4 Perspectives

Microorganisms are increasingly exploited to address some of the most challenging global problems such as sustainability and energy security. In many cases, cell factories used for industrial applications require a combination of complex phenotypes such as high tolerance to inhibitors in the raw materials, toxic products at high concentrations, low pH, and high temperature. Directed evolution approaches have been successful in coping with these challenges. Nevertheless, challenges and opportunities still remain in strain development by directed evolution. First of all, novel screening and selection methods should be developed and integrated into directed evolution pipelines. Currently, the phenotypes that directed evolution can cope are mainly limited to those closely related to cellular growth, such as substrate utilization and tolerance to toxic compounds. The development of small molecule biosensors based on TFs [153], G-protein-coupled receptors (GPCRs) [100], and riboswitches [60] can be incorporated to expand the scope of directed evolution. For example, a malonyl-CoA biosensor was developed by Li et al. and then used to screen a cDNA library that increased the intracellular malonyl-CoA levels [86]. The robotic platform [31, 138] and microfluidic system [54, 135] may also be used for the screening of desired phenotypes. Another challenge is the trade-off of directed evolution [21, 51, 97]. For example, the evolved galactose-fermenting strain demonstrated decreased growth in glucose, and the evolved thermotolerant strain showed trade-offs when growing at ancestral temperatures. Evolutionary trade-offs may hinder its industrial applications, in which multiple and complex traits are required. Systems biology tools [132] may help elucidate the molecular mechanisms of evolved phenotypes and minimize the evolutionary trade-offs for cellular factory development. Nevertheless, the development of novel genome engineering tools such as the CRISPR-Cas system provides new dimensions in the generation of strain libraries with improved diversity. Although genome-scale screening based on CRISPR knockout [126], CRISPR interference [43], and CRISPR activation [43, 69] has been demonstrated in mammalian cells, their applications in the construction and optimization of cell factories have yet to be explored.

References

1. Abbott DA, Zelle RM, Pronk JT, Maris AJA (2009) Metabolic engineering of *Saccharomyces cerevisiae* for production of carboxylic acids: current status and challenges. FEMS Yeast Res 9(8):1123–1136
2. Alexeyev MF, Shokolenko IN (1995) Mini-Tn10 transposon derivatives for insertion mutagenesis and gene delivery into the chromosome of gram-negative bacteria. Gene 160(1): 59–62
3. Almario MP, Reyes LH, Kao KC (2013) Evolutionary engineering of *Saccharomyces cerevisiae* for enhanced tolerance to hydrolysates of lignocellulosic biomass. Biotechnol Bioeng 110(10):2616–2623
4. Alper H, Moxley J, Nevoigt E, Fink GR, Stephanopoulos G (2006) Engineering yeast transcription machinery for improved ethanol tolerance and production. Science 314(5805): 1565–1568
5. Alper H, Stephanopoulos G (2007) Global transcription machinery engineering: a new approach for improving cellular phenotype. Metab Eng 9(3):258–267
6. Atsumi S, Liao JC (2008) Directed evolution of *Methanococcus jannaschii* citramalate synthase for biosynthesis of 1-propanol and 1-butanol by *Escherichia coli*. Appl Environ Microbiol 74(24):7802–7808
7. Baba T, Ara T, Hasegawa M, Takai Y, Okumura Y, Baba M, Datsenko KA, Tomita M, Wanner BL, Mori H (2006) Construction of *Escherichia coli* K-12 in-frame, single-gene knockout mutants: the Keio collection. Mol Syst Biol 2:2006.0008
8. Bailey JE (1991) Toward a science of metabolic engineering. Science 252(5013): 1668–1675
9. Bailey JE, Sburlati A, Hatzimanikatis V, Lee K, Renner WA, Tsai PS (1996) Inverse metabolic engineering: a strategy for directed genetic engineering of useful phenotypes. Biotechnol Bioeng 52(1):109–121
10. Bao Z, Xiao H, Liang J, Zhang L, Xiong X, Sun N, Si T, Zhao H (2015) Homology-integrated CRISPR-Cas (HI-CRISPR) system for one-step multigene disruption in *Saccharomyces cerevisiae*. ACS Synth Biol 4(5):585–594
11. Barrick JE, Yu DS, Yoon SH, Jeong H, Oh TK, Schneider D, Lenski RE, Kim JF (2009) Genome evolution and adaptation in a long-term experiment with *Escherichia coli*. Nature 461(7268):1243–1247
12. Biot-Pelletier D, Martin VJ (2014) Evolutionary engineering by genome shuffling. Appl Microbiol Biotechnol 98(9):3877–3887
13. Blank D, Wolf L, Ackermann M, Silander OK (2014) The predictability of molecular evolution during functional innovation. Proc Natl Acad Sci U S A 111(8):3044–3049
14. Brennan TC, Williams TC, Schulz BL, Palfreyman RW, Kromer JO, Nielsen LK (2015) Evolutionary engineering improves tolerance for replacement jet fuels in *Saccharomyces cerevisiae*. Appl Environ Microbiol 81(10):3316–3325

15. Burgard AP, Pharkya P, Maranas CD (2003) Optknock: a bilevel programming framework for identifying gene knockout strategies for microbial strain optimization. Biotechnol Bioeng 84(6):647–657
16. Cadiere A, Aguera E, Caille S, Ortiz-Julien A, Dequin S (2012) Pilot-scale evaluation the enological traits of a novel, aromatic wine yeast strain obtained by adaptive evolution. Food Microbiol 32(2):332–337
17. Cadiere A, Ortiz-Julien A, Camarasa C, Dequin S (2011) Evolutionary engineered *Saccharomyces cerevisiae* wine yeast strains with increased in vivo flux through the pentose phosphate pathway. Metab Eng 13(3):263–271
18. Cakar ZP, Turanli-Yildiz B, Alkim C, Yilmaz U (2012) Evolutionary engineering of *Saccharomyces cerevisiae* for improved industrially important properties. FEMS Yeast Res 12(2):171–182
19. Carr PA, Church GM (2009) Genome engineering. Nat Biotechnol 27(12):1151–1162
20. Caspeta L, Chen Y, Ghiaci P, Feizi A, Buskov S, Hallstrom BM, Petranovic D, Nielsen J (2014) Altered sterol composition renders yeast thermotolerant. Science 346(6205):75–78
21. Caspeta L, Nielsen J (2015) Thermotolerant yeast strains adapted by laboratory evolution show trade-off at ancestral temperatures and preadaptation to other stresses. MBio 6(4):e00431
22. Chen Y, Sheng J, Jiang T, Stevens J, Feng X, Wei N (2016) Transcriptional profiling reveals molecular basis and novel genetic targets for improved resistance to multiple fermentation inhibitors in *Saccharomyces cerevisiae*. Biotechnol Biofuels 9:9
23. Chou HH, Keasling JD (2013) Programming adaptive control to evolve increased metabolite production. Nat Commun 4:2595
24. Cobb RE, Wang Y, Zhao H (2015) High-efficiency multiplex genome editing of *Streptomyces* species using an engineered CRISPR/Cas system. ACS Synth Biol 4(6):723–728
25. Cong L, Ran FA, Cox D, Lin SL, Barretto R, Habib N, Hsu PD, Wu XB, Jiang WY, Marraffini LA, Zhang F (2013) Multiplex genome engineering using CRISPR/Cas systems. Science 339(6121):819–823
26. Dean AM, Thornton JW (2007) Mechanistic approaches to the study of evolution: the functional synthesis. Nat Rev Genet 8(9):675–688
27. Demeke MM, Foulquie-Moreno MR, Dumortier F, Thevelein JM (2015) Rapid evolution of recombinant *Saccharomyces cerevisiae* for Xylose fermentation through formation of extra-chromosomal circular DNA. PLoS Genet 11(3):e1005010
28. DiCarlo JE, Conley AJ, Penttila M, Jantti J, Wang HH, Church GM (2013) Yeast oligo-mediated genome engineering (YOGE). ACS Synth Biol 2(12):741–749
29. Dietrich JA, McKee AE, Keasling JD (2010) High-throughput metabolic engineering: advances in small-molecule screening and selection. Annu Rev Biochem 79:563–590
30. Dietrich JA, Shis DL, Alikhani A, Keasling JD (2013) Transcription factor-based screens and synthetic selections for microbial small-molecule biosynthesis. ACS Synth Biol 2(1):47–58
31. Dörr M, Fibinger MPC, Last D, Schmidt S, Santos-Aberturas J, Böttcher D, Hummel A, Vickers C, Voss M, Bornscheuer UT (2016) Fully automatized high-throughput enzyme library screening using a robotic platform. Biotechnol Bioeng. doi:10.1002/bit.25925
32. Dragosits M, Mattanovich D (2013) Adaptive laboratory evolution – principles and applications for biotechnology. Microb Cell Fact 12:64
33. Dymond JS, Richardson SM, Coombes CE, Babatz T, Muller H, Annaluru N, Blake WJ, Schwerzmann JW, Dai J, Lindstrom DL, Boeke AC, Gottschling DE, Chandrasegaran S, Bader JS, Boeke JD (2011) Synthetic chromosome arms function in yeast and generate phenotypic diversity by design. Nature 477(7365):471–476
34. Ellis HM, Yu D, Di Tizio T, Court DL (2001) High efficiency mutagenesis, repair, and engineering of chromosomal DNA using single-stranded oligonucleotides. Proc Natl Acad Sci U S A 98(12):6742–6746
35. Enquist-Newman M, Faust AM, Bravo DD, Santos CN, Raisner RM, Hanel A, Sarvabhowman P, Le C, Regitsky DD, Cooper SR, Peereboom L, Clark A, Martinez Y, Goldsmith J, Cho MY, Donohoue PD, Luo L, Lamberson B, Tamrakar P, Kim EJ, Villari JL, Gill A, Tripathi SA,

Karamchedu P, Paredes CJ, Rajgarhia V, Kotlar HK, Bailey RB, Miller DJ, Ohler NL, Swimmer C, Yoshikuni Y (2014) Efficient ethanol production from brown macroalgae sugars by a synthetic yeast platform. Nature 505(7482):239–243

36. Esvelt KM, Wang HH (2013) Genome-scale engineering for systems and synthetic biology. Mol Syst Biol 9:641
37. Flikweert MT, Swaaf M, Dijken JP, Pronk JT (1999) Growth requirements of pyruvate-decarboxylase-negative *Saccharomyces cerevisiae*. FEMS Microbiol Lett 174(1):73–79
38. Fondi M, Liò P (2015) Multi-omics and metabolic modelling pipelines: challenges and tools for systems microbiology. Microbiol Res 171:52–64
39. Fong SS, Burgard AP, Herring CD, Knight EM, Blattner FR, Maranas CD, Palsson BO (2005) In silico design and adaptive evolution of *Escherichia coli* for production of lactic acid. Biotechnol Bioeng 91(5):643–648
40. Fullwood MJ, Wei CL, Liu ET, Ruan Y (2009) Next-generation DNA sequencing of paired-end tags (PET) for transcriptome and genome analyses. Genome Res 19(4):521–532
41. Geertman JM, Maris AJ, Dijken JP, Pronk JT (2006) Physiological and genetic engineering of cytosolic redox metabolism in *Saccharomyces cerevisiae* for improved glycerol production. Metab Eng 8(6):532–542
42. Giaever G, Chu AM, Ni L, Connelly C, Riles L, Veronneau S, Dow S, Lucau-Danila A, Anderson K, Andre B, Arkin AP, Astromoff A, Bakkoury M, Bangham R, Benito R, Brachat S, Campanaro S, Curtiss M, Davis K, Deutschbauer A, Entian K-D, Flaherty P, Foury F, Garfinkel DJ, Gerstein M, Gotte D, Guldener U, Hegemann JH, Hempel S, Herman Z, Jaramillo DF, Kelly DE, Kelly SL, Kotter P, LaBonte D, Lamb DC, Lan N, Liang H, Liao H, Liu L, Luo C, Lussier M, Mao R, Menard P, Ooi SL, Revuelta JL, Roberts CJ, Rose M, Ross-Macdonald P, Scherens B, Schimmack G, Shafer B, Shoemaker DD, Sookhai-Mahadeo S, Storms RK, Strathern JN, Valle G, Voet M, Volckaert G, Wang C-Y, Ward TR, Wilhelmy J, Winzeler EA, Yang Y, Yen G, Youngman E, Yu K, Bussey H, Boeke JD, Snyder M, Philippsen P, Davis RW, Johnston M (2002) Functional profiling of the *Saccharomyces cerevisiae* genome. Nature 418(6896):387–391
43. Gilbert LA, Horlbeck MA, Adamson B, Villalta JE, Chen Y, Whitehead EH, Guimaraes C, Panning B, Ploegh HL, Bassik MC, Qi LS, Kampmann M, Weissman JS (2014) Genome-scale CRISPR-mediated control of gene repression and activation. Cell 159(3):647–661
44. Greener A, Callahan M, Jerpseth B (1997) An efficient random mutagenesis technique using an *E. coli* mutator strain. Mol Biotechnol 7(2):189–195
45. Gresham D, Dunham MJ (2014) The enduring utility of continuous culturing in experimental evolution. Genomics 104(6 Pt A):399–405
46. Guadalupe-Medina V, Metz B, Oud B, Graaf CM, Mans R, Pronk JT, Maris AJ (2014) Evolutionary engineering of a glycerol-3-phosphate dehydrogenase-negative, acetate-reducing *Saccharomyces cerevisiae* strain enables anaerobic growth at high glucose concentrations. J Microbial Biotechnol 7(1):44–53
47. Guimaraes PM, Berre V, Sokol S, Francois J, Teixeira JA, Domingues L (2008) Comparative transcriptome analysis between original and evolved recombinant lactose-consuming *Saccharomyces cerevisiae* strains. Biotechnol J 3(12):1591–1597
48. Hasunuma T, Sakamoto T, Kondo A (2016) Inverse metabolic engineering based on transient acclimation of yeast improves acid-containing xylose fermentation and tolerance to formic and acetic acids. Appl Microbiol Biotechnol 100(2):1027–1038
49. Hawkins GM, Doran-Peterson J (2011) A strain of *Saccharomyces cerevisiae* evolved for fermentation of lignocellulosic biomass displays improved growth and fermentative ability in high solids concentrations and in the presence of inhibitory compounds. Biotechnol Biofuels 4(1):49
50. Ho CH, Magtanong L, Barker SL, Gresham D, Nishimura S, Natarajan P, Koh JL, Porter J, Gray CA, Andersen RJ, Giaever G, Nislow C, Andrews B, Botstein D, Graham TR, Yoshida M, Boone C (2009) A molecular barcoded yeast ORF library enables mode-of-action analysis of bioactive compounds. Nat Biotechnol 27(4):369–377

51. Hong KK, Nielsen J (2013) Adaptively evolved yeast mutants on galactose show trade-offs in carbon utilization on glucose. Metab Eng 16:78–86
52. Horinouchi T, Minamoto T, Suzuki S, Shimizu H, Furusawa C (2014) Development of an automated culture system for laboratory evolution. J Lab Autom 19(5):478–482
53. Hosaka T, Ohnishi-Kameyama M, Muramatsu H, Murakami K, Tsurumi Y, Kodani S, Yoshida M, Fujie A, Ochi K (2009) Antibacterial discovery in actinomycetes strains with mutations in RNA polymerase or ribosomal protein S12. Nat Biotechnol 27(5):462–464
54. Huang M, Bai Y, Sjostrom SL, Hallstrom BM, Liu Z, Petranovic D, Uhlen M, Joensson HN, Andersson-Svahn H, Nielsen J (2015) Microfluidic screening and whole-genome sequencing identifies mutations associated with improved protein secretion by yeast. Proc Natl Acad Sci U S A 112(34):E4689–E4696
55. Hutchison CA, Peterson SN, Gill SR, Cline RT, White O, Fraser CM, Smith HO, Venter JC (1999) Global transposon mutagenesis and a minimal *Mycoplasma* genome. Science 286(5447):2165–2169
56. Jakiela S, Kaminski TS, Cybulski O, Weibel DB, Garstecki P (2013) Bacterial growth and adaptation in microdroplet chemostats. Angew Chem Int Ed 52(34):8908–8911
57. Jakociunas T, Bonde I, Herrgard M, Harrison SJ, Kristensen M, Pedersen LE, Jensen MK, Keasling JD (2015) Multiplex metabolic pathway engineering using CRISPR/Cas9 in *Saccharomyces cerevisiae*. Metab Eng 28:213–222
58. Jiang W, Bikard D, Cox D, Zhang F, Marraffini LA (2013) RNA-guided editing of bacterial genomes using CRISPR-Cas systems. Nat Biotechnol 31(3):233–239
59. Keasling JD (2010) Manufacturing molecules through metabolic engineering. Science 330(6009):1355–1358
60. Kim HJ, Ha S, Lee HY, Lee KJ (2015) ROSics: chemistry and proteomics of cysteine modifications in redox biology. Mass Spectrum Rev 34(2):184–208
61. Kim SJ, Seo SO, Jin YS, Seo JH (2013) Production of 2,3-butanediol by engineered *Saccharomyces cerevisiae*. Bioresour Technol 146:274–281
62. Kim SJ, Seo SO, Park YC, Jin YS, Seo JH (2014) Production of 2,3-butanediol from xylose by engineered *Saccharomyces cerevisiae*. J Biotechnol 192:376–382
63. Kim SR, Skerker JM, Kang W, Lesmana A, Wei N, Arkin AP, Jin YS (2013) Rational and evolutionary engineering approaches uncover a small set of genetic changes efficient for rapid xylose fermentation in *Saccharomyces cerevisiae*. PLoS One 8(2):e57048
64. Kim SR, Xu H, Lesmana A, Kuzmanovic U, Au M, Florencia C, Oh EJ, Zhang G, Kim KH, Jin YS (2015) Deletion of *PHO13*, encoding haloacid dehalogenase type IIA phosphatase, results in upregulation of the pentose phosphate pathway in *Saccharomyces cerevisiae*. Appl Environ Microbiol 81(5):1601–1609
65. Kinnersley M, Wenger J, Kroll E, Adams J, Sherlock G, Rosenzweig F (2014) Ex uno plures: clonal reinforcement drives evolution of a simple microbial community. PLoS Genet 10(6):e1004430
66. Klein-Marcuschamer D, Stephanopoulos G (2008) Assessing the potential of mutational strategies to elicit new phenotypes in industrial strains. Proc Natl Acad Sci U S A 105(7):2319–2324
67. Klimacek M, Kirl E, Krahulec S, Longus K, Novy V, Nidetzky B (2014) Stepwise metabolic adaption from pure metabolization to balanced anaerobic growth on xylose explored for recombinant *Saccharomyces cerevisiae*. Microb Cell Fact 13(1):37
68. Koffas M (2005) Evolutionary metabolic engineering. Metab Eng 7(1):1–3
69. Konermann S, Brigham MD, Trevino AE, Joung J, Abudayyeh OO, Barcena C, Hsu PD, Habib N, Gootenberg JS, Nishimasu H, Nureki O, Zhang F (2015) Genome-scale transcriptional activation by an engineered CRISPR-Cas9 complex. Nature 517(7536):583–588
70. Koppram R, Albers E, Olsson L (2012) Evolutionary engineering strategies to enhance tolerance of xylose utilizing recombinant yeast to inhibitors derived from spruce biomass. Biotechnol Biofuels 5(1):32

71. Koren S, Phillippy AM (2015) One chromosome, one contig: complete microbial genomes from long-read sequencing and assembly. Curr Opin Microbiol 23:110–120
72. Kryazhimskiy S, Rice DP, Jerison ER, Desai MM (2014) Global epistasis makes adaptation predictable despite sequence-level stochasticity. Science 344(6191):1519–1522
73. Kucukgoze G, Alkim C, Yilmaz U, Kisakesen HI, Gunduz S, Akman S, Cakar ZP (2013) Evolutionary engineering and transcriptomic analysis of nickel-resistant *Saccharomyces cerevisiae*. FEMS Yeast Res 13(8):731–746
74. Lang GI, Botstein D, Desai MM (2011) Genetic variation and the fate of beneficial mutations in asexual populations. Genetics 188(3):647–661
75. Lang GI, Desai MM (2014) The spectrum of adaptive mutations in experimental evolution. Genomics 104(6):412–416
76. Lang GI, Murray AW (2008) Estimating the per-base-pair mutation rate in the yeast *Saccharomyces cerevisiae*. Genetics 178(1):67–82
77. Lang GI, Rice DP, Hickman MJ, Sodergren E, Weinstock GM, Botstein D, Desai MM (2013) Pervasive genetic hitchhiking and clonal interference in fourty evolving yeast populations. Nature 500(7464):571–574
78. Lee H, Popodi E, Tang H, Foster PL (2012) Rate and molecular spectrum of spontaneous mutations in the bacterium *Escherichia coli* as determined by whole-genome sequencing. Proc Natl Acad Sci U S A 109(41):E2774–E2783
79. Lee HJ, Kim SJ, Yoon JJ, Kim KH, Seo JH, Park YC (2015) Evolutionary engineering of *Saccharomyces cerevisiae* for efficient conversion of red algal biosugars to bioethanol. Bioresour Technol 191:445–451
80. Lee JW, Na D, Park JM, Lee J, Choi S, Lee SY (2012) Systems metabolic engineering of microorganisms for natural and non-natural chemicals. Nat Chem Biol 8(6):536–546
81. Lee KS, Hong ME, Jung SC, Ha SJ, Yu BJ, Koo HM, Park SM, Seo JH, Kweon DH, Park JC, Jin YS (2011) Improved galactose fermentation of *Saccharomyces cerevisiae* through inverse metabolic engineering. Biotechnol Bioeng 108(3):621–631
82. Lee SM, Jellison T, Alper HS (2014) Systematic and evolutionary engineering of a xylose isomerase-based pathway in *Saccharomyces cerevisiae* for efficient conversion yields. Biotechnol Biofuels 7(1):122
83. Lee SY, Kim HU (2015) Systems strategies for developing industrial microbial strains. Nat Biotechnol 33(10):1061–1072
84. Lenski RE, Mongold JA, Sniegowski PD, Travisano M, Vasi F, Gerrish PJ, Schmidt TM (1998) Evolution of competitive fitness in experimental populations of *E. coli*: what makes one genotype a better competitor than another? Antonie Van Leeuwenhoek 73(1):35–47
85. Li P, Sun H, Chen Z, Li Y, Zhu T (2015) Construction of efficient xylose utilizing Pichia pastoris for industrial enzyme production. Microb Cell Fact 14:22
86. Li S, Si T, Wang M, Zhao H (2015) Development of a synthetic Malonyl-CoA sensor in *Saccharomyces cerevisiae* for intracellular metabolite monitoring and genetic screening. ACS Synth Biol 4(12):1308–1315
87. Lian J, Chao R, Zhao H (2014) Metabolic engineering of a *Saccharomyces cerevisiae* strain capable of simultaneously utilizing glucose and galactose to produce enantiopure (2R,3R)-butanediol. Metab Eng 23:92–99
88. Lind PA, Andersson DI (2008) Whole-genome mutational biases in bacteria. Proc Natl Acad Sci U S A 105(46):17878–17883
89. Ling H, Pratomo Juwono NK, Teo WS, Liu R, Leong SS, Chang MW (2015) Engineering transcription factors to improve tolerance against alkane biofuels in *Saccharomyces cerevisiae*. Biotechnol Biofuels 8:231
90. Lopez-Malo M, Garcia-Rios E, Melgar B, Sanchez MR, Dunham MJ, Guillamon JM (2015) Evolutionary engineering of a wine yeast strain revealed a key role of inositol and mannoprotein metabolism during low-temperature fermentation. BMC Genomics 16:537

91. Lou DI, Hussmann JA, McBee RM, Acevedo A, Andino R, Press WH, Sawyer SL (2013) High-throughput DNA sequencing errors are reduced by orders of magnitude using circle sequencing. Proc Natl Acad Sci U S A 110(49):19872–19877
92. Luan G, Cai Z, Li Y, Ma Y (2013) Genome replication engineering assisted continuous evolution (GREACE) to improve microbial tolerance for biofuels production. Biotechnol Biofuels 6(1):137
93. Lynch M, Sung W, Morris K, Coffey N, Landry CR, Dopman EB, Dickinson WJ, Okamoto K, Kulkarni S, Hartl DL, Thomas WK (2008) A genome-wide view of the spectrum of spontaneous mutations in yeast. Proc Natl Acad Sci U S A 105(27):9272–9277
94. Lynch MD, Warnecke T, Gill RT (2007) SCALEs: multiscale analysis of library enrichment. Nat Methods 4(1):87–93
95. Maharjan R, Seeto S, Notley-McRobb L, Ferenci T (2006) Clonal adaptive radiation in a constant environment. Science 313(5786):514–517
96. Mali P, Yang L, Esvelt KM, Aach J, Guell M, DiCarlo JE, Norville JE, Church GM (2013) RNA-guided human genome engineering via Cas9. Science 339(6121):823–826
97. Martinez JL, Bordel S, Hong KK, Nielsen J (2014) Gcn4p and the Crabtree effect of yeast: drawing the causal model of the Crabtree effect in *Saccharomyces cerevisiae* and explaining evolutionary trade-offs of adaptation to galactose through systems biology. FEMS Yeast Res 14(4):654–662
98. Michener JK, Smolke CD (2012) High-throughput enzyme evolution in *Saccharomyces cerevisiae* using a synthetic RNA switch. Metab Eng 14(4):306–316
99. Mitsumasu K, Liu ZS, Tang YQ, Akamatsu T, Taguchi H, Kida K (2014) Development of industrial yeast strain with improved acid- and thermo-tolerance through evolution under continuous fermentation conditions followed by haploidization and mating. J Biosci Bioeng 118(6):689–695
100. Mukherjee K, Bhattacharyya S, Peralta-Yahya P (2015) GPCR-based chemical biosensors for medium-chain fatty acids. ACS Synth Biol 4(12):1261–1269
101. Nan H, Seo SO, Oh EJ, Seo JH, Cate JH, Jin YS (2014) 2,3-butanediol production from cellobiose by engineered *Saccharomyces cerevisiae*. Appl Microbiol Biotechnol 98(12): 5757–5764
102. Nielsen J, Keasling Jay D (2016) Engineering cellular metabolism. Cell 164(6):1185–1197
103. Nijland JG, Shin HY, Jong RM, Waal PP, Klaassen P, Driessen AJ (2014) Engineering of an endogenous hexose transporter into a specific D-xylose transporter facilitates glucose-xylose co-consumption in *Saccharomyces cerevisiae*. Biotechnol Biofuels 7(1):168
104. Notley-McRobb L, King T, Ferenci T (2002) rpoS mutations and loss of general stress resistance in *Escherichia coli* populations as a consequence of conflict between competing stress responses. J Bacteriol 184(3):806–811
105. Novick A, Szilard L (1950) Description of the chemostat. Science 112(2920):715–716
106. Ochi K (2007) From microbial differentiation to ribosome engineering. Biosci Biotechnol Biochem 71(6):1373–1386
107. Oud B, Flores CL, Gancedo C, Zhang X, Trueheart J, Daran JM, Pronk JT, Maris AJ (2012) An internal deletion in *MTH1* enables growth on glucose of pyruvate-decarboxylase negative, non-fermentative *Saccharomyces cerevisiae*. Microb Cell Fact 11:131
108. Oud B, Maris AJ, Daran JM, Pronk JT (2012) Genome-wide analytical approaches for reverse metabolic engineering of industrially relevant phenotypes in yeast. FEMS Yeast Res 12(2):183–196
109. Park K-S, Lee D-K, Lee H, Lee Y, Jang Y-S, Kim YH, Yang H-Y, Lee S-I, Seol W, Kim J-S (2003) Phenotypic alteration of eukaryotic cells using randomized libraries of artificial transcription factors. Nat Biotechnol 21(10):1208–1214
110. Park KS, Jang YS, Lee H, Kim JS (2005) Phenotypic alteration and target gene identification using combinatorial libraries of zinc finger proteins in prokaryotic cells. J Bacteriol 187(15):5496–5499

111. Patzschke A, Steiger MG, Holz C, Lang C, Mattanovich D, Sauer M (2015) Enhanced glutathione production by evolutionary engineering of *Saccharomyces cerevisiae* strains. Biotechnol J 10(11):1719–1726
112. Pereira SR, Sanchez INV, Frazao CJ, Serafim LS, Gorwa-Grauslund MF, Xavier AM (2015) Adaptation of *Scheffersomyces stipitis* to hardwood spent sulfite liquor by evolutionary engineering. Biotechnol Biofuels 8:50
113. Pinel D, Colatriano D, Jiang H, Lee H, Martin VJ (2015) Deconstructing the genetic basis of spent sulphite liquor tolerance using deep sequencing of genome-shuffled yeast. Biotechnol Biofuels 8:53
114. Pinel D, D'Aoust F, Cardayre SB, Bajwa PK, Lee H, Martin VJ (2011) *Saccharomyces cerevisiae* genome shuffling through recursive population mating leads to improved tolerance to spent sulfite liquor. Appl Environ Microbiol 77(14):4736–4743
115. Qi X, Zha J, Liu GG, Zhang W, Li BZ, Yuan YJ (2015) Heterologous xylose isomerase pathway and evolutionary engineering improve xylose utilization in *Saccharomyces cerevisiae*. Front Microbiol 6:1165
116. Quandt EM, Deatherage DE, Ellington AD, Georgiou G, Barrick JE (2014) Recursive genome-wide recombination and sequencing reveals a key refinement step in the evolution of a metabolic innovation in *Escherichia coli*. Proc Natl Acad Sci U S A 111(6): 2217–2222
117. Rabinovitch-Deere CA, Oliver JWK, Rodriguez GM, Atsumi S (2013) Synthetic biology and metabolic engineering approaches to produce biofuels. Chem Rev 113(7): 4611–4632
118. Reyes LH, Gomez JM, Kao KC (2014) Improving carotenoids production in yeast via adaptive laboratory evolution. Metab Eng 21:26–33
119. Rodriguez-Verdugo A, Carrillo-Cisneros D, Gonzalez-Gonzalez A, Gaut BS, Bennett AF (2014) Different tradeoffs result from alternate genetic adaptations to a common environment. Proc Natl Acad Sci U S A 111(33):12121–12126
120. Rosenzweig F, Sherlock G (2014) Experimental evolution: prospects and challenges. Genomics 104(6, Part A):v–vi
121. Sanchez BJ, Nielsen J (2015) Genome scale models of yeast: towards standardized evaluation and consistent omic integration. Integr Biol 7(8):846–858
122. Santos CNS, Stephanopoulos G (2008) Combinatorial engineering of microbes for optimizing cellular phenotype. Curr Opin Chem Biol 12(2):168–176
123. Sauer U (2001) Evolutionary engineering of industrially important microbial phenotypes. Adv Biochem Eng Biotechnol 73:129–169
124. Scalcinati G, Otero JM, Vleet JR, Jeffries TW, Olsson L, Nielsen J (2012) Evolutionary engineering of *Saccharomyces cerevisiae* for efficient aerobic xylose consumption. FEMS Yeast Res 12(5):582–597
125. Serero A, Jubin C, Loeillet S, Legoix-Ne P, Nicolas AG (2014) Mutational landscape of yeast mutator strains. Proc Natl Acad Sci U S A 111(5):1897–1902
126. Shalem O, Sanjana NE, Hartenian E, Shi X, Scott DA, Mikkelsen TS, Heckl D, Ebert BL, Root DE, Doench JG, Zhang F (2014) Genome-scale CRISPR-Cas9 knockout screening in human cells. Science 343(6166):84–87
127. Sharan SK, Thomason LC, Kuznetsov SG, Court DL (2009) Recombineering: a homologous recombination-based method of genetic engineering. Nat Protoc 4(2):206–223
128. Shen Y, Stracquadanio G, Wang Y, Yang K, Mitchell LA, Xue Y, Cai Y, Chen T, Dymond JS, Kang K, Gong J, Zeng X, Zhang Y, Li Y, Feng Q, Xu X, Wang J, Wang J, Yang H, Boeke JD, Bader JS (2016) SCRaMbLE generates designed combinatorial stochastic diversity in synthetic chromosomes. Genome Res 26(1):36–49
129. Shi S, Liang Y, Zhang MM, Ang EL, Zhao H (2016) A highly efficient single-step, markerless strategy for multi-copy chromosomal integration of large biochemical pathways in *Saccharomyces cerevisiae*. Metab Eng 33:19–27

130. Shima J, Hesketh A, Okamoto S, Kawamoto S, Ochi K (1996) Induction of actinorhodin production by rpsL (encoding ribosomal protein S12) mutations that confer streptomycin resistance in *Streptomyces lividans* and *Streptomyces coelicolor* A3(2). J Bacteriol 178(24):7276–7284
131. Shin HY, Nijland JG, Waal PP, Jong RM, Klaassen P, Driessen AJ (2015) An engineered cryptic Hxt11 sugar transporter facilitates glucose-xylose co-consumption in *Saccharomyces cerevisiae*. Biotechnol Biofuels 8:176
132. Shui W, Xiong Y, Xiao W, Qi X, Zhang Y, Lin Y, Guo Y, Zhang Z, Wang Q, Ma Y (2015) Understanding the mechanism of thermotolerance distinct from heat shock response through proteomic analysis of industrial strains of *Saccharomyces cerevisiae*. Mol Cell Proteomics 14(7):1885–1897
133. Si T, Luo Y, Bao Z, Zhao H (2015) RNAi-assisted genome evolution in *Saccharomyces cerevisiae* for complex phenotype engineering. ACS Synth Biol 4(3):283–291
134. Si T, Xiao H, Zhao H (2015) Rapid prototyping of microbial cell factories via genome-scale engineering. Biotechnol Adv 33(7):1420–1432
135. Sjostrom SL, Bai Y, Huang M, Liu Z, Nielsen J, Joensson HN, Andersson Svahn H (2014) High-throughput screening for industrial enzyme production hosts by droplet microfluidics. Lab Chip 14(4):806–813
136. Skretas G, Kolisis FN (2012) Combinatorial approaches for inverse metabolic engineering applications. Comput Struct Biotechnol J 3:e201210021
137. Smith KM, Liao JC (2011) An evolutionary strategy for isobutanol production strain development in *Escherichia coli*. Metab Eng 13(6):674–681
138. Snoek T, Picca Nicolino M, Bremt S, Mertens S, Saels V, Verplaetse A, Steensels J, Verstrepen KJ (2015) Large-scale robot-assisted genome shuffling yields industrial *Saccharomyces cerevisiae* yeasts with increased ethanol tolerance. Biotechnol Biofuels 8:32
139. Sonderegger M, Sauer U (2003) Evolutionary engineering of *Saccharomyces cerevisiae* for anaerobic growth on xylose. Appl Environ Microbiol 69(4):1990–1998
140. Standage-Beier K, Zhang Q, Wang X (2015) Targeted large-scale deletion of bacterial genomes using CRISPR-nickases. ACS Synth Biol 4(11):1217–1225
141. Steensels J, Snoek T, Meersman E, Picca Nicolino M, Voordeckers K, Verstrepen KJ (2014) Improving industrial yeast strains: exploiting natural and artificial diversity. FEMS Microbiol Rev 38(5):947–995
142. Stovicek V, Borodina I, Forster J (2015) CRISPR–Cas system enables fast and simple genome editing of industrial *Saccharomyces cerevisiae* strains. Metab Eng Commun 2:13–22
143. Sung W, Ackerman MS, Miller SF, Doak TG, Lynch M (2012) Drift-barrier hypothesis and mutation-rate evolution. Proc Natl Acad Sci U S A 109(45):18488–18492
144. Tenaillon O, Rodriguez-Verdugo A, Gaut RL, McDonald P, Bennett AF, Long AD, Gaut BS (2012) The molecular diversity of adaptive convergence. Science 335(6067):457–461
145. Teoh ST, Putri S, Mukai Y, Bamba T, Fukusaki E (2015) A metabolomics-based strategy for identification of gene targets for phenotype improvement and its application to 1-butanol tolerance in *Saccharomyces cerevisiae*. Biotechnol Biofuels 8:144
146. Tilloy V, Cadiere A, Ehsani M, Dequin S (2015) Reducing alcohol levels in wines through rational and evolutionary engineering of *Saccharomyces cerevisiae*. Int J Food Microbiol 213:49–58
147. Toprak E, Veres A, Michel JB, Chait R, Hartl DL, Kishony R (2012) Evolutionary paths to antibiotic resistance under dynamically sustained drug selection. Nat Genet 44(1):101–105
148. Torres EM, Dephoure N, Panneerselvam A, Tucker CM, Whittaker CA, Gygi SP, Dunham MJ, Amon A (2010) Identification of aneuploidy-tolerating mutations. Cell 143(1):71–83
149. Maris AJ, Geertman JM, Vermeulen A, Groothuizen MK, Winkler AA, Piper MD, Dijken JP, Pronk JT (2004) Directed evolution of pyruvate decarboxylase-negative *Saccharomyces cerevisiae*, yielding a C_2-independent, glucose-tolerant, and pyruvate-hyperproducing yeast. Appl Environ Microbiol 70(1):159–166

150. Vanee N, Fisher AB, Fong SS (2012) Evolutionary engineering for industrial microbiology. Subcell Biochem 64:43–71
151. Vilela Lde F, de Araujo VP, Paredes Rde S, Bon EP, Torres FA, Neves BC, Eleutherio EC (2015) Enhanced xylose fermentation and ethanol production by engineered *Saccharomyces cerevisiae* strain. AMB Express 5:16
152. Wang HH, Isaacs FJ, Carr PA, Sun ZZ, Xu G, Forest CR, Church GM (2009) Programming cells by multiplex genome engineering and accelerated evolution. Nature 460(7257):894–U133
153. Wang M, Li S, Zhao H (2016) Design and engineering of intracellular-metabolite-sensing/regulation gene circuits in *Saccharomyces cerevisiae*. Biotechnol Bioeng 113(1):206–215
154. Wang T, Wei JJ, Sabatini DM, Lander ES (2014) Genetic screens in human cells using the CRISPR-Cas9 system. Science 343(6166):80–84
155. Wang Y, Zhang S, Liu H, Zhang L, Yi C, Li H (2015) Changes and roles of membrane compositions in the adaptation of *Saccharomyces cerevisiae* to ethanol. J Basic Microbiol 55(12):1417–1426
156. Wargacki AJ, Leonard E, Win MN, Regitsky DD, Santos CN, Kim PB, Cooper SR, Raisner RM, Herman A, Sivitz AB, Lakshmanaswamy A, Kashiyama Y, Baker D, Yoshikuni Y (2012) An engineered microbial platform for direct biofuel production from brown macroalgae. Science 335(6066):308–313
157. Warner JR, Reeder PJ, Karimpour-Fard A, Woodruff LB, Gill RT (2010) Rapid profiling of a microbial genome using mixtures of barcoded oligonucleotides. Nat Biotechnol 28(8):856–862
158. Wei N, Quarterman J, Jin YS (2013) Marine macroalgae: an untapped resource for producing fuels and chemicals. Trends Biotechnol 31(2):70–77
159. Winkler JD, Kao KC (2014) Recent advances in the evolutionary engineering of industrial biocatalysts. Genomics 104(6 Pt A):406–411
160. Wisselink HW, Cipollina C, Oud B, Crimi B, Heijnen JJ, Pronk JT, Maris AJ (2010) Metabolome, transcriptome and metabolic flux analysis of arabinose fermentation by engineered *Saccharomyces cerevisiae*. Metab Eng 12(6):537–551
161. Woodruff LBA, Gill RT (2011) Engineering genomes in multiplex. Curr Opin Biotechnol 22(4):576–583
162. Wright J, Bellissimi E, Hulster E, Wagner A, Pronk JT, Maris AJ (2011) Batch and continuous culture-based selection strategies for acetic acid tolerance in xylose-fermenting *Saccharomyces cerevisiae*. FEMS Yeast Res 11(3):299–306
163. Xiao H, Zhao H (2014) Genome-wide RNAi screen reveals the E3 SUMO-protein ligase gene SIZ1 as a novel determinant of furfural tolerance in *Saccharomyces cerevisiae*. Biotechnol Biofuels 7:78
164. Xu H, Kim S, Sorek H, Lee Y, Jeong D, Kim J, Oh EJ, Yun EJ, Wemmer DE, Kim KH, Kim SR, Jin YS (2016) PHO13 deletion-induced transcriptional activation prevents sedoheptulose accumulation during xylose metabolism in engineered *Saccharomyces cerevisiae*. Metab Eng 34:88–96
165. Yan D, Wang C, Zhou J, Liu Y, Yang M, Xing J (2014) Construction of reductive pathway in *Saccharomyces cerevisiae* for effective succinic acid fermentation at low pH value. Bioresour Technol 156:232–239
166. Zelle RM, Hulster E, Winden WA, Waard P, Dijkema C, Winkler AA, Geertman JM, Dijken JP, Pronk JT, Maris AJ (2008) Malic acid production by *Saccharomyces cerevisiae*: engineering of pyruvate carboxylation, oxaloacetate reduction, and malate export. Appl Environ Microbiol 74(9):2766–2777
167. Zha J, Shen M, Hu M, Song H, Yuan Y (2014) Enhanced expression of genes involved in initial xylose metabolism and the oxidative pentose phosphate pathway in the improved xylose-utilizing *Saccharomyces cerevisiae* through evolutionary engineering. J Ind Microbiol Biotechnol 41(1):27–39

168. Zhang Y, Liu G, Engqvist MK, Krivoruchko A, Hallstrom BM, Chen Y, Siewers V, Nielsen J (2015) Adaptive mutations in sugar metabolism restore growth on glucose in a pyruvate decarboxylase negative yeast strain. Microb Cell Fact 14:116
169. Zhou H, Cheng JS, Wang BL, Fink GR, Stephanopoulos G (2012) Xylose isomerase overexpression along with engineering of the pentose phosphate pathway and evolutionary engineering enable rapid xylose utilization and ethanol production by *Saccharomyces cerevisiae*. Metab Eng 14(6):611–622
170. Zhu L, Li Y, Cai Z (2015) Development of a stress-induced mutagenesis module for autonomous adaptive evolution of *Escherichia coli* to improve its stress tolerance. Biotechnol Biofuels 8:93
171. Ziv N, Brandt NJ, Gresham D (2013) The use of chemostats in microbial systems biology. J Vis Exp 80:18

Back to Basics: Creating Genetic Diversity

8

Kang Lan Tee and Tuck Seng Wong

Abstract

Directed evolution has emerged as a key enabling technology for improving the properties of biomolecules, biochemical pathways, and microorganisms to satisfy a wide range of biotechnological applications, from synthetic biology through to industrial biocatalysis. Laboratory evolution is an iterative process, alternating between creating genetic diversity and selection/screening to identify improved variants. This book chapter focuses on genetic diversity only. We describe and critically review recent advances in the methods for DNA assembly, random mutagenesis, focused mutagenesis, and DNA recombination. We also identify trends in these areas and highlight commercial kits that are developed to streamline and expedite these molecular biology techniques.

8.1 Introduction

Charles Darwin has taught us an invaluable lesson on how to engineer a biological system, using the principle of genetic diversity coupled with selection. Built upon the concept of Darwinian evolution, directed evolution has transformed the field of biological engineering. Not only is it an indispensable tool in the academic arena, it is also a key enabling technology in the commercial sector. Directed evolution is now a popular method of choice for tailoring the properties of nucleic acids, proteins, pathways, and organisms to suit various applications, including biocatalysis, bioremediation, biosensing, and synthetic biology.

K.L. Tee • T.S. Wong (✉)
ChELSI Institute and Advanced Biomanufacturing Centre (ABC), Department of Chemical and Biological Engineering, University of Sheffield,
Sir Robert Hadfield Building, Mappin Street, Sheffield S1 3JD, UK
e-mail: t.wong@sheffield.ac.uk

M. Alcalde (ed.), *Directed Enzyme Evolution: Advances and Applications*,
DOI 10.1007/978-3-319-50413-1_8

Creating a high-quality mutant library is the first step in a successful directed evolution campaign. The phrase *Back to Basics* in the title refers to this fundamental step in directed evolution. It involves choosing the right mutagenesis strategy and an efficient cloning method. In 2006, we wrote a comprehensive review on the methods for creating genetic diversity [1]. This survey was subsequently updated in 2013 and published in *Biotechnology Advances* [2]. Encouraged by the positive feedback from peers and the attention received for these two reviews, we decide to extend our previous work to survey methods published since 2013. In other words, this book chapter is a sequel to the previous two papers [1, 2]; the organization and the method categorization are identical to the systems used before to facilitate reading and understanding. Those methods that have already been covered previously will not be repeated here, and interested readers are kindly asked to refer to these papers [1, 2]. We have endeavored to be inclusive in this chapter, and we apologize if there is a method that we miss.

The aim of this chapter is twofold: (1) to provide a method selection guide for those wanting to create a gene library, particularly newcomers, and (2) to stimulate the development of many more novel and exciting techniques to advance the field of directed evolution. This chapter starts with a description of DNA assembly methods. This is followed by an update and critical review of the methodologies in random mutagenesis, focused mutagenesis, and DNA recombination. We also provide an overview of the commercial kits developed for each application and compare the principles of these kits. This chapter is concluded with a perspective on the future developments in the field of genetic diversity creation.

8.2 Molecular Cloning and DNA Assembly

Molecular cloning is a necessary step in most directed evolution experiments, where the gene of interest (GOI) is introduced into a vector for subsequent mutagenesis or the mutant library is subcloned for expression in an appropriate host organism. Further to directed evolution of a single protein, we have seen increased application of directed evolution in pathway engineering, where multiple genes in a pathway are engineered to achieve a greater final output [3, 4]. Consequentially, more methods are now adapted for introducing multiple GOIs into a vector backbone simultaneously or sequentially, instead of just a single GOI. As molecular cloning is traditionally used for introducing a GOI into a vector, the term *DNA assembly* is used in this chapter to better encompass all the methods developed to insert a single or multiple GOI(s) into a vector backbone.

We have previously categorized molecular cloning methods into four main strategies back in the year 2013 [2]. Development of new methods, however, necessitates the introduction of a new category. DNA assembly methods are now described in the following categories: (1) complementary overhangs, (2) homologous sequences, (3) overlapping polymerase chain reaction (PCR), (4) megaprimers, and (5) bridged ligation. The principles of these five strategies are illustrated in Fig. 8.1, and an overview is provided in Table 8.1.

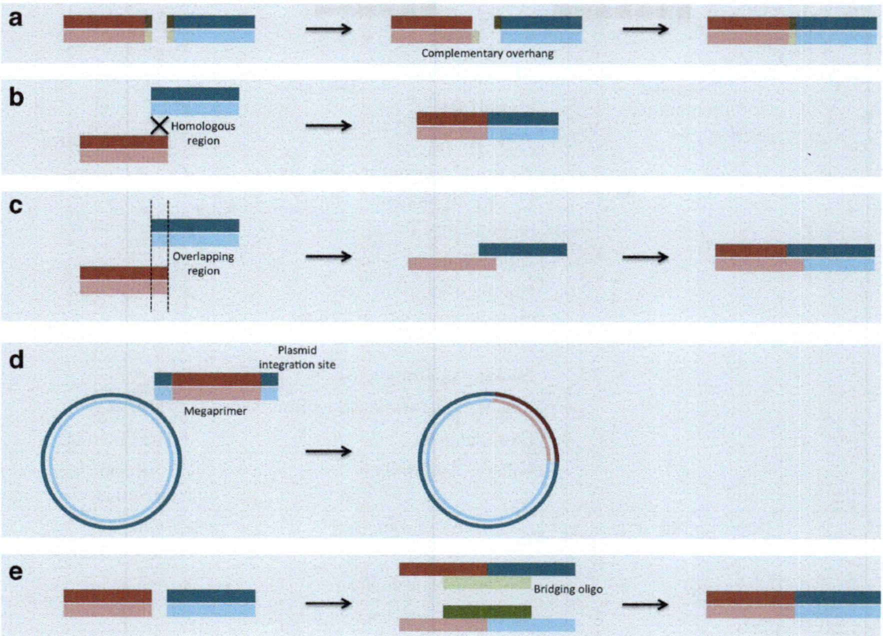

Fig. 8.1 Five strategies for DNA assembly: (**a**) complementary overhangs, (**b**) homologous sequences, (**c**) overlapping PCR, (**d**) megaprimers, and (**e**) bridged ligation

8.2.1 DNA Assembly Based on Complementary Overhangs

DNA assembly based on complementary overhangs resembles the conventional restriction–ligation cloning, but varies in the ways complementary overhangs between the GOI(s) and vector are generated. Short complementary overhangs [2–4 nucleotides (nts)] are commonly created using Type II restriction enzymes (REs), terminal transferase activity of some thermophilic DNA polymerases, asymmetric PCRs, or zinc finger nucleases, before GOI(s) and vector are covalently linked using a DNA ligase. Alternatively, long complementary overhangs (12–14 nts) allow formation of a stable nicked plasmid that can be directly transformed without the need of a DNA ligase. The nicks are then repaired by the host mechanisms. Examples of the methods used to generate long overhangs include phosphorothioate bond cleavage with iodine/ethanol solution or the use of nicking DNA endonucleases. New methods described since year 2013 continue to use the above methods to generate complementary overhangs, with increasing effort dedicated to adapting these methods to assemble multiple DNA fragments.

In an attempt to increase the cloning efficiency, Gao *et al.* used an additional gel extraction followed by a second ligation, after the first ligation of the linearized vector and the DNA insert. This additional gel purification step was aimed to isolate the right-sized ligation product to reduce self-ligation of digested vectors [5]. Using the

Table 8.1 Methods for DNA assembly

Category	Strategy	Mechanism	Availability of cloning kit
A	Complementary overhangs	Type II RE	✗
		Type IIS RE	✓ (Golden Gate)
		Terminal transferase activity of DNA polymerase	✓ (TA cloning)
		Asymmetric PCR	✗
		Zinc finger nuclease	✗
		Chemical method (e.g., phosphorothioate + iodine/ethanol)	✗
		Nicking endonuclease	✗
		Uracil excision-based technique	✓ (USER)
		Exonuclease	✓ (In-Fusion)
		Gibson Assembly	✓ (Gibson Assembly)
B	Homologous sequences	*In vitro* recombination	✓ (Gateway, GeneArt Seamless)
		In vivo recombination	✗
		Yeast homologous recombination	✗
C	Overlapping PCR	PCR-based assembly	✗
D	Megaprimers	GOI as primer for whole-plasmid amplification	✗
E	Bridged ligation	Ligase cycling reaction	✗

same principle, a sticky-/blunt-end ligation was used to clone two DNA fragments into a vector [6]. simpleUSER cloning was developed to simplify the USER cloning procedure by modifying the plasmid preparation step [7]. Instead of using a combination of one Type II RE (e.g., XbaI) and one nicking endonuclease (e.g., Nt.BbvCI) to create a linearized vector with two different overhangs [8], simpleUSER cloning made use of a single nicking endonuclease (i.e., Nb.BtsI). The same authors also reported nicking cloning [7], in which the overhangs on the DNA insert were created using a single nicking endonuclease (i.e., Nb.BbvCI), thereby abolishing the need of a USER enzyme mix [uracil-DNA glycosylase (UDG) and the DNA glycosylase-lyase endonuclease VIII] and uracil-containing primers. Nicking cloning shares identical principle with the Nicking DNA Endonuclease (NiDE) method [9]. In comparison, NiDE is more elegant as the same nicking endonuclease was used to prepare both the plasmid and the DNA insert.

Complementary Annealing Mediated by Exonuclease (CAME) uses exonuclease activity to expose the complementary overhangs on the DNA insert and the vector [10]. Enzymes that could be used to recess the DNA ends include *Pfu* DNA polymerase (3′→5′ exonuclease activity), T4 DNA polymerase (3′→5′ exonuclease activity), and λ exonuclease (5′→3′ exonuclease activity). This method is, in principle, identical to In-Fusion cloning.

There are an increasing number of publications in the area of multiple DNA fragment assembly. Gibson Assembly, developed in 2009, is an isothermal reaction (50 °C) that uses the concerted action of three enzymes. T5 exonuclease generates long overhangs, Phusion/*Taq* DNA polymerase fills in the gaps of the annealed single-strand regions, and *Taq* DNA ligase seals the nicks of the annealed DNA [11]. The simplicity of this method has made it the method of choice for many researchers [12]. Variations of Gibson Assembly mainly revolve around the enzyme mix used. Hot Fusion omits the *Taq* DNA ligase in its enzyme mix to achieve a lower self-ligation of the base vector and, thus, a higher assembly efficiency compared to Gibson Assembly (92% vs 70%) for a seven-DNA fragment assembly [13]. The Multiple Patch Cloning (MUPAC) uses T5 exonuclease, Klenow fragment (exo-), and T4 DNA ligase to lower the isothermal reaction temperature to 37 °C, which facilitates the assembly of fragments with shorter overhang (16 nts) [14].

Single Strand Assembly (SSA) is another variation of Gibson Assembly. Instead of assembling double-stranded DNA fragments, the authors used long single-stranded oligonucleotides with a homology sequence of 20 bp to enable Gibson Assembly. This approach is particularly useful when investigating transcriptional element (e.g., promoter) and translational element [e.g., ribosome-binding site (RBS)]. Synthetic libraries of these elements can be easily synthesized as long single-stranded oligonucleotides.

Further to Gibson Assembly, USER cloning has also been used for multiple DNA fragment assembly to create expression vector for mammalian cell engineering [15], integrative vector set for *Saccharomyces cerevisiae* [16], and vector for insect cell-baculovirus expression system [17].

Biopart Assembly Standard for Idempotent Cloning (BASIC) relies on multiple restrictions/ligations of the DNA fragments and linkers with BsaI (a Type IIS RE) and T4 DNA ligase to assemble up to seven fragments [18]. The authors were able to recapitulate the BsaI sites for subsequent rounds of assembly by methylation of the specified linker oligonucleotides to avoid cleavage.

8.2.2 DNA Assembly Based on Homologous Sequences

Methods that use homologous sequences eliminate the need of complementary overhangs between the GOI(s) and vector. This group of methods does not use REs or ligases but instead utilizes DNA recombination of homologous sequences. Recombination was demonstrated *in vitro* using *E. coli* cell extract or recombinases and *in vivo* in *E. coli* or yeast. Recombination efficiency can further be improved by including the λ prophage Red recombination for both *in vivo* and *in vitro* systems.

Advanced Quick Assembly (AQUA) is an *in vivo* homologous recombination method for DNA assembly of up to four fragments using a 32 bp homologous sequence [19]. The authors demonstrated the recombination using chemically prepared *E. coli* TOP10 cells and compared its efficiency to that of five different commercially available competent cells [i.e., TOP10, NEB5α, NEB10β, BL21(DE3), and JM109]. The efficiencies across all *E. coli* strains for two DNA fragment assembly were between 83–100%. Jacobus *et al.* investigated the optimal conditions for *in vivo* homologous recombination of two DNA fragments in *E. coli* DH5α [20]. When changing stoichiometry of vector to insert (from 1:1 to 1:4), the authors did not observe significant difference in the number of colonies obtained. The authors did, however, observe an optimal number of colonies were obtained when they used 150 ng of linearized pUC19 vector at a vector to insert ratio of 1:2. The cloning efficiency increases from 25% to 70% and to 100%, when the homology sequence increases from 10 bp to 20 bp and 30 bp, respectively. *In vivo* recombination was also tested for *E. coli* XL10-Gold [21], pointing to the fact that most of the laboratory *E. coli* strains can in fact be used for DNA assembly via recombination.

In an attempt to make Seamless Ligation Cloning Extract (SLiCE) [22], an *in vitro* homologous recombination method, more economically accessible, Motohashi and coworkers investigated a homemade bacterial cell extract prepared from *E. coli* JM109 using a Tris-HCl/Triton X-100 cell lytic buffer [23] instead of a commercial cell lytic buffer (CelLytic B Cell Lysis Reagent from Sigma) as in the original protocol [22]. Both the homemade extract and the commercial In-Fusion HD Cloning Kit provided comparable cloning efficiency, but the commercial kit generated about twice the number of clones compared to the homemade extract [24].

HomeRun Vector Assembly System (HVAS) utilized both Gateway cloning (a recombination system) and homing endonucleases for DNA assembly [25]. Despite having the potential for high-throughput automation in cloning, this approach is limited by the number of homing endonucleases available. Homing endonucleases are double-stranded DNases that have large, asymmetric recognition sites (12–40 bps). Currently, there are only four commercialized by New England Biolabs, i.e., I-CeuI, I-SceI, PI-PspI, and PI-SceI.

Yeast-based homology recombination has been used since 1987 for DNA assembly [26]. This powerful cloning method is not more widely adopted as it requires a yeast-compatible shuttle vector. To increase the versatility of yeast homologous recombination, the groups of Belden [27] and Andréasson [28] have both adapted yeast homologous recombination cloning for use with bacteria plasmids. Belden and coworkers achieved this by including a DNA fragment that contains both the 2-micron origin of replication (2 μm *ori*) and the *ura3* gene for selection in yeast, during a recombination of up to four DNA fragments [27]. Alternatively, Andréasson and coworkers overcame the problem by adding to the bacteria plasmid a short autonomous replication sequence (*ARS*) and a centromere sequence (*CEN*) for stable single-copy replication and mitotic segregation in yeast. Further, the authors cloned the yeast *TEF* promoter of the kanMX cassette (a resistance marker cassette for G418) upstream of the kanamycin resistance gene and were able to use the *kan* gene for selection in both yeast and *E. coli*. The final plasmid became a shuttle vector that the authors used to demonstrate yeast homologous recombination of up to seven DNA fragments. Yeast-based cloning method is proven an efficient way of making large or complex DNA constructs, as demonstrated by the assembly of the entire *Mycoplasma genitalium* genome [29]. Kilaru *et al.* also applied yeast-based recombination as a convenient way to construct vectors for fungus *Zymoseptoria tritici* [30].

8.2.3 DNA Assembly Based on Overlapping PCR

In overlapping PCR, DNA fragments that share overlapping regions at both ends are assembled using PCR. In other words, only DNA polymerase is required. Cao *et al.* demonstrated assembly of three PCR fragments and a linear vector in a single PCR using PrimeSTAR HS DNA polymerase [31]. Prior to assembly, all fragments and vector were amplified in high-fidelity PCRs to incorporate 20–50 bp overlapping regions at both termini.

8.2.4 DNA Assembly Based on Megaprimers

Cloning based on megaprimer strategy is, in principle, similar to QuikChange mutagenesis. Following the initial amplification of GOI, the generated PCR products served as megaprimers for the subsequent linear or exponential amplification of the whole plasmid.

QuickStep-Cloning, developed in our group, is a sequence-independent ligation-free method for directional cloning of a GOI into any plasmid at any position [32]. The method uses megaprimers with 3′ overhangs generated from asymmetric PCRs, thus allowing exponential amplification of the plasmid. The generated nicked plasmids can be transformed directly into expression host without DNA ligation. Based on the time requirement and the cloning efficiency, we are confident to say that QuickStep-Cloning fares exceptionally well in comparison to other megaprimer-based cloning methods.

Recombination-Assisted Megaprimer (RAM) [33] was designed to complement Restriction-Free (RF) cloning [34] by revising the method into an exponential amplification. RAM cloning involves three steps, i.e., megaprimer synthesis, integration of megaprimer into the target vector, and *in vivo* homologous recombination for end joining using *E. coli* [33].

8.2.5 DNA Assembly Based on Bridged Ligation

Ligase Cycling Reaction (LCR) is a one-step, scarless DNA assembly method [35]. It uses single-stranded bridging oligos complementary to the ends of adjoining DNA fragments, thermostable ligase, and multiple cycles of denaturation–annealing–ligation to assemble complex DNA constructs. The authors demonstrated one-step assembly of up to 20 DNA parts and up to 20 kb DNA constructs with very low single-nucleotide polymorphisms (<1 per 25 kb) and insertion/deletion (<1 per 50 kb). Zhao group combined this automation-friendly LCR with *in vivo* yeast-based DNA assembly for the construction of large biochemical pathways [36].

8.2.6 "De-cloning"

In the Sects. 8.2.1, 8.2.2, 8.2.3, 8.2.4, and 8.2.5, we have mainly focused on methods that "add" or "insert" a DNA into a construct. Equally important, we need the capability to "remove" or "delete" a DNA sequence.

Krishnamurthy *et al.* described a multiplex gene removal protocol using a two-step PCR [37]. They demonstrated the method through removing three genes from a plasmid. In the first PCR, segments to be kept are amplified. These PCR fragments are subsequently assembled in a PCR by incorporating single-stranded oligonucleotides of 40 bases in length, sharing a 20-nt complementarity with the two fragments to be connected. In fact, most DNA assembly methods can be adapted for gene removal. AQUA cloning (described in Sect. 8.2.2), for instance, was shown to be a good tool for deleting a DNA sequence [19].

8.2.7 Trends in DNA Assembly Methods

Through following the development of methods for DNA assembly, we observe a few trends in the development of DNA assembly methods: (1) The ability to synthesize long oligonucleotides with high accuracy and at a low cost has opened up many possibilities for new method development. (2) DNA polymerases (e.g., Q5 and Phusion) that display high fidelity, high processivity, and capability of amplifying DNA with high GC content are increasingly used to enhance efficiency and avoid unwanted mutations. (3) Many methods now involve whole-plasmid amplification or preparing plasmid with a PCR, partly owing to the availability of high-fidelity DNA polymerases. (4) Many variations of QuikChange protocol and megaprimer-based integration are continuously being developed. (5) Intermediate DNA purification steps (e.g., gel extraction, reaction cleanup) are undesirable. (6) Sequence-independent and scarless insertion or assembly of DNA fragments (commonly known as seamless) is emphasized. (7) Directional cloning or assembly in a desired order is important. (8) Homologous recombination is gaining momentum as a mainstream method. (9) Multiple fragment insertion or deletion in a single reaction (i.e., multiplexing) is preferred. (10) The use of combination of methods is getting more common to create complex and large DNA constructs. (11) Many DNA assembly methods have been adapted for focused mutagenesis, through incorporating primers containing desired modifications. (12) Many methods for assembling DNA can, in principle, be modified for "de-cloning." (13) Automation-friendly methods (e.g., LCR) will continue to receive more attention.

8.3 Genetic Diversity

"Strategies and applications of *in vitro* mutagenesis," written by Botstein and Shortle, is perhaps one of the earliest comprehensive summaries of genetic diversity creation methods [38]. Despite being published over 30 years ago, some of the strategies described in this review (e.g., oligonucleotide-directed mutagenesis, transposon mutagenesis) are still widely used in today's laboratories.

Broadly, genetic diversity creation methods can be classified into three categories, i.e., random mutagenesis, focused mutagenesis, and DNA recombination. With random mutagenesis, point mutations are introduced into the GOI at random positions, typically through PCR employing an error-prone DNA polymerase. Focused mutagenesis is the method of choice for altering a GOI at a selected position. Point mutation, insertion, or deletion can be introduced by incorporating mutagenic primer(s) containing the desired modification. Contrary to random mutagenesis and focused mutagenesis in which the mutant library is prepared from a single parental template, DNA recombination involves assembly of DNA fragments derived from more than one parental sequence encoding proteins of identical/similar function.

Noteworthy, commercial kits are now available for each category to streamline and standardize the experimental procedure. These kits contain all the components required (e.g., buffers, chemicals, enzymes, control DNA, and competent cells) for creating a gene library. They are also accompanied with a comprehensive manual, describing the background/principle of the method, kit components, recipes for media/buffers, step-by-step protocol, result expected, and troubleshooting guide. This is particularly helpful for newcomers who are yet to develop their competency in molecular biology and protein engineering. In this section, we discuss recent development in each category focusing on those methods reported after 2013 and provide a summary of commercial kits available in the market.

8.3.1 Random Mutagenesis

Ever since Leung described the protocol of error-prone polymerase chain reaction (epPCR) in 1989, in which the fidelity of *Taq* DNA polymerase was intentionally reduced by the use of suboptimal reaction conditions (e.g., high $MgCl_2$ concentration in the presence of $MnCl_2$ and high number of PCR cycles starting with a low template concentration) [39], this particular genetic diversity creation method remains the front-runner in terms of its usage frequency. The popularity of epPCR is attributed to its technical simplicity and cost-effectiveness [2]. Not surprising, we continue to see modifications being made to epPCR and diversification of its applications.

AXM mutagenesis was developed to eliminate the need of subcloning an epPCR library [40]. GOI is amplified under error-prone conditions (70 mM $MgCl_2$, 5 mM $MnCl_2$, and unbalanced nucleotide concentrations), using a reverse primer containing phosphorothioate linkages on its 5′ end. The double-stranded PCR product is treated with T7 exonuclease to selectively degrade the unmodified strand. The resulting ssDNA then acts as a megaprimer, which anneals to a uracilated, circular, single-stranded phagemid DNA and primes DNA polymerization, similar to a Kunkel mutagenesis [41]. The ligated heteroduplex product is then transformed into *E. coli*, where the uracilated strand is cleaved *in vivo* by uracil-*N*-glycosylase. Although the subcloning step is bypassed, the method necessitates a laborious preparation of DNA template, which is less appealing. The same group also develops an *E. coli* strain expressing *Eco*29kI restriction-methylation system to restrict incoming parental DNA in the transformed cells [42].

Instead of accumulating mutations on the GOI, Schwaneberg group extended the application of epPCR to mutagenize vector backbone to increase recombinant protein expression, in a method called epMEGAWHOP [43]. GOI is first amplified in a PCR. The PCR product serves as a megaprimer for the subsequent whole-plasmid amplification that is conducted under error-prone condition (0.05 mM Mn^{2+}). After DpnI digestion to remove the methylated or hemimethylated parental template, the resulting DNA is transformed into *E. coli* DH5α. Following plasmid isolation and transformation into an expression host, the library is screened for enhanced protein expression. The group demonstrated the applicability by enhancing the expression of cellulose (2× improvement), lipase (2×), and protease (4×), which were cloned into pET28a(+) (5369 bp), pET22b(+) (5493 bp), and pHY-300PLK (4870 bp), respectively. This strategy does not require *a priori* knowledge of the bacterial transcription/translation machinery and is applicable to all enzymes, expression vectors, and bacterial hosts. Potentially, the same strategy could be used to increase the stability and the copy number of a plasmid.

Building on the previously reported TriNEx method [44], Jones group reported an approach to introduce in-frame codon replacement with an amber stop codon (TAG), as a means to incorporate noncanonical amino acid into protein [45]. The method was verified by incorporating *p*-azido-L-phenylalanine or *p*-iodo-L-phenylalanine into cytochrome *b562* and keratinocyte growth factor. The possibility to expand the genetic code would mean that the protein sequence space can further be broadened from 20^N to $(20 + AA_{NC})^N$, in which N represents the number of amino acids within a protein and AA_{NC} is the number of noncanonical amino acids added to the pool of 20 naturally occurring amino acids.

8.3.1.1 Commercial Kits for Random Mutagenesis

To provide a selection guide to readers, particularly newcomers, we subdivide the commercial kits for random mutagenesis into four categories (Table 8.2). Kits in RI category are PCR kits to conduct epPCR. Briefly, GOI is amplified under error-prone conditions (e.g., high Mg^{2+} concentration, addition of Mn^{2+}, unbalanced dNTP concentration, or combination of these factors) or using an error-prone DNA polymerase (e.g., Mutazyme I and/or Taq). In RII category, mutations are introduced via the use of nucleotide analogues (e.g., dPTP, 8-oxo-dGTP) or DNA-modifying chemicals. Kit in RIII contains *E. coli* XL1-Red competent cells ready for plasmid transformation. This is a mutator strain that is deficient in three DNA repair pathways: *mutS* (error-prone mismatch repair), *mutD* (deficient in 3′→5′ exonuclease of DNA polymerase III), and *mutT* (unable to hydrolyze 8-oxo-dGTP). Kits in category RIV are more recent compared to the other three categories, and they involve the use of a transposase (e.g., MuA or Tn5) for *in vitro* transposition. Through proper design of a transposon cassette, random insertion or deletion is possible with these kits.

8.3.1.2 Mutational Spectrum and Quality of a Random Mutagenesis Library

While applying a random mutagenesis method or developing a new one for directed evolution, key considerations include mutation frequency (i.e., mutations/kb), distribution of mutations, mutational spectrum (e.g., transition, transversion, insertion,

Table 8.2 Commercially available random mutagenesis kits

Category	Kit	Company	Principle	Type of mutations	Mutation frequency	Mutational spectrum
RI	JBS Error-Prone Kit	Jena Bioscience	Addition of Mn^{2+} and unbalanced dNTP concentration	Point mutation	6–20 nt/kb	Not reported
	Diversify PCR Random Mutagenesis Kit	Clontech	Addition of Mn^{2+} and unbalanced dNTP concentration	Point mutation	2–8 nt/kb	T_s, 47–80% T_v, 20–53%
	GeneMorph II	Agilent	Blend of Mutazyme I and Taq DNA polymerase mutant	Point mutation	3–16 nt/kb	T_s, 43% T_v, 51.4% Indel, 5.5%
RII	AquaMutant Random Mutagenesis Reagent Kit	MoBiTec	Chemical mutagen	Point mutation	50 nt/kb	T_s, 50–60% T_v, 30–40%
	JBS dNTP-Mutagenesis Kit	Jena Bioscience	Nucleotide analogues (dPTP, 8-oxo-dGTP)	Point mutation	up to 200 nt/bp	T_s if dPTP is used T_v if 8-oxodGTP is used
RIII	XL1-Red	Agilent	*E. coli* strain deficient in three DNA repair pathways: *mutS* (error-prone mismatch repair), *mutD* (deficient in 3′–5′ exonuclease of DNA polymerase III), and *mutT* (unable to hydrolyze 8-oxo-dGTP)	Point mutation	Not reported	Not reported
RIV	Mutation Generation System Kit	Thermo Scientific	In vitro transposition using MuA transposase	15 bp insertion	Transposition frequency = 0.5–20%	N/A
	EZ-Tn5 In-Frame Insertion Kit	Epicentre	In vitro transposition using Tn5 transposase	In-frame 19 codon insertion	Not reported	N/A
	Stop Generation System Kit	Thermo Scientific	In vitro transposition using MuA transposase	C-terminal deletion	Not reported	N/A

deletion, consecutive nucleotide substitution), amino acid substitution pattern, percentage of unique sequences (or percentage of duplicated sequences), and percentage of wild-type sequences. These parameters are often used to assess the quality of the resultant mutant library and have direct impact on the screening effort one needs to invest. These are also the factors one needs to contemplate when choosing a commercial kit to do the job (Table 8.2), besides its price and the reliability of the kit.

Quantifying the quality of a mutant library requires sequencing large number of clones to obtain statistically significant data. Schwaneberg group attempted this by preparing three random mutagenesis libraries of lipase (549 bp) and sequencing 1000 mutations per library [46]. These three libraries (epPCR-low, epPCR-high, and SeSaM-Tv P/P) were prepared with standard epPCR protocols with a low mutation frequency (0.1 mM $MnCl_2$) or a high mutation frequency (0.3 mM $MnCl_2$), as well as with the Sequence Saturation Mutagenesis (SeSaM) [47]. Transitions were predominant in both epPCR libraries (>72%), while transversions were enriched in the SeSaM library (43%). Further, consecutive nucleotide substitutions occurred at a higher frequency (30%) in the SeSaM library, while retaining high fraction of clones expressing active lipase (52%). Interestingly, the amino acid substitution pattern of both epPCR libraries complements that of the SeSaM library in terms of the amino acid's side-chain property. This important piece of work has conveyed a few key messages: (1) the field of directed evolution can accommodate more genetic diversity creation methods and should encourage their developments, (2) combination of methods allows exploring larger diversity, and (3) libraries created by affordable methodology (e.g., epPCR, SeSaM) are often sufficient to identify improved variants.

8.3.2 Focused Mutagenesis

First described by Michael Smith in 1978 at the University of British Columbia, Site-Directed Mutagenesis (SDM) has quickly become an invaluable tool for protein engineering and for studying structure-function relationships [48]. In Smith's oligonucleotide-directed mutagenesis, a short synthetic oligonucleotide of 12 bases containing a mismatch was used to amplify a single-stranded phage DNA, using combination of *E. coli* DNA polymerase I and T4 DNA ligase. Smith and his team demonstrated that it was possible to introduce a permanent mutation in the circular phage DNA, resulting in a phenotypic change. This seminar work has inspired the development of QuikChange [49] and related techniques that are widely used for focused mutagenesis [2].

Important to note, a lot of the techniques described in Sect. 8.2 for molecular cloning or DNA assembly have now been adapted for focused mutagenesis. This further supports the presentation of both DNA assembly and mutagenesis together in this chapter.

QuikChange is a straightforward method for focused mutagenesis, using a pair of overlapping mutagenic primers to linearly amplify the whole plasmid harboring the GOI. Further to using *Pfu* DNA polymerase or its derivatives for QuikChange mutagenesis, Q5 DNA polymerase (New England Biolabs) has been demonstrated to be a suitable alternative [50]. Q5 is a high-fidelity, thermostable DNA polymerase with 3′→5′ exonuclease activity, fused to a processivity-enhancing Sso7d domain to support robust DNA amplification. Q5 is reported to have an error rate >100-fold lower than that of *Taq* and 12-fold lower than that of *Pfu*. The other attractive feature is its high polymerization rate of 10–40 s/kb depending on the template used. Xia *et al.* attempted to revise the QuikChange protocol by using partially overlapping primer and Phusion DNA polymerase [51], which has an error rate of >50-fold lower than that of *Taq* and 6-fold lower than that of *Pfu*. The authors provided new insights into the molecular mechanism of QuikChange. They proposed that their revised protocol is an exponential amplification process. The resultant linear DNA products with homologous ends are joined to generate circular plasmids within *E. coli* via homologous recombination. From what we have gathered so far, Q5 and Phusion have been popular DNA polymerases for development of molecular biology techniques.

In our experience, the efficacy of QuikChange mutagenesis also depends on the plasmid itself. The method does occasionally fail in cases involving amplifying a large plasmid or a plasmid with high GC content. To overcome this, a Cut-and-Paste-based Cloning Strategy (CPCS) was described for focused mutagenesis of a large gene [52]. Target gene is split into two segments at the site for mutagenesis. Each segment is amplified with one flanking primer and one mutagenic primer. The two PCR fragments are individually cloned into vectors using TA cloning. Subsequently, one fragment is cut out with a pair of restriction enzymes and inserted into the vector containing the other fragment. Comparatively, the method employing Type IIS RE for mutagenesis at multiple sites is far more elegant [53]. This approach is essentially identical to Golden Gate Assembly, which exploits the ability of Type IIS RE [e.g., BsaI, BbsI, BsmBI/Esp3I] to cleave DNA outside of its recognition sequence. The fragments to be assembled are designed in such a way that the Type IIS recognition site is distal to the cleavage site, such that the Type IIS RE can remove its own recognition sequence completely from the assembly. Key advantages of such an arrangement are threefold: (1) the overhang sequence created is not dictated by the RE, and therefore, no scar sequence is introduced, (2) the fragment-specific sequence of the overhangs allows assembly of multiple fragments simultaneously in the desired order, and (3) the restriction site is eliminated from the ligated product. The net result is an ordered and seamless assembly of DNA fragments in a one-pot reaction. MUPAC, a Gibson Assembly derivative described in Sect. 8.2.1, is another technique that can be used for molecular cloning, DNA assembly, and focused mutagenesis [14]. As the overlapping sequence of adjoining fragments is much longer in Gibson Assembly than those used in Golden Gate Assembly (typically an overhang of 4 nts), higher percentage of correct assemblies is expected with a Gibson Assembly approach. The authors of MUPAC reported >90% efficiency in assembly of six fragments to introduce five mutations. Overlapping Extension Polymerase Chain Reaction (OE-PCR) is perhaps a more economical method for introducing multiple changes in a DNA sequence [54], compared to Gibson Assembly and Golden Gate Assembly. The principle of OE-PCR is essentially the same, which involves assembly of PCR fragments sharing overlapping regions, using DNA polymerase such as Phusion. Another alternative is *in vitro* or *in vivo* DNA recombination. Motohashi extended the use of SLiCE from *E. coli* to site-directed mutagenesis, in a process termed SLiCE-mediated PCR-based site-directed mutagenesis (SLiP) [55]. It is also possible to carry out focused mutagenesis with *in vivo* recombination using *E. coli* [e.g., AQUA cloning (19)] or yeast [e.g., plasmid constructed by Belden's group (27)]. In all the seven methods described above (CPCS, Type IIS assisted, MUPAC, OE-PCR, SLiP, AQUA, and yeast-based recombination), mutations are introduced into the primers used for generating the PCR fragments.

Protein fusion with variable linker insertion (P-Link) is classified as a focused mutagenesis method here, as it was designed to create fusion protein library with linkers of variable length [56]. In other words, the linker is the localized variable region. The method capitalizes on creating complementary ends using iodine cleavage of phosphorothioate linkages within the primers. In P-Link, the two genes encoding the proteins to be fused are first cloned into a vector, and this construct is served as the template DNA for subsequent PCR to insert linker region. Two partially overlapping primers, containing the linker sequence, are used to amplify the entire plasmid. Upon DpnI digestion to remove the template DNA and iodine cleavage to expose complementary ends and enable annealing, the product is transformed into *E. coli* for protein expression. Recombineering of Ends of Linearised Plasmids after PCR (REPLACR) is a method for creating substitution, insertion, and deletion, which relies on *in vivo* recombineering [57]. Partially overlapping primers containing the desired mutation are used to amplify the whole plasmid. This generates a linear PCR product with both the ends contain overlapping sequences for recombination. Following DpnI digestion of template DNA, the product is transformed into *E. coli* expressing the recombineering proteins (Redγ, β, α, and RecA). P-Link and REPLACR are conceptually similar, and mutations are designed into the primers like methods in the previous paragraph. Their main difference is how the ends of the linear DNA product are joined.

Contrary to all the abovementioned methods that are developed for creating genetic diversity using *E. coli*, Mutagenic Organized Recombination Process by Homologous *In Vivo* Grouping (MORPHING) is a method to generate diversity at specific regions by taking advantage of the efficient DNA recombination in *S. cerevisiae* [58]. GOI is divided into segments, and each segment is amplified individually using a PCR. Only the segments targeted for mutagenesis are amplified under error-prone conditions. As all the adjoining fragments share homologous sequences, this pool of fragments can be co-transformed into *S. cerevisiae* along with a linearized plasmid.

8.3.2.1 Randomization Scheme

Having decided on which residues to target for mutation, the next immediate task is to define a randomization scheme for primer design. Through the use of a specific codon or a degenerate codon, a residue can be substituted to a specific amino acid, a set of amino acids, or all 20 canonical amino acids. The choice of a degenerate codon is important as it determines the library size and, therefore, the screening effort.

In Table 8.3, we summarize the most commonly used degenerate codons for focused mutagenesis [60–63]. Traditionally, one would use the codon NNN for

Table 8.3 Comparison of randomization schemes

Randomization scheme[a,b]	Number of oligonucleotide syntheses required[c]	Mixing ratio of oligonucleotides	Number of codons	Probability of stop codons	Amino acids represented in the pool[d]	Redundant codons	Amino acid occurrence
NNN (64)	1	1	64	3/64	All (20]	41	Biased
NNK (32)	1	1	32	1/32	All (20)	11	Biased
NNS (32)	1	1	32	1/32	All (20)	11	Biased
NDT (12)	1	1	12	0	*Mixture of polar, nonpolar, positive, and negative*: G, V, L, I, C, N, H, S, D, R, F, Y (12)	0	Unbiased
NTN (16)	1	1	16	0	*Nonpolar*: M, F, V, L, I (5)	11	Biased
NAN (16)	1	1	16	2/16	*Charged, large side chains*: Y, H, Q, N, K, D, E (7)	7	Unbiased
NCN (16)	1	1	16	0	*Smaller side chains, polar and nonpolar*: S, P, T, A (4)	12	Unbiased
RST (4)	1	1	4	0	*Small side chains*: G, A, S, T (4)	0	Unbiased
NDT (12)	4	12/20	20	0	All (20)	0	Unbiased
VMA (6)		6/20					
ATG (1)		1/20					
TGG (1)		1/20					
NDT (12)	3	12/22	22	0	All (20)	2	Almost unbiased
VHG (9)		9/22					
TGG (1)		1/22					

(continued)

Table 8.3 (continued)

Randomization scheme[a,b]	Number of oligonucleotide syntheses required[c]	Mixing ratio of oligonucleotides	Number of codons	Probability of stop codons	Amino acids represented in the pool[d]	Redundant codons	Amino acid occurrence
HAT (3)	4	3/20	29	0	All (20)	9	Unbiased[e]
VMR (12)		6/20					
WKK (8)		8/20					
GDY (6)		3/20					
SYN (16)	2	4/5	17	0	*Aliphatic*: A, V, L, I, P (5)	12	Unbiased[e]
ATA (1)		1/5					
YAY (4)	3	2/4	6	0	*Aromatic*: F, Y, W, H (4)	2	Unbiased[e]
TGG (1)		1/4					
TTT (1)		1/4					
RAM (4)	2	4/6	6	0	*Polar*: R, K, D, E, N, Q (6)	0	Unbiased
CRA (2)		2/6					

[a]IUPAC nomenclatures: N = A/T/G/C, V = A/C/G, H = A/C/T, D = A/G/T, B = C/G/T, M = A/C, K = G/T, W = A/T, S = G/C, Y = C/T, R = A/G
[b]Number in the bracket indicates number of codons
[c]Number of oligonucleotide syntheses required for mutating a single strand
[d]Number in the bracket indicates number of amino acids
[e]Bias is removed through adjusting the ratio of oligonucleotides

saturation mutagenesis. Despite encoding all 20 possible amino acids, this set of 64 (4 × 4 × 4) codons also contains 3 stop codons and 41 redundant codons. Moreover, the occurrence of each amino acid is not identical. Amino acids encoded by six codons (e.g., S, L, R) occur at a higher probability, compared to those encoded by one codon (e.g., M, W) or two codons (e.g., F, Y, H, Q, N, K, D, E, C). NNK and NNS (each 32 codons) are proposed to reduce the redundancy of the mutant library. Instead of using a single degenerate oligonucleotide, one can consider synthesizing a set of oligonucleotides and mix them in a specific ratio to achieve an unbiased (or less biased) occurrence of the desired amino acids. As an example, if NDT, VMA, ATG, and TGG are mixed in the ratio of 12:6:1:1, all 20 amino acids are expected to occur with the same probability. That being said, this ideal situation is only possible if (1) all oligonucleotides are synthesized perfectly (e.g., purify, sequence accuracy, mixing of phosphoramidite building blocks), (2) the concentration of each oligonucleotide is measured accurately, and (3) each oligonucleotide binds to its template with the same affinity and is elongated by DNA polymerase in an unbiased manner. In other words, the quality of the oligonucleotide is essential. Acevedo-Rocha *et al.* explored the economics of focused mutagenesis libraries, by considering various randomization schemes, purity of oligonucleotide, and suppliers of oligonucleotide [64].

Recently, our group reported OneClick, a user-friendly web-based program, developed specifically for quick-and-easy design of focused mutagenesis experiments [59]. To our best knowledge, OneClick is the only tool that offers a step-by-step experimental design, from mutagenic primer design through to analysis of a mutant library. Upon input of GOI sequence, OneClick designs the mutagenic primers according to user input, e.g., amino acid position to mutate, type of amino acid substitutions (e.g., substitution to a group of amino acids with similar chemical property), and type of mutagenic primers. OneClick has incorporated libraries of commercially available plasmids and of DNA polymerases suitable for focused mutagenesis. Therefore, OneClick also provides information such as PCR mixture preparation, thermal cycling condition, the expected size of PCR product, and the type of agar plate to use during bacterial transformation. Importantly, OneClick also carries out a statistical analysis of the resultant mutant library, information of which is important for selection/screening.

8.3.2.2 Commercially Available Kits

Similar to random mutagenesis, a host of commercial kits are available for focused mutagenesis (Table 8.4). Again, to facilitate the selection of a kit, we classify these kits into three subcategories (FI, FII, and FIII; Fig. 8.2).

Eighty percent of the commercial kits fall under category FI. This is perhaps not a surprising figure, considering the ease of using a pair of primers to amplify the whole plasmid. This primer pair could either be (FIA) two overlapping mutagenic primers, (FIB) one non-mutagenic primer and one mutagenic primer that are non-overlapping, and (FIC) one mutagenic primer and one selection primer. Among these kits, the only subtle difference is the way of removing the parental DNA carrying the wild-type sequence. In Phusion Site-Directed Mutagenesis Kit, there is no

Table 8.4 Commercially available site-directed mutagenesis kits

Category	Kit	Supplier	Principle	Primer design[a]	Primer modification	Efficiency	Enzyme	Requirement of special strain	Template
FI	QuikChange Lightning	Agilent	Whole plasmid amplification	Two overlapping primers	Nil	>80% (one site) >55% (three sites)	Derivative of *PfuUltra*, DpnI	No	dsDNA
	ATB	Ameritech Biomedicines	Whole plasmid amplification	Two overlapping primers	Nil	Not reported	*Pfu* Ultimate, DpnI	No	dsDNA
	Muta-Direct	SBS Gentech	Whole plasmid amplification	Two overlapping primers	Nil	100% (one site)	Muta-direct, Mutazyme	No	dsDNA
	TagMaster	GM Biosciences	Whole plasmid amplification	Two overlapping primers	Nil	>95% (one site) 50% (two sites)	Blend of Pfu and Phusion	TagMaster cells	dsDNA
	AMAP	MBL	Whole plasmid amplification	One mutagenic primer	5′-P	>80% (one site) >80% (two sites) >60% (three sites) 40% (four/five sites)	*Pfu*, *Taq* DNA ligase, DpnI	No	dsDNA
	Q5	New England Biolabs	Whole plasmid amplification	Two non-overlapping primers	Nil	>90% (one site)	Q5, kinase, ligase, DpnI	No	dsDNA
	Phusion	Thermo Scientific	Whole plasmid amplification	Two non-overlapping primers	5′-P	>80% (one site)	Phusion Hot Start II, T4 DNA ligase	No	dsDNA
	KOD Plus	Toyobo	Whole plasmid amplification	Two non-overlapping primers	Nil	95% (one site)	KOD Plus, T4 polynucleotide kinase, T4 DNA ligase, DpnI	No	dsDNA
	Change-IT	Affymetrix	Whole plasmid amplification	Two non-overlapping primers	5′-P	90% (one site) 90% (two sites) 80% (three sites)	FideliTaq, DNA ligase, DpnI	No	dsDNA
	EZchange	Enzynomics	Whole plasmid amplification	Two non-overlapping primers	Nil	Not reported	*n*Pfu-Forte, kinase, ligase, DpnI	No	dsDNA
	Transformer	Clontech	Whole plasmid amplification	One mutagenic primer and one selection primer	5′-P	>70% (one site)	T4 DNA polymerase, T4 DNA ligase, restriction enzyme	*E. coli mut*S	dsDNA
	Mutan-Super Express Km	TaKaRa	Whole plasmid amplification	One mutagenic primer and one selection primer	5′-P	>80% (one site)	TaKaRa LA *Taq*	*E. coli sup*^{0}	dsDNA

FII	Gibson Assembly	Synthetic Genomics	Assembly of PCR fragments	Four partially overlapping primers	Nil	>90% (five sites)	Non-disclosed enzyme mix, DpnI	No	dsDNA
	PickMutant	Canvax	Assembly of PCR fragments and cloning vector	Four partially overlapping primers	Nil	Not reported	Non-disclosed enzyme mix	No	dsDNA
FIII	GeneArt (Plus)	Thermo Scientific	Assembly of PCR fragments	Two overlapping primers	Nil	>90% (one site) >90% (two sites) >90% (three sites)	DNA methylase, non-disclosed enzyme mix	DH5α™-T1^R	dsDNA

[a]Primer design for mutagenesis at one site

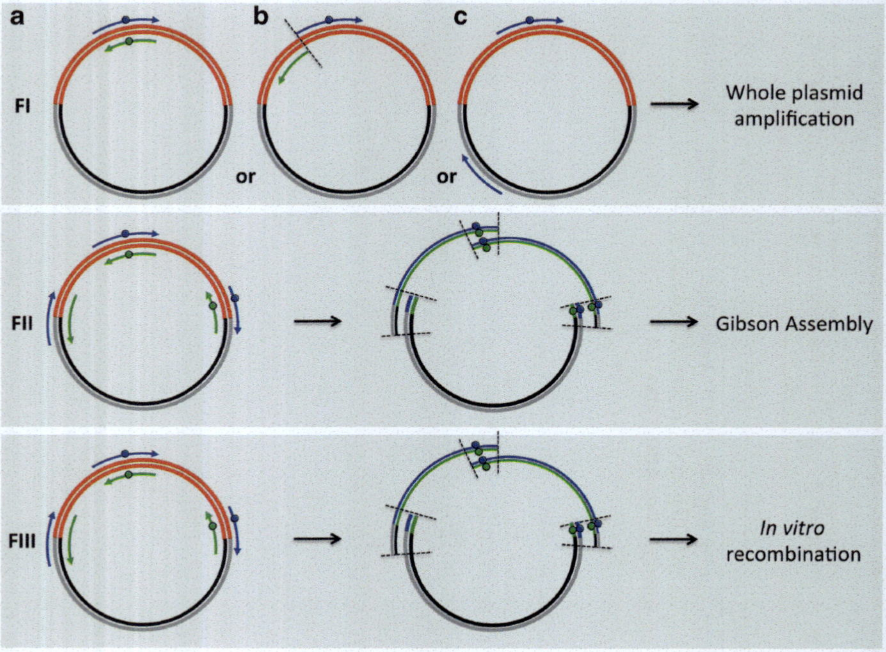

Fig. 8.2 Principles of commercial kits for focused mutagenesis: (FI) whole-plasmid amplification using (**a**) two overlapping mutagenic primers, (**b**) one non-mutagenic primer and one mutagenic primer that are non-overlapping, and (**c**) one mutagenic primer and one selection primer, (FII) Gibson Assembly, and (FIII) in vitro recombination

template removal step recommended. The supplier has argued that minute amount of parental DNA is exponentially amplified in this method, and therefore, the fraction of non-mutated template is minimal. Most kits apply DpnI to degrade methylated or hemimethylated parent DNA, which are usually plasmids isolated from *dam+ E. coli* strains. Instead of relying on a RE, TagMaster cell can discriminate parent DNA and novel synthesized daughter DNA. Therefore, parent plasmid template DNA would be eliminated, and only daughter DNA synthesized in the mutagenesis reaction can be enriched. In Transformer Site-Directed Mutagenesis Kit, one mutagenic primer is used to introduce the desired mutation. It also employs an additional selection primer containing a mutation in the recognition sequence for a unique restriction enzyme site. The two primers simultaneously anneal to one strand of the denatured double-stranded plasmid. After standard DNA elongation, ligation, and a primary selection by restriction digest, the mixture of mutated and unmutated plasmids is transformed into a mutS *E. coli* strain defective in mismatch repair. Transformants are pooled, and plasmid DNA is prepared from the mixed bacterial population. The isolated DNA is then subjected to a second selective restriction enzyme digestion. Since the mutated DNA lacks the restriction enzyme recognition site, it is resistant to digestion. A second transformation is then used to isolate the individual transformant carrying the mutated DNA. Mutan-Super Express Km Site-Directed Mutagenesis System also depends on a selection system. PCR is performed using an oligonucleotide containing the desired mutation and a selection primer to revert the amber mutations on the kanamycin-resistant gene present in the plasmid. When the nicked DNA is transformed into *sup^o^* *E. coli* strain, the nick is repaired, and then transformants containing a site-specific mutation can be obtained after culturing in the media containing kanamycin.

Kits in category FII and FIII rely on assembling PCR fragments, using either Gibson Assembly (FII) or *in vitro* recombination (FIII). These kits are more efficient for performing mutagenesis at multiple sites.

Recently, Agilent Technologies introduced a new product termed QuikChange HT Protein Engineering System. This system capitalizes on Agilent's capability of synthesizing long oligonucleotides (~150 bases). The workflow consists of three steps: (1) select one or more 50-amino acid mutational region in a protein sequence, design a set of oligonucleotides containing desired modifications (random mutations, site-specific mutations, or combinatorial mutations), and place the oligo order; (2) amplify each oligo set with a pair of primers; and (3) perform QuikChange mutagenesis with each amplified oligo set. This system does offer multiple advantages, including elimination of codon redundancy and bias. Further, this system is more economical compared to a synthetic library.

8.3.3 DNA Recombination

Invented by Stemmer, preparation of a shuffled DNA library involves DNA fragmentation by DNase I followed by fragment assembly via PCR [65]. Besides Stemmer shuffling (or DNA shuffling) [65], there is a host of other DNA recombination methods, which include Staggered Extension Process (StEP) [66], Incremental Truncation for the Creation of Hybrid Enzymes (ITCHY) [67], Sequence Homology-Independent Protein Recombination (SHIPREC) [68], *In Vivo* Shuffling [69], Combinatorial Libraries Enhanced by Recombination in Yeast (CLERY) [70], Nonhomologous Random Recombination (NRR) [71] etc. StEP is frequently used to combine beneficial mutations found in random mutagenesis experiments. ITCHY and SHIPREC are, in principle, identical. Both methods allow creating chimeras of one crossover from two nonhomologous sequences. Contrary to most DNA recombination methods, *In Vivo* Shuffling and CLERY rely on recombination within *E. coli* and yeast, respectively. NRR, even though a more laborious process, is homology independent and enables creating chimeras of more than one crossovers.

In comparison with random mutagenesis and focused mutagenesis, we see far less molecular techniques being developed for DNA recombination. Lehtonen *et al.* reported a minor change to Stemmer protocol [71] by using the Gateway cloning procedure directly after the gene reassembly reaction, without additional purification and amplification.

8.3.3.1 Commercial Kits for DNA Recombination

The only commercial kit available for DNA recombination is the JBS DNA-Shuffling Kit (Jena Biosciences), which in essence adopted the principle of the Stemmer protocol [65]. It can be applied for recombination of gene fragments originating from one or several related genes. When used for single-gene shuffling, only one gene is digested and subsequently reassembled resulting in point mutations at a mutation frequency of 7 nt/kb.

8.3.4 Key Milestones and the Trends in the Development of Genetic Diversity Creation Methods

We consider it appropriate to end this book chapter with a reflection on the key milestones in the field of genetic diversity creation. In Fig. 8.3, we cherry-pick some methods that, in our opinion, are transformative or disruptive technologies for creating gene libraries. In the 1960s, Sol Spiegelman and coworkers demonstrated how RNA molecules could be evolved in the test tube [72]. This seminar work is widely acknowledged as the first *in vitro* Darwinian evolution experiment that paved the way for the many directed evolution experiments that followed [73]. In the 1970s, Michael Smith demonstrated oligonucleotide-directed mutagenesis, laying the groundwork for focused mutagenesis [48]. It is fair to say that all DNA assembly methods and methods for creating genetic diversity, applied today, rely on PCR and Kary Mullis's contribution shall not be forgotten [74]. The Nobel Prize in Chemistry 1993 was awarded "for contributions to the developments of methods within DNA-based chemistry" jointly with one half to Kary Mullis "for his invention of the PCR method" and with one half to Michael Smith "for his fundamental contributions to the establishment of oligonucleotide-based, site-directed mutagenesis and its development for protein studies." Leung's epPCR protocol [39] and Stemmer's DNA shuffling [65] are still popular and widely used in many laboratories, despite being reported more than 20 years ago. In 1996, QuikChange was published [49], a highly influential technique that has inspired the development of many focused mutagenesis methods and commercial kits [2]. Huimin Zhao and Frances Arnold reported StEP in 1998 [66], an extremely convenient approach to recombine point mutations. Stemmer and Arnold were awarded the Draper Prize 2011 for their contribution to directed evolution. ITCHY was reported in 1999 to create combinatorial fusion libraries between genes in a manner that is independent of DNA homology [67]. In 2000, the Mutazyme kit was commercialized, and it remains a popular kit to complement epPCR [75]. Two years later, David Liu and coworkers published the NRR that enabled DNA fragments to be randomly recombined in a length-controlled manner without the need for sequence homology [76]. In 2004, we reported SeSaM method to enrich the occurrence of transversion mutations and consecutive nucleotide substitutions in a random mutagenesis experiment [47]. TriNEx was published 4 years later for trinucleotide exchange using an *in vitro* transposition approach [44]. In 2015, Agilent released its QuikChange HT Protein Engineering System.

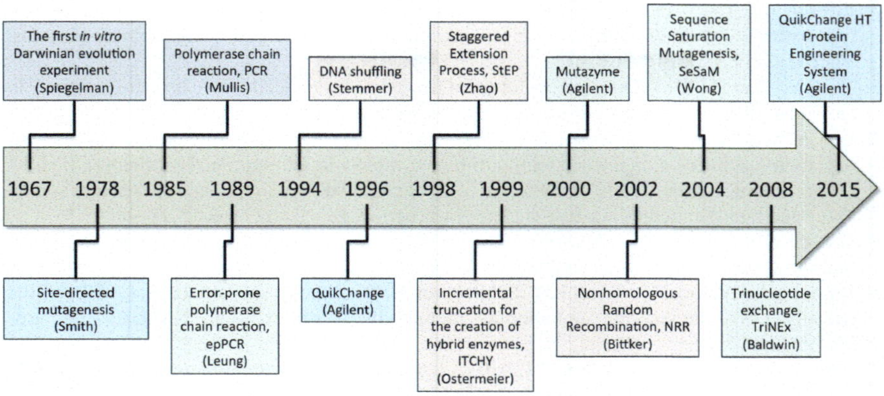

Fig. 8.3 Key milestones in the development of genetic diversity creation methods. Methods for random mutagenesis are colored in *light green*, focused mutagenesis in *light blue*, and DNA recombination in *light orange*

The timeline in Fig. 8.3 and the survey we have conducted allow us to spot a few trends: (1) The pace of new method development will continue to be dictated by the capability of oligonucleotide synthesis (length, accuracy, trimer phosphoramidite), the discovery of new enzymes for molecular biology, and the creativity of researchers. (2) Genetic diversity creation methods will constantly be modified to be compatible with more efficient cloning methods (e.g., Gibson Assembly, Gateway cloning, In-Fusion). (3) Methods for single-gene mutagenesis will be replaced by those for mutating multiple genes or DNA elements (e.g., promoter, RBS). (4) *E. coli*-based methods will continue to dominate, but yeast-based methods will emerge to be equally powerful. (5) High-throughput mutagenesis methods (e.g., QuikChange HT Protein Engineering System) will eventually catch up in terms of its usage frequency. (6) We envisage to see more automation in the preparation of a gene library.

Conclusion

Genetic diversity creation is a rigorous field of research. It has never been stagnant, and we are flooded with a constant stream of new developments, making it extremely difficult to keep up to date. This chapter has, hopefully, provided some convenience and useful guides for readers.

References

1. Wong TS, Zhurina D, Schwaneberg U (2006) The diversity challenge in directed protein evolution. Comb Chem High Throughput Screen 9:271–288
2. Tee KL, Wong TS (2013) Polishing the craft of genetic diversity creation in directed evolution. Biotechnol Adv 31:1707–1721

3. Eriksen DT, Hsieh PC, Lynn P, Zhao H (2013) Directed evolution of a cellobiose utilization pathway in *Saccharomyces cerevisiae* by simultaneously engineering multiple proteins. Microb Cell Factories 12:61
4. Yuan Y, Zhao H (2013) Directed evolution of a highly efficient cellobiose utilizing pathway in an industrial *Saccharomyces cerevisiae* strain. Biotechnol Bioeng 110:2874–2881
5. Gao S, Li Y, Zhang J, Chen H, Ren D, Zhang L, An Y (2014) A modified version of the digestion-ligation cloning method for more efficient molecular cloning. Anal Biochem 453:55–57
6. Gao S, Zhang J, Miao T, Ma D, Su Y, An Y, Zhang Q (2015) A simple and convenient sticky/blunt-end ligation method for fusion gene construction. Biochem Genet 53:42–48
7. Hansen NB, Lubeck M, Lubeck PS (2014) Advancing USER cloning into simpleUSER and nicking cloning. J Microbiol Methods 96:42–49
8. Bitinaite J, Rubino M, Varma KH, Schildkraut I, Vaisvila R, Vaiskunaite R (2007) USER friendly DNA engineering and cloning method by uracil excision. Nucleic Acids Res 35:1992–2002
9. Yang J, Zhang Z, Zhang XA, Luo Q (2010) A ligation-independent cloning method using nicking DNA endonuclease. BioTechniques 49:817–821
10. Sun S, Huang H, Qi YB, Qiu M, Dai ZM (2015) Complementary annealing mediated by exonuclease: a method for seamless cloning and conditioning site-directed mutagenesis. Biotechnol Biotechnol Equip 29:105–110
11. Gibson DG, Young L, Chuang RY, Venter JC, Hutchison CA 3rd, Smith HO (2009) Enzymatic assembly of DNA molecules up to several hundred kilobases. Nat Methods 6:343–345
12. Kahl LJ, Endy D (2013) A survey of enabling technologies in synthetic biology. J Biol Eng 7:13
13. Fu C, Donovan WP, Shikapwashya-Hasser O, Ye X, Cole RH (2014) Hot Fusion: an efficient method to clone multiple DNA fragments as well as inverted repeats without ligase. PLoS ONE 9:e115318
14. Taniguchi N, Nakayama S, Kawakami T, Murakami H (2013) Patch cloning method for multiple site-directed and saturation mutagenesis. BMC Biotechnol 13:91
15. Lund AM, Kildegaard HF, Petersen MB, Rank J, Hansen BG, Andersen MR, Mortensen UH (2014) A versatile system for USER cloning-based assembly of expression vectors for mammalian cell engineering. PLoS ONE 9:e96693
16. Jensen NB, Strucko T, Kildegaard KR, David F, Maury J, Mortensen UH, Forster J, Nielsen J, Borodina I (2014) EasyClone: method for iterative chromosomal integration of multiple genes in *Saccharomyces cerevisiae*. FEMS Yeast Res 14:238–248
17. Zhang Z, Yang J, Barford D (2015) Recombinant expression and reconstitution of multiprotein complexes by the USER cloning method in the insect cell-baculovirus expression system. Methods 95:13–25
18. Storch M, Casini A, Mackrow B, Fleming T, Trewhitt H, Ellis T, Baldwin GS (2015) BASIC: a new biopart assembly standard for idempotent cloning provides accurate, single-tier DNA assembly for synthetic biology. ACS Synth Biol 4:781–787
19. Beyer HM, Gonschorek P, Samodelov SL, Meier M, Weber W, Zurbriggen MD (2015) AQUA cloning: a versatile and simple enzyme-free cloning approach. PLoS ONE 10:e0137652
20. Jacobus AP, Gross J (2015) Optimal cloning of PCR fragments by homologous recombination in *Escherichia coli*. PLoS ONE 10:e0119221
21. Wang Y, Liu Y, Chen J, Tang MJ, Zhang SL, Wei LN, Li CH, Wei DB (2015) Restriction-ligation-free (RLF) cloning: a high-throughput cloning method by in vivo homologous recombination of PCR products. Genet Mol Res: GMR 14:12306–12315
22. Zhang Y, Werling U, Edelmann W (2012) SLiCE: a novel bacterial cell extract-based DNA cloning method. Nucleic Acids Res 40:e55
23. Okegawa Y, Motohashi K (2015) Evaluation of seamless ligation cloning extract preparation methods from an *Escherichia coli* laboratory strain. Anal Biochem 486:51–53
24. Okegawa Y, Motohashi K (2015) A simple and ultra-low cost homemade seamless ligation cloning extract (SLiCE) as an alternative to a commercially available seamless DNA cloning kit. Biochem Biophys Rep 4:148–151

25. Li MV, Shukla D, Rhodes BH, Lall A, Shu J, Moriarity BS, Largaespada DA (2014) HomeRun Vector Assembly System: a flexible and standardized cloning system for assembly of multi-modular DNA constructs. PLoS ONE 9:e100948
26. Ma H, Kunes S, Schatz PJ, Botstein D (1987) Plasmid construction by homologous recombination in yeast. Gene 58:201–216
27. Joska TM, Mashruwala A, Boyd JM, Belden WJ (2014) A universal cloning method based on yeast homologous recombination that is simple, efficient, and versatile. J Microbiol Methods 100:46–51
28. Holmberg MA, Gowda NKC, Andreasson C (2014) A versatile bacterial expression vector designed for single-step cloning of multiple DNA fragments using homologous recombination. Protein Expr Purif 98:38–45
29. Gibson DG, Benders GA, Andrews-Pfannkoch C, Denisova EA, Baden-Tillson H, Zaveri J, Stockwell TB, Brownley A, Thomas DW, Algire MA et al (2008) Complete chemical synthesis, assembly, and cloning of a Mycoplasma genitalium genome. Science 319:1215–1220
30. Kilaru S, Steinberg G (2015) Yeast recombination-based cloning as an efficient way of constructing vectors for Zymoseptoria tritici. Fungal Genet Biol: FG & B 79:76–83
31. Cao P, Wang L, Zhou G, Wang Y, Chen Y (2014) Rapid assembly of multiple DNA fragments through direct transformation of PCR products into *E. coli* and Lactobacillus. Plasmid 76C:40–46
32. Jajesniak P, Wong TS (2015) QuickStep-Cloning: a sequence-independent, ligation-free method for rapid construction of recombinant plasmids. J Biol Eng 9
33. Mathieu J, Alvarez E, Alvarez PJ (2014) Recombination-assisted megaprimer (RAM) cloning. MethodsX 1:23–29
34. van den Ent F, Lowe J (2006) RF cloning: a restriction-free method for inserting target genes into plasmids. J Biochem Biophys Methods 67:67–74
35. de Kok S, Stanton LH, Slaby T, Durot M, Holmes VF, Patel KG, Platt D, Shapland EB, Serber Z, Dean J et al (2014) Rapid and reliable DNA assembly via ligase cycling reaction. ACS Synth Biol 3:97–106
36. Yuan Y, Andersen E, Zhao H (2016) Flexible and versatile strategy for the construction of large biochemical pathways. ACS Synth Biol 5:46–52
37. Krishnamurthy VV, Khamo JS, Cho E, Schornak C, Zhang K (2015) Multiplex gene removal by two-step polymerase chain reactions. Anal Biochem 481:7–9
38. Botstein D, Shortle D (1985) Strategies and applications of in vitro mutagenesis. Science 229:1193–1201
39. Leung DW, Chen EY, Goeddel DV (1989) Techniques 1:11–15
40. Holland EG, Buhr DL, Acca FE, Alderman D, Bovat K, Busygina V, Kay BK, Weiner MP, Kiss MM (2013) AXM mutagenesis: an efficient means for the production of libraries for directed evolution of proteins. J Immunol Methods 394:55–61
41. Kunkel TA (1985) Rapid and efficient site-specific mutagenesis without phenotypic selection. Proc Natl Acad Sci U S A 82:488–492
42. Holland EG, Acca FE, Belanger KM, Bylo ME, Kay BK, Weiner MP, Kiss MM (2015) In vivo elimination of parental clones in general and site-directed mutagenesis. J Immunol Methods 417:67–75
43. Jakob F, Lehmann C, Martinez R, Schwaneberg U (2013) Increasing protein production by directed vector backbone evolution. AMB Express 3:39
44. Baldwin AJ, Busse K, Simm AM, Jones DD (2008) Expanded molecular diversity generation during directed evolution by trinucleotide exchange (TriNEx). Nucleic Acids Res 36:e77
45. Arpino JA, Baldwin AJ, McGarrity AR, Tippmann EM, Jones DD (2015) In-frame amber stop codon replacement mutagenesis for the directed evolution of proteins containing non-canonical amino acids: identification of residues open to bio-orthogonal modification. PLoS ONE 10:e0127504
46. Zhao J, Kardashliev T, Joelle Ruff A, Bocola M, Schwaneberg U (2014) Lessons from diversity of directed evolution experiments by an analysis of 3000 mutations. Biotechnol Bioeng 111:2380–2389

47. Wong TS, Tee KL, Hauer B, Schwaneberg U (2004) Sequence saturation mutagenesis (SeSaM): a novel method for directed evolution. Nucleic Acids Res 32:e26
48. Hutchison CA 3rd, Phillips S, Edgell MH, Gillam S, Jahnke P, Smith M (1978) Mutagenesis at a specific position in a DNA sequence. J Biol Chem 253:6551–6560
49. Papworth C, Bauer JC, Braman J, Wright DA (1996) Site-directed mutagenesis in one day with >80% efficiency. Strategies 9:3–4
50. Liu H, Ye R, Wang YY (2015) Highly efficient one-step PCR-based mutagenesis technique for large plasmids using high-fidelity DNA polymerase. Genet Mol Res: GMR 14:3466–3473
51. Xia Y, Chu W, Qi Q, Xun L (2015) New insights into the QuikChange process guide the use of Phusion DNA polymerase for site-directed mutagenesis. Nucleic Acids Res 43:e12
52. Wang C, Wang TY, Zhang LY, Gao XJ, Wang XW, Jin CJ (2015) Cut-and-paste-based cloning strategy for large gene site-directed mutagenesis. Genet Mol Res: GMR 14:5585–5591
53. Zhang Z, Xu K, Xin Y, Zhang Z (2015) An efficient method for multiple site-directed mutagenesis using type IIs restriction enzymes. Anal Biochem 476:26–28
54. Waneskog M, Bjerling P (2014) Multi-fragment site-directed mutagenic overlap extension polymerase chain reaction as a competitive alternative to the enzymatic assembly method. Anal Biochem 444:32–37
55. Motohashi K (2015) A simple and efficient seamless DNA cloning method using SLiCE from *Escherichia coli* laboratory strains and its application to SLiP site-directed mutagenesis. BMC Biotechnol 15:47
56. Belsare KD, Ruff AJ, Martinez R, Shivange AV, Mundhada H, Holtmann D, Schrader J, Schwaneberg U (2014) P-LinK: a method for generating multicomponent cytochrome P450 fusions with variable linker length. BioTechniques 57:13–20
57. Trehan A, Kielbus M, Czapinski J, Stepulak A, Huhtaniemi I, Rivero-Muller A (2016) REPLACR-mutagenesis, a one-step method for site-directed mutagenesis by recombineering. Sci Rep 6:19121
58. Gonzalez-Perez D, Molina-Espeja P, Garcia-Ruiz E, Alcalde M (2014) Mutagenic Organized Recombination Process by Homologous IN vivo Grouping (MORPHING) for directed enzyme evolution. PLoS ONE 9:e90919
59. Warburton M, Omar Ali H, Liong WC, Othusitse AM, Abdullah Zubir AZ, Maddock S, Wong TS (2015) OneClick: a program for designing focused mutagenesis experiments. AIMS Bioeng 2:126–143
60. Currin A, Swainston N, Day PJ, Kell DB (2015) Synthetic biology for the directed evolution of protein biocatalysts: navigating sequence space intelligently. Chem Soc Rev 44:1172–1239
61. Kille S, Acevedo-Rocha CG, Parra LP, Zhang ZG, Opperman DJ, Reetz MT, Acevedo JP (2013) Reducing codon redundancy and screening effort of combinatorial protein libraries created by saturation mutagenesis. ACS Synth Biol 2:83–92
62. Nov Y, Segev D (2013) Optimal codon randomization via mathematical programming. J Theor Biol 335:147–152
63. Tang L, Gao H, Zhu X, Wang X, Zhou M, Jiang R (2012) Construction of "small-intelligent" focused mutagenesis libraries using well-designed combinatorial degenerate primers. BioTechniques 52:149–158
64. Acevedo-Rocha CG, Reetz MT, Nov Y (2015) Economical analysis of saturation mutagenesis experiments. Sci Rep 5:10654
65. Stemmer WP (1994) DNA shuffling by random fragmentation and reassembly: in vitro recombination for molecular evolution. Proc Natl Acad Sci U S A 91:10747–10751
66. Zhao H, Giver L, Shao Z, Affholter JA, Arnold FH (1998) Molecular evolution by staggered extension process (StEP) in vitro recombination. Nat Biotechnol 16:258–261
67. Ostermeier M, Shim JH, Benkovic SJ (1999) A combinatorial approach to hybrid enzymes independent of DNA homology. Nat Biotechnol 17:1205–1209
68. Sieber V, Martinez CA, Arnold FH (2001) Libraries of hybrid proteins from distantly related sequences. Nat Biotechnol 19:456–460

69. Xu S, Ju J, Misono H, Ohnishi K (2006) Directed evolution of extradiol dioxygenase by a novel in vivo DNA shuffling. Gene 368:126–137
70. Abecassis V, Pompon D, Truan G (2000) High efficiency family shuffling based on multi-step PCR and in vivo DNA recombination in yeast: statistical and functional analysis of a combinatorial library between human cytochrome P450 1A1 and 1A2. Nucleic Acids Res 28:E88
71. Lehtonen SI, Taskinen B, Ojala E, Kukkurainen S, Rahikainen R, Riihimaki TA, Laitinen OH, Kulomaa MS, Hytonen VP (2015) Efficient preparation of shuffled DNA libraries through recombination (Gateway) cloning. Protein Eng Des Select: PEDS 28:23–28
72. Mills DR, Peterson RL, Spiegelman S (1967) An extracellular Darwinian experiment with a self-duplicating nucleic acid molecule. Proc Natl Acad Sci U S A 58:217–224
73. Joyce GF (2007) Forty years of in vitro evolution. Angew Chem 46:6420–6436
74. Saiki RK, Scharf S, Faloona F, Mullis KB, Horn GT, Erlich HA, Arnheim N (1985) Enzymatic amplification of beta-globin genomic sequences and restriction site analysis for diagnosis of sickle cell anemia. Science 230:1350–1354
75. Cline J, Hogrefe H (2000) Randomize gene sequences with new PCR mutagenesis kit. Strategies 13:157–162
76. Bittker JA, Le BV, Liu DR (2002) Nucleic acid evolution and minimization by nonhomologous random recombination. Nat Biotechnol 20:1024–1029

Resurrected Ancestral Proteins as Scaffolds for Protein Engineering

9

Valeria A. Risso and Jose M. Sanchez-Ruiz

Abstract

High stability and enhanced promiscuity (likely linked to conformational flexibility/diversity) contribute to evolvability and are advantageous features in protein scaffolds for laboratory-directed evolution and molecular design. Furthermore, the two features are not necessarily incompatible, and proteins may simultaneously be promiscuous/flexible and highly stable. In fact, it appears plausible that the combination of the two features was not uncommon among the most ancient proteins because (i) ancient life was likely thermophilic and (ii) ancient proteins were likely promiscuous generalists with broad functionalities. Phylogenetic analyses allow the reconstruction of ancestral sequences and provide an approach to explore the properties of ancient proteins. High stability and promiscuity have been often found for proteins encoded by reconstructed ancestral sequences, i.e., for "resurrected" ancestral proteins. The combination of the two features, i.e., the ancestral hyperstable generalist phenotype, has actually been obtained in recent studies. Ancestral protein resurrection thus emerges as a useful source of scaffolds for protein engineering.

V.A. Risso • J.M. Sanchez-Ruiz (✉)
Facultad de Ciencias, Departamento de Química Física, Universidad de Granada, 18071 Granada, Spain

M. Alcalde (ed.), *Directed Enzyme Evolution: Advances and Applications*, DOI 10.1007/978-3-319-50413-1_9

9.1 A Very Brief Introduction to Ancestral Protein Resurrection

The possibility of deriving plausible approximations to ancient protein sequences from the known sequences of modern proteins was originally proposed by Pauling and Zuckerkandl in the 1960s [106]. It remained, however, a theoretical idea until the 1990s, when (i) advances in bioinformatics and the increasing availability of protein sequences facilitated the reliable reconstruction of ancestral sequences and (ii) advances in molecular biology methodologies allowed the preparation in the laboratory (i.e., the "resurrection") of the corresponding encoded proteins.

Derivation of ancestral sequences involves a phylogenetic analysis based on an alignment of modern protein sequences and maximum likelihood or Bayesian estimation of the sequences at the nodes of the phylogenetic tree. The reader is referred to recent excellent reviews for detailed accounts of the procedures involved [13, 27, 91, 94, 127]. Nevertheless, an intuitive feeling about the process can be gained by drawing an analogy with the reconstructions of extinct languages performed by historical linguists since the late eighteenth century (Fig. 9.1). Knowledge of a given word in several modern related languages can be used, together with plausible models of word evolution, to derive a plausible reconstruction of the corresponding word in the common ancestor languages [6, 21]. Extinct languages (known as protolanguages) have been thus reconstructed, including Proto-Germanic and Proto-Indo-European (the common ancestor of Indo-European languages). Likewise, a protein sequence can be viewed as a word written using an alphabet of 20 words, and, therefore, the sequences ("words") for a set of homologous proteins, together with a model of protein evolution, can be used to reconstruct the sequence of the common ancestor protein (i.e., the "ancestral word").

9.2 Experimental Validation of Ancestral Protein Resurrection: Phenotypic Robustness and Evolutionary Narratives

Ancestral sequence reconstruction is an unavoidably uncertain task due to a number of reasons: the evolutionary models used are necessarily simple (for instance, time-invariant amino acid substitution matrices are typically used and site coevolution is typically neglected), the set of modern sequences used as input for the analysis might not be a fair representation of the extant (modern) sequence diversity, horizontal gene transfer processes may complicate phylogenetic analysis, gene duplication events may add uncertainty by preventing a direct comparison of the tree derived from the sequence alignment with known organisms phylogenies, etc.

It appears reasonable, therefore, to ask how reconstructed ancestral sequences and the corresponding "resurrected" ancestral proteins (i.e., the proteins encoded by the reconstructed sequences) can be validated. In this context, two important points must be noted: (1) bioinformatics procedures used for ancestral sequence reconstruction do not provide just a single reconstructed sequence for each phylogenetic node. Rather, they return for each position a vector with the probabilities of being

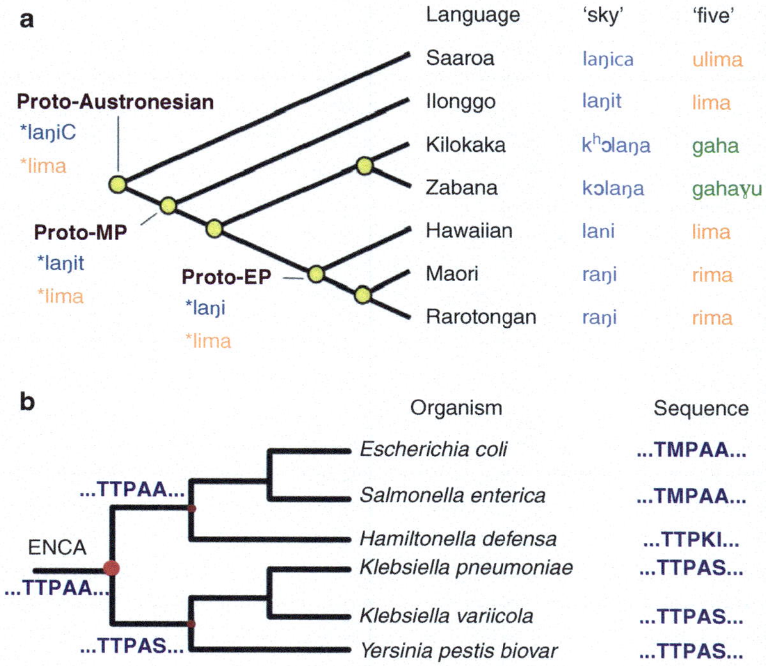

Fig. 9.1 Reconstructing words from extinct languages versus reconstructing sequences of proteins from extinct organisms. (**a**) Language tree for Austronesian languages. The words for "sky" and "five" in modern Austronesian languages were used [21] to reconstruct the corresponding words in the extinct languages Proto-Austronesian, Proto-Malayo-Polynesian (Proto-MP), and Proto-Eastern-Polynesian (Proto-EP) (Reprinted, with permission, from Atkinson [6]). (**b**) Section of a tree derived from the phylogenetic analysis of an alignment of sequences of extant (i.e., modern) class A β-lactamases. The complete tree is shown in Fig. 9.2, while only the section corresponding to enterobacteria is shown here. Sequences of lactamases from modern organisms were used [109] to reconstruct the sequences of the lactamases at the internal nodes. These internal (ancestral) nodes correspond to extinct organisms (ENCA stands for the common ancestor of enterobacteria). For illustration, only the sequence for residues 181–185 is shown (the complete sequences of 262 residues were reconstructed at all the internal nodes). Note that ancestral sequence reconstruction is a broad-scale phylogenetic procedure that simultaneously and consistently reconstructs the sequences of all the nodes in a phylogenetic tree from the information contained in all the extant sequences. Therefore, the reconstructed sequence at the ENCA node is "affected" by the sequences at all the nodes in the tree (not only by the sequences "below" the node). Note also that reconstructed ancestral sequences may substantially differ from consensus sequences [110]. The residue at position 185 is reconstructed as A in ENCA lactamase, while the consensus of the six modern lactamase sequences below ENCA is S at position 185

the ancestral for the 20 amino acids. While for each node a "most probabilistic" sequence can thus be defined as the sequence with the most probable amino acid at each position, alternative representations can be easily constructed by, for instance, Monte Carlo sampling of the amino acid probabilities. In a certain sense, ancestral sequence reconstruction procedures lead to an ensemble of statistically plausible sequences at each phylogenetic node. (2) Obviously, the claim in the field is not that the existed sequences of the proteins that existed millions or billions years ago can be recovered but, rather, that the properties of the proteins encoded by the reconstructed sequences (the laboratory resurrected proteins) may provide a useful approximation to the ancestral protein properties. In short, the claim in the field is that the ancestral protein phenotype (the protein properties) may be to some extent recovered. Therefore, as elaborated below in some detail, validation is performed at the protein phenotype level.

The outcomes of ancestral sequence reconstruction are routinely tested for phenotypic robustness. Briefly, several sequence representations of the protein at a given node are "resurrected" in the laboratory and studied in terms of relevant properties (stability, catalysis, etc.). The most probabilistic sequence and several alternative sequences are typically studied. The desired result is, of course, that the same or similar properties are obtained for different members of the ancestral sequence ensemble. At a more general level, the properties of the resurrected proteins for different phylogenetic nodes must "tell" a convincing evolutionary story or, to use the common jargon in the field, a convincing evolutionary narrative. A specific example will suffice to illustrate the point. In one of the first experimental ancestral resurrection studies [69], Benner and coworkers addressed the laboratory resurrection of ancestral ribonucleases for the artiodactyl lineage (the order of mammals to which sheep, deer, camel, pig, and ox belong). They found the resurrected proteins to display the substrate scope expected for digestive ribonucleases, but only up to the node corresponding to the common ancestors of ruminants. "Older" ribonucleases, in fact, showed enhanced activity toward non-digestive substrates. The ancestral resurrection exercise of Jermann et al. [69] thus supported that digestive ribonucleases originated at about 40 million years when a non-digestive ribonuclease was recruited for ruminant digestion, a scenario fully consistent with the lowering of temperatures at the end of the Eocene, the concomitant widespread emergence of grasses as a source of food, and the appearance of ruminant herbivores. This correlation between the molecular and paleontological records provides clear support for the ancestral ribonuclease resurrection exercise at the phenotype level [13, 69].

In view of the above, one ideal scenario for ancestral sequence reconstruction would involve that the following results hold: (1) the tree derived from the analysis of the alignment of modern sequences is reasonably close to accepted organismal phylogenies, and, in case that paralogous proteins are included, the tree is also consistent with reasonable hypotheses regarding the corresponding gene duplication events; (2) phenotypic robustness is demonstrated by the congruence of the biophysical properties of the proteins encoded by several alternative resurrections at key nodes (obtained from random sampling of the posterior probability, alternative tree topologies below the node, etc.); (3) the biophysical properties determined for the resurrected proteins provide convincing evolutionary narratives that correlate the paleontological and molecular records of life and reveal relevant evolutionary adaptations.

A recently reported [109] laboratory resurrection of ancestral forms of the antibiotic resistance enzyme β-lactamase approaches to some extent the ideal scenario described in the preceding paragraph. A set of sequences providing a uniform coverage of the phyla in Bacteria was used as starting point of the ancestral sequence reconstruction exercise (Fig. 9.2). Sequences from clinical isolates were excluded

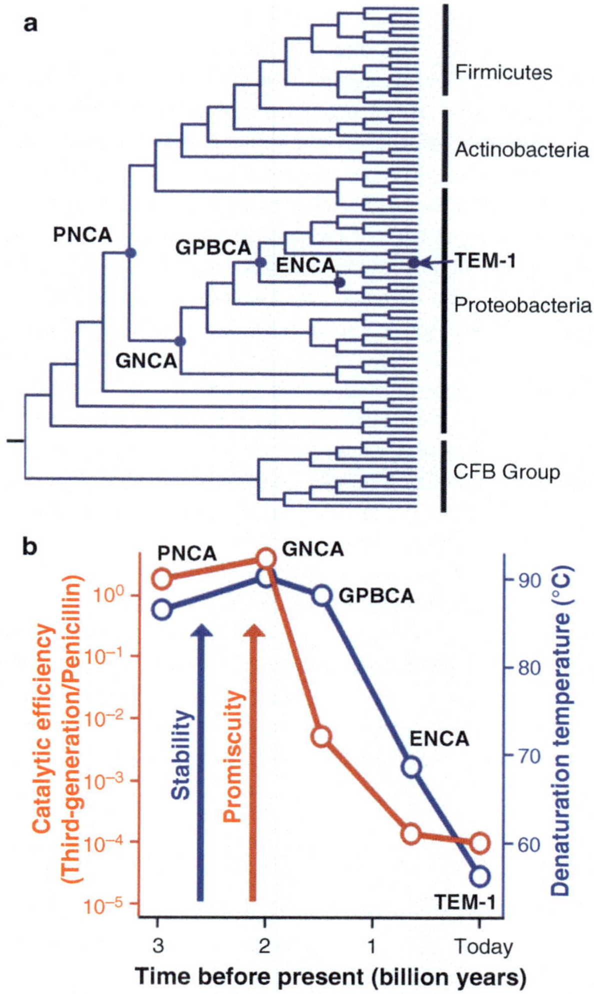

Fig. 9.2 Hyperstability and promiscuity in laboratory resurrections of Precambrian β-lactamases [109]. (**a**) Tree derived from the phylogenetic analysis of an alignment of 75 sequences of modern lactamases. The tree is close to an accepted phylogeny of the organisms and, therefore, sequence reconstruction targeted well-defined Precambrian nodes: ENCA (common ancestor of enterobacteria), GPBCA (common ancestor of gammaproteobacteria), GNCA (common ancestor of various Gram-negative bacteria), and PNCA (common ancestor of various Gram-positive and Gram-negative bacteria). (**b**) Biophysical properties of laboratory resurrection of Precambrian lactamases. Denaturation temperature and the ratio of catalytic efficiencies for the degradation of cefotaxime (a third-generation antibiotic) and penicillin are plotted against the age of the targeted Precambrian nodes. Note that the ancestral PNCA and GNCA lactamases display efficient catalysis of both antibiotics (k_{cat}/K_M around 10^5 M^{-1} s^{-1} for penicillin and cefotaxime), while the modern TEM-1 lactamase is a penicillin specialist

from the set to avoid interference in the reconstruction process from recent evolution during the antibiotic era. The inferred phylogenetic tree was reasonably close to an accepted organismal phylogeny, indicating limited interference from horizontal gene transfer. Phenotypic robustness was supported by stability and antibiotic degradation determinations on the proteins encoded by several alternative reconstructions of the lactamase corresponding to the last common ancestor of Gram-negative bacteria (an organism that inhabited Earth about 2 billion years ago). Finally, the biophysical properties of the resurrected ancestral lactamases provided convincing evolutionary narratives: (1) lactamase stability, as probed by the value of the denaturation temperature, increased by many degrees (30–35 °C) upon "traveling back in time" 2–3 billion years. Furthermore, such change in stability was found to correlate with estimates of ancestral ocean temperatures derived from isotopic composition of cherts, supporting lactamase adaptation to environmental temperatures over geological time scales. (2) Patterns of antibiotic degradation activities for the resurrected ancestral lactamases conformed to the expected evolutionary transition from promiscuous generalists to specialist enzymes, and the estimated time for the emergence of penicillin specialists roughly matched the divergence time of fungi (about 1.2 billion years before present).

Certainly, ancestral sequence reconstruction does not need to conform to "ideal scenarios" to be useful. This would be particularly true if the purpose of the ancestral reconstruction exercise is to obtain scaffolds for protein engineering. The goal in this case would be the preparation of proteins with useful properties from an engineering point of view (such as high stability or catalytic promiscuity), and, therefore, the degree of phenotypic correspondence of the laboratory resurrected proteins with the proteins that actually existed millions or billions of years ago would be a secondary concern. We will return to this issue in several sections of this chapter.

9.3 Ancestral Protein Resurrection Has Been Used in the Last ~20 Years to Address Important Issues in Molecular Evolution

Resurrected ancestral proteins have been used in the last 20 years or so mostly as "tools" to explore evolutionary hypotheses and processes that would have been difficult to address otherwise. Early work has been summarized in excellent reviews [13, 52, 53, 94]. Here we will only provide a list of recent applications to convey to the reader the "flavor" of the field. Ancestral protein resurrection has been recently used to study molecular adaptations to changing environmental conditions over planetary time scales [2, 42, 107, 109]; the origin of thermophily [57]; the evolution of complexity in biomolecular machines [36]; the evolution of divergent DNA specificities in paralogous transcription factors [60]; the evolutionary effects of epistatic interactions across molecular interfaces [4]; the molecular mechanisms of evolution of new functions [105]; the degree of conservation of protein structure over long geological time scales [64]; the origin of detoxifying enzymes [11]; the role of gene duplication in evolutionary innovation [130]; the characterization of the evolutionary events leading to gene silencing [85]; the determination of the time at which our ancestors were exposed to alcohol in the diet [23]; the degree of conservation of site-specific amino acid preferences over evolutionary history [111]; the evolution of protein conformational dynamics [145]; the historical trajectory, timing, and mechanisms of evolution of a new protein function important to organized multicellularity in diverse animal phyla [5]; the mechanism for the de novo emergence of a functional lectin β-propeller from short motifs [122]; the molecular origin and adaptive changes in the visual system of mammals [16]; and the adaptive evolution of binding specificity in solute-binding proteins [26].

9.4 Recent Work Suggests the Biotechnological Potential of Resurrected Ancestral Proteins

As described in the preceding section, resurrected ancestral proteins have been mostly used as tools in molecular evolution studies. It has emerged from very recent studies, however, that resurrected ancestral may often display the biophysical features (high stability, substrate and catalytic promiscuity linked to enhanced conformational flexibility/diversity) that are likely to contribute to protein evolvability and that are, therefore, advantageous in scaffolds for molecular design and laboratory-directed evolution. These useful ancestral properties should not come as a surprise. High stability in resurrected ancestral proteins is likely a consequence of the probable thermophilic character of ancient life. Likewise, promiscuity upon ancestral resurrection is to be expected if ancient proteins, unlike many modern proteins, were generalists with broad specificities. In the following sections, we expound on the evolutionary origin and usefulness in engineering of the typical ancestral phenotype.

9.5 High Stability Should Contribute to Protein Evolvability

Protein stability is often described as being marginal. The prevalent view in modern literature is that marginal protein stability is not adaptive. That is, high stability does not necessarily impair function (by, for instance, making the protein more "rigid"). In fact, the simplest and widely proposed explanation of marginal protein stability involves two straightforward assumptions: (1) evolution of protein stability can be described in terms of an evolutionary stability threshold, in the sense that mutations that bring stability below the threshold compromise organism survival are rejected (purifying selection), while mutations that keep stability above the threshold are essentially neutral. (2) Most mutations in a protein are destabilizing, and, consequently, available protein sequences become scarcer as stability is increased. The combined effect of these two factors should make protein stability to fluctuate slightly above the threshold during evolution [15, 18, 47, 120, 126].

In a protein engineering scenario, marginal stability obviously limits protein evolvability because many functionally useful mutations are destabilizing (since most mutations are destabilizing anyway) and would compromise proper folding. On the other hand, a substantial number of those mutations would become available to molecular design or laboratory-directed evolution if the background scaffold has been stabilized [17, 77, 90, 95].

9.6 Why We Should Expect to Obtain Proteins with Enhanced Stability upon Ancestral Resurrection Targeting Very Ancient Phylogenetic Nodes

The reason is simply that ancient life was likely thermophilic and proteins from thermophilic organisms are expected to display enhanced stability. The several proposed scenarios and experimental analyses that are consistent with ancient life being thermophilic are summarized below:

1. Evidence from ribosomal RNA sequences suggested the thermophilic character of the earliest branches of the tree of life [135].
2. Estimates of ancestral ocean temperatures derived from the isotopic composition of cherts suggest that late Hadean/Archaean ancestral oceans that hosted life were hot [79–81, 112].
3. High levels of CO_2 may have contributed to make the Archean climate much warmer than today [72]. See, however, point 4 below.
4. While some analyses [123] disfavor high levels of traditional greenhouse gasses in the primitive Earth, recent work [73, 140] has demonstrated that nitrogen and hydrogen may have served as greenhouse gasses under particular conditions and would have helped to maintain high Archaean temperatures.
5. Primordial life may have flourished in hydrothermal vents on the ocean floor [86, 93].
6. Perhaps, only the "tougher," thermophilic organisms could survive catastrophic impact events in the young planet ("impact bottleneck" scenarios) [100, 121].
7. Experimental studies on the temperature dependence of the rates of nonenzymatic reactions suggest that a massive thermal acceleration of the emergence of primordial chemistry was required to set the stage for enzyme evolution to get started [125, 138, 139].

9.7 Ancestral Protein Resurrection Targeting Precambrian Phylogenetic Nodes Often Leads in Fact to Highly Stable Proteins

If ancient life was thermophilic, as suggested by the evidence summarized in the preceding section, ancient proteins should have been highly stable. Consequently, ancestral sequence reconstruction targeting "old" Precambrian nodes (in particular "very old" Archaean nodes) could be expected to lead to proteins with substantially enhanced stability as compared with their modern counterparts. Indeed, several recent ancestral resurrection exercises are consistent with this expectation. Denaturation temperatures for laboratory resurrections of Precambrian elongation factors [42], thioredoxins [107], β-lactamases [109], and nucleoside diphosphate kinases [2] are up to about 30–35° higher than their modern homologs. Furthermore, for these four different families, plots of denaturation temperature versus age of the targeted phylogenetic nodes define a trend that mirrors the cooling of the oceans as inferred from the isotopic composition of cherts [80, 112]. These results are clearly consistent with the proposal that Archaean temperatures were higher than present average temperatures, that oceans cooled over geological time scales, and that protein stability, at least for the four systems mentioned above, followed the cooling trend. The overall congruence is, in fact, remarkable, given that the four protein systems included display different sizes, functions, and structures.

It must be recognized, of course, that some ancestral resurrection studies targeting Precambrian nodes have failed to achieve large increments in denaturation temperature [54]. This is not too surprising, however, because, as we have elaborated in detail in Sect. 9.2, ancestral sequence reconstruction is an unavoidably uncertain task that should be convincingly validated at the phenotypic level (as, for instance, by the congruence of denaturation temperatures with estimated ancestral ocean temperatures). Also, in some cases, denaturation temperatures for resurrected enzymes may correspond to local environments that do not always reflect global environmental temperatures. Furthermore, it is plausible that natural selection operates in many cases on the basis of protein kinetic stability [116], and enhanced kinetic stability may not necessarily translate into increased values of the equilibrium denaturation temperature or other metrics of thermodynamic stability (for a recent example, see [101]). In view of all this, it would perhaps be unrealistic to expect that *all* ancestral resurrection efforts lead to proteins with very high denaturation temperatures. On the other hand, if ancient life was thermophilic, we should expect a statistically significant trend toward substantially stabilized proteins when targeting Precambrian nodes in ancestral sequence reconstruction. This is in fact supported by the several experimental studies summarized in the preceding paragraph.

9.8 Is the Enhanced Stability of Resurrected Precambrian Proteins an Artifact of Ancestral Sequence Reconstruction?

About 10 years ago, Goldstein and coworkers reported an insightful computational analysis into the accuracy of ancestral reconstruction methods [132]. They performed population evolution simulations using an off-lattice protein model with a fitness function related to protein stability. Their results supported that ancestral reconstruction methodologies (maximum parsimony and maximum likelihood, in particular) that reconstruct the "best guess" amino acid at each position may substantially overestimate ancestral protein stability (the problem appeared to be less acute with Bayesian methodologies). The work of Goldstein and coworkers [132] has been sometimes used to argue that the stability enhancements often observed for resurrected proteins are artifacts inherent to ancestral sequence reconstruction procedures. This interpretation, however, hardly applies to the large stabilizations observed upon Precambrian protein resurrection [2, 42, 107, 109], as the corresponding increments in denaturation temperature are very large. They are, in fact, much larger than (1) the increments in denaturation temperature typically obtained in protein engineering studies aimed at preparing stabilized variants and (2) the increments in denaturation temperature predicted by energetic biases reported by Williams et al. [132]. We elaborate these two points below in some detail:

1. Figure 9.1 in the review paper of Wijma et al. [133] shows a statistics of the denaturation temperature increments obtained through protein engineering of enzymes. Original data are taken from literature reports for the 2007–2012 period. It is clear from the inspection of the figure that T_m enhancements by mutagenesis are typically in the 2–15° range. Only two specific studies, with $\Delta T_m = 35°$ and $\Delta T_m = 45°$, report increments comparable to those obtained by Precambrian protein resurrection. Actually, the $\Delta T_m = 35°$ data point in Fig. 9.1c corresponds to resurrected Precambrian thioredoxins [107]. The $\Delta T_m = 45°$ increment was reported [31] for an engineered variant of terpene synthase that originally displayed a very low denaturation temperature, a fact that may have facilitated stabilization by mutation. Overall, the ΔT_m values often obtained through Precambrian protein resurrection are much larger than those typically achieved through engineering. It is difficult to envision how simple methodology biases could produce stability enhancements that are larger than those obtained on the basis of sophisticated computational approaches and/or efficient directed evolution.
2. In their computational analyses into the accuracy of ancestral reconstruction methods, Williams et al. [132] reported a bias of 1.5 kcal/mol in unfolding free energy for reconstructions based on maximum likelihood (biases for reconstructions based on maximum parsimony and, in particular, Bayesian inference were reported to be smaller). As a first approximation, changes in unfolding free energy ($\Delta\Delta G$) can be translated into changes in denaturation temperature by using the well-known Shellman equation [117], $\Delta T_m = \Delta\Delta G/\Delta S$, where ΔS is the unfolding entropy change. ΔS values can be estimated from the size of the protein from known structure-energetic correlations [113]. Specifically, for a temperature of 60 °C, the unfolding entropy change scales with protein size according to $\Delta S \approx 2.1 \cdot N_{RES}$ $cal \cdot K^{-1} \cdot mol^{-1}$, where N_{RES} is the number of amino acid residues. This gives ΔS values of 0.21, 0.42, and 0.63 kcal $K^{-1} \cdot mol^{-1}$ for proteins of 100, 200, and 300 residues, respectively. Application of Shellman equation with $\Delta\Delta G = 1.5$ kcal/mol then gives denaturation temperature increments of 7.1, 3.6, and 2.4 for 100, 200, and 300 residues. These values are consistent with the estimate (about 6°) given by Williams et al. [132] using the thermodynamic data for T4 lysozyme unfolding. While these are obviously "back-of-the-envelope" calculations, it is clear that the stability overestimation in ancestral reconstruction discussed by Williams et al. [132] is expected to lead to increments in denaturation temperature of a few degrees, substantially smaller

than the increments of 30–35° found in several Precambrian resurrection studies [2, 42, 107, 109].

9.9 Enzyme Promiscuity Is a Convenient Feature in Biotechnological Application Scenarios

Textbooks sometimes picture enzymes as being highly specific and enormously efficient catalysts. This description, while valid in general terms, is perhaps an excessive simplification. Certainly, enzymes can achieve rate enhancements of many orders of magnitude with respect to the non-catalyzed reactions [136, 137]. Still, the catalytic efficiency of most enzymes is substantially lower than the maximum attainable rate, the so-called diffusion limit [10]. Regarding specificity, many proteins can actually carry out several functions [7, 20, 28, 35, 74–76, 78, 102, 142]. Often, an enzyme shows one clear primary activity, together with some other low-level "secondary" activities involving substrates and/or chemical transformations similar to those involved in the primary activity. Furthermore, some enzymes can efficiently perform a variety of related functions (the detoxification enzyme glutathione transferase is a paradigmatic example: [144]) or even clearly differentiated functions linked to different active sites, a type of promiscuity often referred to as "moonlighting."

However, even the most common low-level, secondary promiscuous activities are interesting from a biotechnological point of view [62, 74, 102]. Enzymes are being increasingly employed in the industry (Chap. 19 in [48, 114]) since their use as technological catalysts often replaces traditional processes based on chemicals that are typically inefficient in terms of the use of energy and natural resources. Unfortunately, biotechnological applications of enzymes are, in principle, severely limited to their natural activities, while many technological applications require catalysis of nonnatural reactions. However, some "nonnatural" reactions can actually be subject to enzyme catalysis, albeit as promiscuous activities. Certainly, in most cases, these secondary catalysis levels are very low. Still they can be used as "seeds" or backgrounds for the preparation of catalytically efficient engineered variants through rational design or, more often, laboratory-directed evolution. In this way, enzymes can be engineered to catalyze chemical reactions that, while related to their primary activities, differ from them in terms of the substrate degraded, the product obtained, and the chemical transformation achieved. In short, promiscuity expands the scope of technological applications of enzymes.

9.10 Why We Should Expect Many Proteins Encoded by Reconstructed Ancestral Sequences to Be Promiscuous

Two lines of reasoning suggest that promiscuity should be a common outcome of ancestral protein resurrection: (1) primordial enzymes were likely promiscuous generalists, and such ancient promiscuity is likely to be recovered upon sequence reconstruction targeting very ancient phylogenetic nodes; (2) most models of gene duplication evolution predict that the ancestral single-copy gene had more functions than the evolved daughter copies, and, therefore, sequence reconstruction targeting pre-duplication nodes is expected to often yield promiscuous proteins. We elaborate on these two scenarios below:

1. A substantial number of modern enzymes are highly efficient specialists, capable to enhance the rate of a very specific chemical reaction by many orders of magnitude [136, 137]. Such high degree of specialization is clearly the product of natural selection operating over evolutionary time scales. There is, therefore, wide agreement [9, 30, 32, 41, 67, 68, 75, 76, 104, 145] that many ancient enzymes were promiscuous generalists with broad functionalities and that many proteins have evolved from generalists into specialist during evolution. The fundamental idea was most clearly expressed by Roy Jensen in a highly influential article [68]. He pointed out that primordial life was likely restricted to limited genetic information that encoded a small number of proteins. Therefore, primitive enzymes likely possessed broad functionalities, thus maximizing the catalytic versatility of the cell under limited enzyme resources. Subsequently, gene duplication (see below) provided the opportunity for the "luxury of increased specialization and the improved metabolic efficiency" [68]. According to this view, the low-level, secondary promiscuous activities detected for many modern enzymes would be a vestige of the broad functionality of their protein ancestors, and sequence reconstruction targeting very old nodes should often recover enzymes with enhanced promiscuity.
2. Gene duplication is the origin of most new genes and likely underlies the origin of most new functions. Not unexpectedly, many models of evolution of gene duplication, when applied to enzymes, suggest multifunctional (promiscuous) ancestors. As famously discussed by Ohno [103], duplication generates redundant gene copies, and if gene dosage is not crucial, one copy can evolve free from functional constraints, while the ancestral function is maintained in the other copy through purifying selection. The copy that is free to evolve could occasionally accept mutations that generate a new function. This simple neo-functionalization model faces, however, several problems. The most obvious one is that accumulation of neutral mutations is likely to render nonfunctional the gene copy that is freed from constraint before it can acquire a new function with adaptive value. That is, in the simple model, nonfunctionalization (pseudogenization) would be far more likely than neo-functionalization, and it would not be clear how the large numbers of functional paralogous genes found in living organisms could have originated. Several alternative models have been proposed to solve this conundrum. In the escape from adaptive conflict (EAC) model [29, 56], the single-copy gene is supposed to be selected to perform a new function while maintaining the ancestral function. While moderate increments in the new function might perhaps occur without seriously compromising the ancestral function [1, 41], large improvements may not be possible, however, because of the concomitant detrimental effects on the ancestral function. Gene duplication provides an escape from this conflict, as one copy is free to improve the new function while the other can improve the ancestral function. In this way, each

copy specializes in one of the functions, while the ancestral gene prior to duplication was bifunctional. In the duplication-degeneration-complementation (DDC) model [39], both genes resulting from duplication experience degenerating mutations that cause loss of function, i.e., sub-functionalization. In the quantitative sub-functionalization version of the model [50], there is one single function, and the duplicates are maintained because their lower function levels impose that the two copies are required to reach the level of the ancestral gene. In the qualitative sub-functionalization version of the DDC model [50, 61], on the other hand, the ancestral gene has two different functions that are independently mutable. Degeneration after duplication causes complementary loss of function in the duplicate genes. Each copy thus becomes specialized in one function through degeneration of the other function, while the pre-duplication ancestral gene had the two functions (i.e., the functions present in the ancestral gene are partitioned between the two descendant duplicates). By contrast with the EAC and DDC models, the innovation-amplification-divergence (IAD) model (or adaptive radiation model) [14, 40] proposes that the duplication step is not neutral. This model still assumes more than one function in the parent gene with, perhaps, one major original function and one (or several) minor side activities. At a certain point one of the side functions may become valuable (due to a change in ecological niche, for instance). Since duplication and amplification events are far more common than improvement by point mutations [14], the requirement for increased levels of the side activity will be more likely met in the first place by amplification of the original gene. Subsequently, however, the possibility of improvement by point mutation is increased because of the presence of several extra copies of the gene that act as mutational targets. Improvement in one copy brings about relaxation on the other copies, and these could then be removed from the population in such a way that the required level of the new function is eventually provided by one copy of the gene. Since optimization of the new function is likely detrimental for the original one, selection will maintain one copy with the original major function.

The three well-known models we have described above (EAC, DDC with qualitative sub-functionalization, and IAD) display one common feature: the ancestral pre-duplication gene has more functions than the evolved daughter copies. Certainly, we have not discussed all possible models of gene duplication evolution (for a more exhaustive catalog, see Table 1 in [65]). It is clear from the above discussion, however, that ancestral sequence reconstruction targeting pre-duplication nodes is likely to yield promiscuous proteins.

9.11 Ancestral Protein Resurrection Has Been Shown to Lead to Substrate Promiscuous Enzymes

In congruence with the proposal summarized in the preceding section, a number of ancestral reconstruction exercises have indeed led to substrate promiscuity in resurrected ancestral enzymes and/or provide evidence supporting the relevance of ancestral promiscuous activities for enzyme evolution. Some of these studies are briefly described below.

Verstrepen and coworkers [130] noted the contrast between the large number of available theoretical models of enzyme evolution upon gene duplication (see Sect. 9.10) and the scarcity of experimental evidence that could be used to decide between the different models. To alleviate this situation, they used ancestral sequence reconstruction to gain experimental insight into the functional properties of the ancestral proteins encoded by pre-duplication genes. As a model system they chose the *MALS* gene family, which encodes modern α-glycosidases with diversified substrate specificities (including a variety of maltose and isomaltose analogues). They indeed found the protein encoded by the sequence reconstructed for the first pre-duplication enzyme to be promiscuous, being primarily active on maltose sugars but also showing significant levels of activity for isomaltose substrates. Duplication events then produced enzymes in which subsequent mutations led to optimization of either maltose or isomaltose activity.

Barkman and coworkers [59] also addressed the fate of ancestral protein activities during evolution, but used as a model system the plant methyltransferases of the SABATH gene family. They used ancestral sequence reconstruction to explore functional divergence after gene duplication with respect to salicylic acid, benzoic acid, and nicotinic acid, the three substrates used by the modern enzymes for production of substances involved in pathogen/herbivore defense and floral scent. Their results provided evidence for evolution through improved catalysis of ancestrally non-preferred substrates and overall supported the relevance of ancestral promiscuous activities for enzyme evolution.

Risso et al. [109] reported a detailed biophysical and functional characterization of proteins encoded by reconstructed sequences corresponding to Precambrian nodes in the evolution of the antibiotic resistance enzyme β-lactamase (Fig. 9.2). While a modern TEM-1 β-lactamase is a penicillin specialist and shows rather low activity levels toward, for instance, third-generation antibiotics, resurrected ancestral β-lactamases for 2–3 billion years nodes were found to be moderately efficient promiscuous generalists, able to degrade both penicillin and third-generation antibiotics. This result does not imply, of course, the existence of third-generation lactam antibiotics (a human invention) in the Precambrian. It rather suggests, however, that ancestral lactamases had to hydrolyze a variety of substances, perhaps because ancient bacteria produced a diversity of antibiotics as a mechanism to obtain nutrients by killing competitors [51] and lactamases arose as a defense device or perhaps because the older Precambrian lactamases were at Jensen's generalist stage [68].

Sterner, Merki, and coworkers [108] have recently used ancestral protein resurrection to study the history of bifunctionality of a sugar isomerase form histidine biosynthesis (see, also, third paragraph in Sect. 9.13). They found evidence that the substrate promiscuity of the ancient HisA enzymes had persisted over 2 billion years at least, apparently without evolutionary pressure. This is a very interesting result because, in our view, it suggests that, for many protein systems, vestiges of Jensen's generalist stage [68] may have "survived" over very long evolutionary periods.

The issue of the persistence of ancestral promiscuity is also germane to the interpretation of a very recent study into the evolution of polar amino acid-binding proteins (AABPs). Clifton and Jackson [26] have characterized proteins encoded by reconstructed sequences corresponding to four nodes at which AABP subfamilies diverged. One of the resurrected proteins appeared to be an inefficient specialist that was only capable of low-affinity binding to one specific amino acid. The authors did not rule out the possibility that this result was due to errors in the sequence reconstruction process. The other three resurrected ancestral AABPs displayed a promiscuous capability to bind several amino acids. Comparison with the binding scope of the modern AABPs revealed two scenarios. In one case, the evolutionary conversion of the ancestral promiscuous protein into a more specialized binder was apparent. In the two other cases, the modern AABP proteins displayed a binding scope similar to that of their resurrected ancestors. In view of these results, the authors [26] concluded that "the evidence that the ancient progenitors of modern proteins had a larger range of physiologically relevant functions in comparison with their descendants remains limited." However, it is plausible (see preceding section) that many ancient proteins were generalists, and this is, in fact, supported by the outcomes of the several ancestral resurrection studies summarized in this section, including the work of Clifton and Jackson [26] on ancestral AABPs. Furthermore, since many modern proteins are specialists, it is reasonable to expect that the generalist to specialist conversion has occurred often during evolution. However, there is no reason to believe that such conversion has occurred in all instances. It is of course conceivable that ancestral promiscuity has been retained (or modulated) in some cases, provided that such retention confers an adaptive advantage. Regardless of interpretation subtleties, however, it is relevant in the context of this chapter that a resurrected ancestral AABP has been used to make a robust genetically encoded sensor for arginine [134].

9.12 Ancestral Protein Resurrection Has Been Recently Shown to Lead to Catalytically Promiscuous Enzymes

The studies summarized in the preceding section provide examples of substrate promiscuity in resurrected ancestral enzymes. Substrate promiscuity in the catalysis of a given chemical reaction involves the capability of accepting substrates of different shape and size while maintaining the same basic structure for the transition state of the process. In contrast, catalytic promiscuity (that is, the catalysis of different reactions by a given enzyme) requires the capability to accept different transition states.

Kazlauskas and coworkers have recently addressed the catalytic promiscuity of resurrected ancestral enzymes [30] using hydroxynitrile lyases as a model system. These enzymes catalyze the elimination of hydrogen cyanide from cyanohydrins and evolved from esterases about 100 million years ago likely as a mechanism of defense against herbivorous insects. The authors thus resurrected ancestral enzymes corresponding to branch nodes in the divergence of hydroxynitrile lyases from esterases. While modern esterases and hydroxynitrile lyases show little catalytic promiscuity, almost all the resurrected ancestral proteins catalyzed both ester hydrolysis and cyanohydrin cleavage. This is a rather remarkable result because, although esterases and hydroxynitrile lyases share the α-/β-hydrolase fold and the catalytic serine-histidine-aspartate triad, the reaction mechanisms for the two catalyzed reactions show substantial differences [30]: (i) ester hydrolysis involves an acyl intermediate, while the elimination reaction catalyzed by hydroxynitrile lyases has no acyl intermediate; (ii) different leaving groups (apolar an polar, respectively) are generated in the hydrolysis and elimination reactions; (iii) there are different requirements regarding the binding of the carbonyl oxygen to the oxyanion hole. The astounding catalytic promiscuity of the resurrected ancestral enzymes was further highlighted by their capability to catalyze, albeit slowly, reactions of enzymes in the α-/β-hydrolase family that are outside the phylogenetic nodes targeted for reconstruction (decarboxylation, Michael addition, γ-lactam hydrolysis, and 1,5-diketone hydrolysis).

9.13 Conformational Flexibility/Diversity Provides a Likely Explanation for Modern and Ancestral Enzyme Promiscuity

Protein promiscuity involves their capability to interact with different binding targets, substrates, or transition states (in the case of catalytic promiscuity). Clearly, the protein needs to be able to "adapt" to a variety of molecules that may differ in size, shape, and interacting groups. A link between promiscuity and conformational flexibility is apparent [12, 67, 98, 145] and often interpreted in terms of induced-fit or conformational diversity models ([19, 131]; see also following section). A few illustrative examples of the relation between promiscuity and flexibility are provided below.

Ubiquitin is recognized by a diversity of proteins, and its flexibility is demonstrated by the conformational heterogeneity in structures of different complexes and by conformational ensemble derived from the analysis of residual dipolar couplings [87]. Likewise, flexibility has been shown to allow the regulatory protein calmodulin to bind and activate a large number of target enzymes [143].

Two related isomerization reactions in histidine and tryptophan biosynthesis are catalyzed in some actinobacteria by a single bifunctional protein, despite a difference of a factor of two between the sizes of the two substrates involved. Structural analysis shows the active site to be highly flexible, in such a way that two different and substrate-specific active-site environments become available [33].

Detoxification enzymes tend to be notoriously promiscuous, as they typically need to degrade a variety of toxic substances. Still, there may be differences in promiscuity profile for related detoxification enzymes. The A1-1 isoform of cytosolic human glutathione S-transferase is highly promiscuous and can degrade a variety of structurally unrelated toxins. By contrast, the A4-4 isoform displays a preference for lipid peroxidation products. X-ray crystallography, hydrogen/deuterium exchange, and fluorescence lifetime distribution analysis support that the local and global conformational flexibility/diversity is responsible for the much higher substrate promiscuity of the A1-1 isoform [58].

The examples summarized above describe modern proteins that display promiscuity linked to conformational flexibility/diversity. Recent computational studies [145] support that conformational flexibility/diversity is also responsible for the fact that resurrected Precambrian β-lactamases are substrate promiscuous [109], as we have already mentioned in a preceding section. Specifically, perturbation-response-scanning analyses based on replica-exchange MD simulations [145] supported that the modern TEM-1 β-lactamase has a rigid active site, plausibly reflecting an adaptation for efficient degradation of a specific antibiotic (penicillin), while enhanced conformational flexibility accounts for the capability of the ancestral resurrected lactamases to bind and degrade antibiotic molecules of different size and shape.

9.14 Conformational Flexibility/Diversity Should Contribute to Protein Evolvability

It is widely accepted that protein promiscuity and the concomitant conformational flexibility contribute to high protein evolvability, i.e., to the capability of the protein to evolve new functions. This point was eloquently made by James and Tawfik [67] on the basis of a conformational diversity description of protein flexibility and will follow here the general argument provided by these authors. There is ample experimental and computational evidence that proteins in solution are best envisioned as ensembles of more or less diverse conformations [18, 19, 24, 43, 44, 67, 71, 98, 128, 131, 145]. Such conformational diversity explains functional diversity, as different conformations could perform different functions (related to the binding of different partners, substrates, or transition states) and furthermore suggest a simple molecular mechanism for the evolution of new functions. For instance, while the major (highly populated) conformation of an enzyme catalyzes the conversion of the "natural" substrate, a minor (scarcely populated) conformation might catalyze the conversion of an alternative substrate. If, at some time, the latter process becomes consequential for organismal fitness, natural selection will bring about an improvement of the new function, likely through mutations that shift the conformational equilibria toward the conformation responsible for the new activity. The process will likely involve gene duplication as necessary step (see Sect. 9.10) to bypass the possible trade-off between the primary function and the new function.

Clearly, conformational flexibility/diversity contributes to protein evolvability in the sense that it provides the molecular basis for the existence of low-level promiscuous activities that can be enhanced during natural evolution and, in biotechnological application scenarios, through laboratory-directed evolution. Furthermore, recent work supports that that conformational flexibility/diversity may facilitate the emergence of totally new enzyme activities (i.e., linked to new active sties) in non-catalytic scaffolds. The only (to our knowledge) reported example [83, 97] of simple, single-mutation generation of de novo activities in non-catalytic scaffolds used calmodulin, a conformationally flexible protein ([143]; see also Sect. 9.13) as background. Screening of combinatorial libraries of proteins designed to fold into four-helix bundles led to proteins able to rescue various *E. coli* auxotrophs [37], and these de novo proteins were shown to be structurally dynamic [99]. An artificial RNA ligase obtained through screening of a very large naive library [119] displays enhanced conformational dynamics [25].

9.15 High Stability and Enhanced Conformational Flexibility, Likely Contributors to Protein Evolvability, Are Not Necessarily Incompatible Features

Both high stability and promiscuity/flexibility should contribute to evolvability (Sects. 9.5 and 9.14). It is important to note, therefore, that these two features are not necessarily incompatible, a statement that is illustrated by the several examples provided below.

As mentioned in Sect. 9.13, ubiquitin has been shown experimentally to display conformational diversity, likely a reflection of its promiscuous capability of being recognized by a large number of different proteins. Still, it is a highly stable protein with a denaturation temperature of about 90 °C at neutral pH [63].

Hydrogen exchange experiments demonstrated substantial conformational flexibility in a highly stable rubredoxin from the hyperthermophile *Pyrococcus furiosus*, as conformational opening for solvent access was shown to occur in the milliseconds or faster time scale [55, 66]. Likewise, hydrogen exchange studies supported a higher degree of flexibility in a thermophilic α-amylase as compared with a mesophilic homolog [38]. Also, NMR studies on homologous mesophilic and thermophilic ribonuclease HI enzymes do not support that stable thermophilic proteins are more rigid [22], a conclusion that is reinforced by a number of computational studies (see [70] and references quoted therein).

Of course, the examples provided above should not be taken to imply that proteins cannot be highly stable and conformationally rigid at the same time. They do support, however, that some highly stable proteins can also be conformationally flexible. This may perhaps come as a surprise to some readers because the notion that protein marginal stability is adaptive can still be found occasionally in the literature (for instance, [118]). According to this interpretation, low protein stability guarantees the degree of flexibility required for function. However, as we have elaborated in Sect. 9.5, marginal protein stability is not likely adaptive, but the outcome of natural purifying selection linked to an evolutionary stability threshold combined with the well-known fact that the number of available sequences decreases with increasing stability. Accordingly, it should be possible to enhance protein stability through laboratory evolution or engineering without compromising function. This, in fact, has been experimentally demonstrated in several studies [45, 96, 129].

9.16 Plausibly, the Combination of High Stability and Promiscuity/Flexibility Was Not Uncommon Among the Most Ancient Proteins, and It Is, in Fact, Found in Several Resurrected Ancestral Proteins

As we have elaborated in preceding sections, enhanced stability and promiscuity/flexibility contribute to protein evolvability and are not incompatible features. Ancient proteins may, therefore, have displayed both features *simultaneously*. This is, in fact, supported by the recent ancestral resurrection studies summarized below.

Some laboratory resurrections of several Precambrian thioredoxins [107] display denaturation temperatures 30–35° higher than their modern mesophilic counterparts and, at the same time, show enhanced activity in single-molecule assays of disulfide reduction. This likely reflects promiscuity because a nonnatural substrate is used in these single-molecule assays.

As previously noted (Fig. 9.2), laboratory resurrections of 2–3-billion-year-old β-lactamases [109] display denaturation temperatures 30–35° higher than their modern mesophilic counterparts and, at the same time, are moderately efficient promiscuous enzymes capable to degrade a variety of antibiotics, including penicillin and third-generation antibiotics (by contrast, the modern TEM-1 β-lactamase is a penicillin specialist). Recent computational studies [145] support that conformational flexibility/diversity is responsible for the substrate promiscuity of these resurrected ancestral enzymes.

Very recently [30], a resurrected ancestral enzyme along the path from esterases to hydroxynitrile lyases has been shown to be able to catalyze both ester hydrolysis and lyase reactions. This catalytic promiscuity (attributed to enhanced conformational flexibility by [30]) did not imply a lower stability, however. Quite the contrary, the promiscuous ancestral protein was reported to display a denaturation temperature (about 80 °C) substantially higher than the denaturation temperatures for its modern descendants (reported as 54–70 °C).

9.17 Ancestral Protein Resurrection Versus Consensus Engineering

It emerges from the models, theoretical scenarios, and experimental results we have discussed so far that ancestral sequence reconstruction should provide a convenient and efficient method to search sequence space for the protein biophysical features that are desirable in protein engineering scenarios.

Certainly, the idea of using sequence analyses as a basis for modulating protein biophysical properties has been around for quite some time, and it is, in fact, readily illustrated by the well-known consensus approach [3, 34, 46, 82, 89, 92, 124]. The most frequent amino acid residue at a given position in a sequence alignment is referred to as the consensus amino acid at that position. Back-to-the-consensus mutations are often found to lead to stability enhancements and function modulations qualitatively similar to those obtained through ancestral sequence reconstruction (see [110] for a recent discussion). This is perhaps not too surprising because the consensus amino acid at a given position may in many cases be the ancestral amino acid. That is, some back-to-the-consensus mutations may actually be back-to-the-ancestor mutations, and their stabilizing effect is consistent with the conservation of energetic preferences over evolutionary history [111]. Still, as elaborated below, ancestral protein resurrection appears to have some clear advantages over consensus engineering.

While comparative analyses are scarce, it appears that consensus engineering captures the useful ancestral properties only to a limited extent [110]. This is no too surprising because consensus is a simple counting procedure, while ancestral sequence reconstruction is a broad-scale phylogenetic procedure that simultaneously and consistently reconstructs the sequences of all the nodes in a phylogenetic tree from the information contained in all the extant sequences. As a result, differences between the ancestral amino acid and the consensus amino acid are likely to occur in particular at sites with a high amount of diversity (see [110] for a more detailed discussion).

While the alignment of a given set of modern protein sequences leads to essentially one consensus sequence, ancestral sequence reconstruction leads to a large number of sequences, a feature that allows for an efficient search of sequence space. For instance, the trend observed in the properties determined for the resurrected proteins corresponding to a given set of phylogenetic nodes may immediately suggest additional nodes at which the properties targeted are likely to be optimized. Likewise, several alternative representations of the protein sequence at a given promising node (from a random sampling of the amino acid probabilities: see Sect. 9.2) can be subjected to laboratory resurrection for property optimization.

It must be recognized, of course, that, compared with ancestral sequence reconstruction, the determination of a consensus sequence is a simple and computationally undemanding procedure. This said, however, it is important to note that there are currently several software packages and online services (for instance, [49, 88, 115, 141]) that can perform ancestral sequence reconstruction. Furthermore, the time-consuming experimental validation procedures required for applications in molecular evolution (see Sect. 9.2) can be substantially relaxed (or even omitted) if the purpose of the ancestral reconstruction exercise is simply to obtain scaffolds with extreme and useful properties in a protein engineering scenario.

9.18 Crystal Ball

"Prediction is very difficult, especially about the future" (commonly attributed to Niels Bohr). Still, we will venture to make two specific predictions about future uses of ancestral resurrection in protein engineering.

Promiscuity is a useful protein property in biotechnological application scenarios (Sect. 9.9). However, it is an accidental property in most modern proteins, in the sense that it is not adaptive (with obvious exceptions, such as some detoxification enzymes). Search for promiscuity in modern proteins is, therefore, inefficient [30].

Ancestral protein resurrection targeting branch points in the divergence of new activities will be used as an efficient method to search for protein promiscuity. This prediction is supported by recent work by Kazlauskas and coworkers [30].

The generation of totally new catalytic sites is, arguably, one of the most important unsolved problems in protein engineering. The development of procedures to engineer new active sites capable to efficiently catalyze reactions totally unrelated with those catalyzed by natural enzymes will have an enormous impact in biotechnology and our understanding of enzyme catalysis. However, previous attempts at the rational design of new catalytic sites have met with limited success [84], as comparatively low catalysis levels are typically achieved despite targeting simple model reactions. The high evolvability of resurrected ancestral proteins will make them the scaffolds of choice for the development of more efficient approaches to engineer new catalytic sites. This prediction is supported by recent (unpublished) work from our group that uses a very simple design approach to generate substantial levels of a nonnatural activity in protein scaffolds derived from ancestral resurrection.

Acknowledgments Work in the authors' lab is supported by FEDER Funds and Grants, CSD2009-00088, and BIO2015-66426-R from the Spanish Ministry of Economy and Competitiveness.

References

1. Aharoni A, Gaidukov L, Khersonsky O et al (2005) The 'evolvability' of promiscuous protein functions. Nat Genet 37(1):73–76
2. Akanuma S, Nakajima Y, Yokobori S et al (2013) Experimental evidence for the thermophilicity of ancestral life. Proc Natl Acad Sci U S A 110:11067–11072
3. Amin N, Liu AD, Ramer S et al (2004) Construction of stabilized proteins by combinatorial consensus mutagenesis. Protein Eng Des Sel 17:787–793
4. Anderson DW, McKeown AN, Thornton JW (2015) Intermolecular epistasis shaped the function and evolution of an ancient transcription factor and its DNA binding sites. Elife 4:e07864
5. Anderson DP, Whitney DS, Hanson-Smith V et al (2016) Evolution of an ancient protein function involved in organized multicellularity in animals. Elife 5:e10147
6. Atkinson QD (2013) The descent of words. Proc Natl Acad Sci U S A 110(11):4159–4160
7. Babtie A, Tokuriki N, Hollfelder F (2010) What makes an enzyme promiscuous? Curr Opin Chem Biol 14:200–207
8. Bahar I, Lezon TR, Yang LW et al (2010) Global dynamics of proteins: bridging between structure and function. Annu Rev Biophys 39:23–42
9. Baier F, Tokuriki N (2014) Connectivity between catalytic landscapes of the metallo-β-lactamase superfamily. J Mol Biol 426:2442–2456
10. Bar-Even A, Noor E, Savir Y et al (2011) The moderately efficient enzyme: evolutionary and physicochemical trends shaping enzyme parameters. Biochemistry 50:4402–4410
11. Bar-Rogovsky H, Hugenmatter A, Tawfik DS (2013) The evolutionary origins of detoxifying enzymes: the mammalian serum paraoxonases (PONs) relate to bacterial homoserine lactonases. J Biol Chem 288(33):23914–23927
12. Ben-David M, Elias M, Filippi JJ et al (2012) Catalytic versatility and backups in enzyme active sites: the case of serum paraoxonase 1. J Mol Biol 418:181–196
13. Benner SA, Sassi SO, Gaucher EA (2007) Molecular paleoscience: systems biology from the past. Adv Enzymol Relat Areas Mol Biol 75:1–132

14. Bergthorsson U, Andersson DI, Roth JR (2007) Ohno's dilemma: evolution of new genes under continuous selection. Proc Natl Acad Sci U S A 104(43):17004–17009
15. Bershtein S, Segal M, Bekerman R et al (2006) Robustness-epistasis link shapes the fitness landscape of a randomly drifting protein. Nature 444:929–932
16. Bickelmann C, Morrow JM, Du J et al (2015) The molecular origin and evolution of dim-light vision in mammals. Evolution 69(11):2995–3003
17. Bloom JD, Labthavikul ST, Otey CR et al (2006) Protein stability promotes evolvability. Proc Natl Acad Sci U S A 103:5869–5874
18. Bloom JD, Arnold FH, Wilke CO (2007) Breaking proteins with mutations: threads and thresholds in evolution. Mol Syst Biol 3:76
19. Boehr DD, Nussinov R, Wright PE (2009) The role of dynamic conformational ensembles in biomolecular recognition. Nat Chem Biol 5:789–796
20. Bornscheuer UT, Kazlauskas RJ (2004) Catalytic promiscuity in biocatalysis: using old enzymes to form new bonds and follow new pathways. Angew Chem Int Ed 43:6032–6040
21. Bouchard-Côte A, Hall D, Griffiths TL et al (2013) Automated reconstruction of ancient languages using probabilistic models of sound change. Proc Natl Acad Sci U S A 110:4224–4229
22. Butterwick JA, Loria JP, Astrof NS et al (2004) Multiple time scale backbone dynamics of homologous thermophilic and mesophilic ribonuclease HI enzymes. J Mol Biol 339(4):855–871
23. Carrigan MA, Uryasev O, Frye CB et al (2015) Hominids adapted to metabolize ethanol long before human-directed fermentation. Proc Natl Acad Sci U S A 112(2):458–463
24. Changeux JP, Edelstein A (2011) Conformational selection or induced fit? 50 years of debate resolved. F1000 Biol Rep 3:19
25. Chao FA, Morelli A, Haugner JC 3rd et al (2013) Structure and dynamics of a primordial catalytic fold generated by in vitro evolution. Nat Chem Biol 9(2):81–83
26. Clifton BE, Jackson CJ (2016) Ancestral protein reconstruction yields insight into adaptive evolution of binding specifity in solute-binding proteins. Cell Chem Biol 23:236–245
27. Cole MF, Gaucher EA (2011) Utilizing natural diversity to evolve protein function: applications towards thermostability. Curr Opin Chem Biol 15(3):399–406
28. Copley SD (2003) Enzymes with extra talents: moonlighting functions and catalytic promiscuity. Curr Opin Chem Biol 7:265–272
29. Des Marais DL, Rausher MD (2008) Escape from adaptive conflict after duplication in an anthocyanin pathway gene. Nature 454:762–765
30. Devamani T, Rauwerdink AM, Lunzer M et al (2016) Catalytic promiscuity of ancestral esterases and hydroxynitrile lyases. J Am Chem Soc 138(3):1046–1056
31. Diaz JE, Lin CS, Kunishiro K et al (2011) Computational design and selections for an engineered, thermostable terpene synthase. Protein Sci 9:1597–1606
32. Duarte F, Amrein BA, Kamerlin SC (2013) Modeling catalytic promiscuity in the alkaline phosphatase superfamily. Phys Chem Chem Phys 15(27):11160–11177
33. Due AV, Kuper J, Geerlof A et al (2011) Bisubstrate specificity in histidine/tryptophan biosynthesis isomerase from Mycobacterium tuberculosis by active site metamorphosis. Proc Natl Acad Sci U S A 108(9):3554–3559
34. Durani V, Magliery TJ (2013) Protein engineering and stabilization from sequence statistics: variation and covariation analysis. Methods Enzymol 523:237–256
35. Erijman A, Aizner Y, Shifman JM (2011) Multispecific recognition: mechanism, evolution, and design. Biochemistry 50:602–611
36. Finnigan GC, Hanson-Smith V, Stevens TH et al (2012) Evolution of increased complexity in a molecular machine. Nature 481:360–364
37. Fisher MA, McKinley KL, Bradley LH et al (2011) De novo designed proteins from a library of artificial sequences function in *Escherichia coli* and enable cell growth. PLoS ONE 6(1):e15364

38. Fitter J, Heberle J (2000) Structural equilibrium fluctuations in mesophilic and thermophilic alpha-amylase. Biophys J 79(3):1629–1636
39. Force A, Lynch M, Pickett FB et al (1999) Preservation of duplicate genes by complementary, degenerative mutations. Genetics 151(4):1531–1545
40. Francino MP (2005) An adaptive radiation model for the origin of new gene functions. Nat Genet 37:573–577
41. Garcia-Seisdedos H, Ibarra-Molero B, Sanchez-Ruiz JM (2012) How many ionizable groups can sit on a protein hydrophobic core? Proteins 80:1–7
42. Gaucher EA, Govindarajan S, Ganesh OK (2008) Palaeotemperature trend for Precambrian life inferred from resurrected proteins. Nature 451:704–707
43. Gerek ZN, Keskin O, Ozkan SB (2009) Identification of specificity and promiscuity of PDZ domain interactions through their dynamic behavior. Proteins 77:796–781
44. Gerek ZN, Ozkan SB (2010) A flexible docking scheme to explore the binding selectivity of PDZ domains. Protein Sci 19:914–928
45. Giver L, Gershenson A, Freskgard PO et al (1998) Directed evolution of a thermostable esterase. Proc Natl Acad Sci U S A 95:12809–12813
46. Godoy-Ruiz R, Perez-Jimenez R, Ibarra-Molero B et al (2004) Relation between protein stability, evolution and structure, as probed by carboxylic acid mutations. J Mol Biol 336:313–318
47. Godoy-Ruiz R, Ariza F, Rodriguez-Larrea D et al (2006) Natural selection for kinetic stability is a likely origin of correlations between mutational effects on protein energetics and frequencies of amino acid occurrences in sequence alignments. J Mol Biol 362:966–997
48. Grunwald P (2009) Use of enzymes in industry. In: Biocatalysis: biochemical fundamentals and applications. Imperial College Press, London, pp 968–992
49. Guindon S, Lethiec F, Duroux P et al (2005) PHYML Online–a web server for fast maximum likelihood-based phylogenetic inference. Nucleic Acids Res 33:W557–W559
50. Hahn MW (2009) Distinguishing among evolutionary models for the maintenance of gene duplicates. J Herpetol 100(5):605–617
51. Hall BG, Barlow M (2004) Evolution of the serine beta-lactamases: past, present and future. Drug Resist Updat 7(2):111–123
52. Harms MJ, Thornton JW (2010) Analyzing protein structure and function using ancestral gene resurrection. Curr Opin Struct Biol 20:260–236
53. Harms MJ, Thornton JW (2013) Evolutionary biochemistry: revealing the historical and physical causes of protein properties. Nat Rev Genet 14:559–571
54. Hart KM, Harms MJ, Schmidt BH et al (2014) Thermodynamic system drift in protein evolution. PLoS Biol 12(11):e1001994
55. Hernandez G, Jenney FE Jr, Adams MW et al (2000) Millisecond time scale conformational flexibility in a hyperthermophile protein at ambient temperature. Proc Natl Acad Sci U S A 97(7):3166–3170
56. Hittinger CT, Carroll SB (2007) Gene duplication and the adaptive evolution of a classic genetic switch. Nature 449:677–681
57. Hobbs JK, Shepherd C, Saul DJ et al (2012) On the origin and evolution of thermophily: reconstruction of functional precambrian enzymes from ancestors of Bacillus. Mol Biol Evol 29:825–835
58. Hou L, Honaker MT, Shireman LM et al (2007) Functional promiscuity correlates with conformational heterogeneity in A-class glutathione S-transferases. J Biol Chem 282(32):23264–23274
59. Huang R, Hippauf F, Rohrbeck D et al (2012) Enzyme functional evolution through improved catalysis of ancestrally nonpreferred substrates. Proc Natl Acad Sci USA 109(8):2966–2971
60. Hudson WH, Kossmann BR, de Vera IM et al (2016) Distal substitutions drive divergent DNA specificity among paralogous transcription factors through subdivision of conformational space. Proc Natl Acad Sci U S A 113(2):326–331

61. Hughes AL (2005) Gene duplication and the origin of novel proteins. Proc Natl Acad Sci U S A 102:8791–8792
62. Hult K, Berglund P (2007) Enzyme promiscuity: mechanism and applications. Trends Biotechnol 25(5):231–238
63. Ibarra-Molero B, Loladze VV, Makhatadze GI et al (1999) Thermal versus guanidine-induced unfolding of ubiquitin. An analysis in terms of the contributions from charge-charge interactions to protein stability. Biochemistry 38(25):8138–8149
64. Ingles-Prieto A, Ibarra-Molero B, Delgado-Delgado A et al (2013) Conservation of protein structure over four billion years. Structure 21:1690–1697
65. Innan H, Kondrashov F (2010) The evolution of gene duplications: classifying and distinguishing between models. Nat Rev Genet 11(2):97–108
66. Jaenicke R (2000) Do ultrastable proteins from hyperthermophiles have high or low conformational rigidity? Proc Natl Acad Sci U S A 97:2962–2964
67. James LC, Tawfik DS (2003) Conformational diversity and protein evolution -a sixty-years-old hypothesis revisited. Trends Biochem Sci 28:361–368
68. Jensen RA (1976) Enzyme recruitment in evolution of new function. Annu Rev Microbiol 30:409–425
69. Jermann TM, Opotz JG, Stackhouse J et al (1995) Reconstructing the evolutionary history of the artiodactyl ribonuclease superfamily. Nature 374:57–59
70. Kalimeri M, Rahaman O, Melchionna S et al (2013) How conformational flexibility stabilizes the hyperthermophilic elongation factor g-domain. J Phys Chem B 117:13775–13785
71. Kar G, Keskin O, Gursoy A et al (2010) Allostery and population shift in drug discovery. Curr Opin Pharmacol 10:715–722
72. Kasting JF (1987) Theoretical constraints on oxygen and carbon dioxide concentrations in the Precambrian atmosphere. Precambrian Res 34:205–229
73. Kasting JF (2013) Atmospheric science. How was early Earth kept warm? Science 339(6115):44–45
74. Kazlauskas RJ (2005) Enhancing catalytic promiscuity for biocatalysis. Curr Opin Chem Biol 2:195–201
75. Khersonsky O, Roodveldt C, Tawfik DS (2006) Enzyme promiscuity: evolutionary and mechanistic aspects. Curr Opin Chem Biol 10:498–508
76. Khersonsky O, Tawfik DS (2010) Enzyme promiscuity: a mechanistic and evolutionary perspective. Annu Rev Biochem 79:471–505
77. Khersonsky O, Kiss G, Röthlisberger D et al (2012) Bridging the gaps in design methodologies by evolutionary optimization of the stability and proficiency of designed Kemp eliminase KE59. Proc Natl Acad Sci U S A 109(26):10358–10363
78. Kim J, Copley SD (2007) Why metabolic enzymes are essential or nonessential for growth of *Escherichia coli* K12 on glucose. Biochemistry 46:12501–12511
79. Knauth LP, Lowe DR (1978) Oxygen Isotope Geochemistry of Cherts from Onverwacht Group (3.4 billion years), Transvaal, South Africa, with implications for secular variations in isotopic composition of cherts. Earth Planet Sci Lett 41:209–222
80. Knauth LP, Lowe DR (2003) High Archean climatic temperature inferred from oxygen isotope geochemistry of cherts in the 3.5 Ga Swaziland Supergroup, South Africa. Geol Soc Am Bull 115:566–580
81. Knauth LP (2005) Temperature and salinity history of the Precambrian ocean: implications for the course of microbial evolution. Palaeogeogr Palaeoclimatol Palaeoecol 219:53–69
82. Kohn A, Binz HK, Forrer P et al (2003) Designed to be stable: crystal structure of a consensus ankyrin repeat protein. Proc Natl Acad Sci U S A 100:1700–1705
83. Korendovych IV, Kulp DW, Wu Y et al (2011) Design of a switchable eliminase. Proc Natl Acad Sci U S A 108:6823–6827
84. Korendovych IV, DeGrado WF (2014) Catalytic efficiency of designed catalytic proteins. Curr Opin Struct Biol 27:113–121
85. Kratzer JT, Lanaspa MA, Murphy MN et al (2014) Evolutionary history and metabolic insights of ancient mammalian uricases. Proc Natl Acad Sci U S A 111(10):3763–3768

86. Lane N, Martin WF (2012) The origin of membrane energetics. Cell 151:1406–1416
87. Lange OF, Lakomek NA, Farès C et al (2008) Recognition dynamics up to microseconds revealed from an RDC-derived ubiquitin ensemble in solution. Science 320(5882):1471–1475
88. Lartillot N, Lepage T, Blanquart S (2009) PhyloBayes 3: a Bayesian software package for phylogenetic reconstruction and molecular dating. Bioinformatics 25:2286–2288
89. Lehman M, Pasamontes L, Lassen SF et al (2000) The consensus concept for thermostability engineering of proteins. Biochim Biophys Acta 1543:408–415
90. Li Y, Drummond DA, Sawayama AM et al (2007) A diverse family of thermostable cytochrome P450s created by recombination of stabilizing fragments. Nat Biotechnol 25(9): 1051–1056
91. Liberles D (2007) Ancestral sequence reconstruction. Oxford University Press, USA
92. Magliery TJ (2015) Protein stability: computation, sequence statistics, and new experimental methods. Curr Opin Struct Biol 33:161–168
93. Martin W, Baross J, Kelley D et al (2008) Hydrothermal vents and the origin of life. Nat Rev Microbiol 6(11):805–814
94. Merkl R, Sterner R (2016) Ancestral protein reconstruction: techniques and applications. Biol Chem 397(1):1–21
95. Merski M, Shoichet BK (2012) Engineering a model protein cavity to catalyze the Kemp elimination. Proc Natl Acad Sci U S A 109:16179–16183
96. Miyazaki K, Wintrode PL, Grayling RA et al (2000) Directed evolution study of temperature adaptation in a psychrophilic enzyme. J Mol Biol 297(4):1015–1026
97. Moroz YS, Dunston TT, Makhlynets OV et al (2015) New tricks for old proteins: single mutations in a nonenzymatic protein give rise to various enzymatic activities. J Am Chem Soc 137(47):14905–14911
98. Münz M, Hein J, Biggin PC (2012) The role of flexibility and conformational selection in the binding promiscuity of PDZ domains. PLoS Comput Biol 8:e1002749
99. Murphy GS, Greisman JB, Hecht MH (2016) De novo proteins with life-sustaining functions are structurally dynamic. J Mol Biol 428(2 Pt A):399–411
100. Nisbet EG, Sleep NH (2001) The habitat and nature of early life. Nature 409(6823):1083–1091
101. Novak MJ, Pattammattel A, Koshmerl B et al (2016) "Stable-on-the-Table" enzymes: engineering the enzyme–graphene oxide interface for unprecedented kinetic stability of the biocatalyst. ACS Catal 6(1):339–347
102. Nobeli I, Favia AD, Thornton JM (2009) Protein promiscuity and its implications for biotechnology. Nat Biotechnol 27:157–167
103. Ohno S (1970) Evolution by gene duplication. Springer, Berlin
104. O'Brien PJ, Herschlag D (1999) Catalytic promiscuity and the evolution of new enzymatic activities. Chem Biol 6:R91–R105
105. Ortlund EA, Bridgham JT, Redimbo MR et al (2007) Crystal structure of an ancient protein: evolution by conformational epistasis. Science 317:1544–1548
106. Pauling L, Zuckerkandl E (1963) Chemical paleogenetics. Molecular 'restoration studies' of extinct forms of life. Acta Chem Scand 17:S9–S16
107. Perez-Jimenez R, Ingles-Prieto A, Zhao ZM et al (2011) Single-molecule paleoenzymology probes the chemistry of resurrected enzymes. Nat Struct Mol Biol 18:592–596
108. Plach MG, Reisinger B, Sterner R et al (2016) Long-term persistence of bi-functionality contributes to the robustness of microbial life through exaptation. PLoS Genet 12(1):e1005836
109. Risso VA, Gavira JA, Mejia-Carmona DF et al (2013) Hyperstability and substrate promiscuity in laboratory resurrections of Precambrian b-lactamases. J Am Chem Soc 135:2899–2902
110. Risso VA, Gavira JA, Gaucher EA et al (2014) Phenotypic comparisons of consensus variants versus laboratory resurrections of Precambrian proteins. Proteins 82(6):887–896
111. Risso VA, Manssour-Triedo F, Delgado-Delgado A et al (2015) Mutational studies on resurrected ancestral proteins reveal conservation of site-specific amino acid preferences throughout evolutionary history. Mol Biol Evol 32(2):440–455

112. Robert F, Chaussidon M (2006) A palaeotemperature curve for the Precambrian oceans based on silicon isotopes in cherts. Nature 443:969–972
113. Robertson AD, Murphy KP (1997) Protein structure and the energetics of protein stability. Chem Rev 97(5):1251–1268
114. Robinson PK (2015) Enzymes: principles and biotechnological applications. Essays Biochem 59:1–41
115. Ronquist F, Huelsenbeck JP (2003) MrBayes 3: Bayesian phylogenetic inference under mixed models. Bioinformatics 19:1572–1574
116. Sanchez-Ruiz JM (2010) Protein kinetic stability. Biophys Chem 148(1–3):1–15
117. Schellman JA (1987) The thermodynamic stability of proteins. Annu Rev Biophys Biophys Chem 16:115–137
118. Schulenburg C, Miller BG (2014) Enzyme recruitment and its role in metabolic expansion. Biochemistry 53(5):836–845
119. Seelig B, Szostak JW (2007) Selection and evolution of enzymes from a partially randomized non-catalytic scaffold. Nature 448(7155):828–831
120. Sikosek T, Chan HS (2014) Biophysics of protein evolution and evolutionary protein biophysics. J R Soc Interface 11:20140419
121. Sleep NH (2010) The Hadean-Archaean environment. Cold Spring Harb Perspect Biol 2:a002527
122. Smock RG, Yadid I, Dym O et al (2016) De novo evolutionary emergence of a symmetrical protein is shaped by folding constraints. Cell 164(3):476–486
123. Som SM, Catling DC, Harnmeijer JP et al (2012) Air density 2.7 billion years ago limited to less than twice modern levels by fossil raindrop imprints. Nature 484(7394):359–362
124. Steipe B, Schiller B, Pluckthun A et al (1994) Sequence statistics reliably predict stabilizing mutations in a protein domain. J Mol Biol 240:188–192
125. Stockbridge RB, Lewis CA Jr, Yuan Y et al (2010) Impact of temperature on the time required for the establishment of primordial biochemistry, and for the evolution of enzymes. Proc Natl Acad Sci U S A 107(51):22102–22105
126. Taverna DR, Goldstein RA (2002) Why are proteins marginally stable? Proteins 46:105–109
127. Thornton JW (2004) Resurrecting ancient genes: experimental analysis of extinct molecules. Nat Rev Genet 5:366–375
128. Tobi D, Bahar I (2005) Structural changes involved in protein binding correlate with intrinsic motions of proteins in the unbound state. Proc Natl Acad Sci U S A 102:18908–18913
129. Van den Burg B, Vriend G, Veltman OR et al (1998) Engineering an enzyme to resist boiling. Proc Natl Acad Sci U S A 95(5):2056–2060
130. Voordeckers K, Brown CA, Vanneste K et al (2012) Reconstruction of ancestral metabolic enzymes reveals molecular mechanisms underlying evolutionary innovation through gene duplication. PLoS Biol 10:e1001446
131. Vogt AD, Di Cera E (2012) Conformational selection or induced fit? A critical appraisal of the kinetic mechanism. Biochemistry 51:5894–5902
132. Williams PD, Pollock DD, Blackburne BP et al (2006) Assessing the accuracy of ancestral protein reconstruction methods. PLoS Comput Biol 2(6):e69
133. Wijma HJ, Floor RJ, Janssen DB (2013) Structure- and sequence-analysis inspired engineering of proteins for enhanced thermostability. Curr Opin Struct Biol 23(4):588–594
134. Whitfield JH, Zhang W, Herde MK et al (2015) Construction of a robust and sensitive arginine biosensor through ancestral protein reconstruction. Protein Sci 24:1412–1422
135. Woese CR (1987) Bacterial evolution. Microbiol Rev 51:221–271
136. Wolfenden R (2006) Degrees of difficulty of water-consuming reactions in the absence of enzymes. Chem Rev 106:3379–3396
137. Wolfenden R (2011) Benchmark reaction rates, the stability of biological molecules in water, and the evolution of the catalytic power in enzymes. Annu Rev Biochem 80:645–647

138. Wolfenden R (2014a) Massive thermal acceleration of the emergence of primordial chemistry, the incidence of spontaneous mutation, and the evolution of enzymes. J Biol Chem 289(44):30198–30204
139. Wolfenden R (2014b) Primordial chemistry and enzyme evolution in a hot environment. Cell Mol Life Sci 71(15):2909–2915
140. Wordsworth R, Pierrehumbert R (2013) Hydrogen-nitrogen greenhouse warming in Earth's early atmosphere. Science 339(6115):64–67
141. Yang Z (1997) PAML: a program package for phylogenetic analysis by maximum likelihood. Comput Appl Biosci 13:555–556
142. Yip SH, Matsumura I (2013) Substrate ambiguous enzymes within the *Escherichia coli* proteome offer different evolutionary solutions to the same problem. Mol Biol Evol 30(9):2001–2012
143. Yamniuk AP, Vogel HJ (2004) Calmodulin's flexibility allows for promiscuity in its interactions with target proteins and peptides. Mol Biotechnol 27(1):33–57
144. Zhang W, Dourado DF, Fernandes PA et al (2012) Multidimensional epistasis and fitness landscapes in enzyme evolution. Biochem J 445(1):39–46
145. Zou T, Risso VA, Gavira JA et al (2015) Evolution of conformational dynamics determines the conversion of a promiscuous generalist into a specialist enzyme. Mol Biol Evol 32:142–143

Molecular Modeling in Enzyme Design, Toward In Silico Guided Directed Evolution

10

Emanuele Monza, Sandra Acebes, M. Fátima Lucas, and Victor Guallar

Abstract

Directed evolution (DE) creates diversity in subsequent rounds of mutagenesis in the quest of increased protein stability, substrate binding, and catalysis. Although this technique does not require any structural/mechanistic knowledge of the system, the frequency of improved mutations is usually low. For this reason, computational tools are increasingly used to focus the search in sequence space, enhancing the efficiency of laboratory evolution. In particular, molecular modeling methods provide a unique tool to grasp the sequence/structure/function relationship of the protein to evolve, with the only condition that a structural model is provided. With this book chapter, we tried to guide the reader through the state of the art of molecular modeling, discussing their strengths, limitations, and directions. In addition, we suggest a possible future template for *in silico* directed evolution where we underline two main points: a hierarchical computational protocol combining several different techniques and a synergic effort between simulations and experimental validation.

E. Monza • S. Acebes
Joint BSC-CRG-IRB Research Program in Computational Biology, Barcelona Supercomputing Center, Jordi Girona 29, 08034 Barcelona, Spain

M.F. Lucas
Joint BSC-CRG-IRB Research Program in Computational Biology, Barcelona Supercomputing Center, Jordi Girona 29, 08034 Barcelona, Spain

Anaxomics Biotech, Balmes 89, 08008 Barcelona, Spain

V. Guallar, PhD (✉)
Joint BSC-CRG-IRB Research Program in Computational Biology, Barcelona Supercomputing Center, Jordi Girona 29, 08034 Barcelona, Spain

ICREA, Passeig Lluís Companys 23, 08010 Barcelona, Spain
e-mail: victor.guallar@bsc.es

M. Alcalde (ed.), *Directed Enzyme Evolution: Advances and Applications*,
DOI 10.1007/978-3-319-50413-1_10

10.1 Introduction

Biotechnology needs catalysts that can work under harsh conditions, catalyze a broad range of substrates, generate maximum amount of product, and tolerate changes in the environment. Enzymes, which are biodegradable and reusable catalysts [1], in addition to remarkable reaction rates, can work in environmentally friendly pH and temperature ranges and display control over stereochemistry and regioselectivity which makes them ideal for many applications [2, 3]. When thinking about enzymes, people normally associate them to expressions such as "perfect catalysts" or "outstanding reaction rate." In fact, there are examples of enzymes that catalyze reactions at extremely high rates such as triose phosphate isomerase, superoxide dismutase, or carbonic anhydrase [4]. These are often limited only by the rate of ligand diffusion into the active site (diffusion-controlled rate). Nevertheless, an extensive analysis by Bar-Even et al., of nearly 2,000 enzymes, showed that the median maximal turnover rate value over all measured enzymes is about 10 s^{-1} nowhere close to the values of 10^5 or 10^6 normally associated with catalysts [5, 6]. So, it would appear that natural enzymes are "just good enough" for the function they must perform in a given organism [7]. One might conclude that if they had evolved to their optimum performance, then trying to improve them (from a kinetic point of view) would be attempting the impossible. On the contrary, as seen by the distribution of reaction rates, k_{cat}, most enzymes function at a lower rate than the diffusion limit, and thus, there is space to further increase their kinetic properties to meet industrial needs. Additionally, we need enzymes capable of catalyzing reactions for which no known enzymes exist, to work with different substrates and for particular conditions that are industrially convenient and economically advantageous. For all these reasons, in most cases, we cannot just use enzymes as they are found, but instead we need to change their physical-chemical and functional properties. This is one of the reasons why engineering enzymes for biocatalysis is an incessantly growing field [8–11].

In nature, enzymes have evolved over millions of years to meet specific demands and operate under tight in vivo regulation. Their degree of adeptness includes diverse criteria such as which substrates they accept, the effective reaction rate, the environment in which they function and how well they tolerate changes in it, inactivation by their own products, etc. These characteristics are precisely the ones that scientists wish to control to their own advantage. Some of the earliest attempts to modify enzymes required a deep knowledge of complex structure/function relationships, and (to the authors' contentment) computer simulations have played an important part in it [12, 13]. Since the pioneering work [10, 14, 15] in computationally designed protein sequences (with experimental validation), many remarkable achievements have been obtained. Interesting work includes predicting sequence changes that alter atomic packing arrangements in buried protein regions or the creation of new metal-binding sites which may have many applications along with potential improvement in protein stability [16–18]. In addition to being able to correctly predict changes in protein structure, there has been, of course, a large interest in altering proteins, through computational techniques, to create new function or adapt them to particular conditions. Rational protein design, which involves modification of specific amino acids in the protein's three-dimensional (3D) structure with previous structural/mechanistic knowledge, can be used to alter specificity, stability, selectivity, and activity. Literature contains a vastness of examples of rationally designed proteins (which we do not presume to cover here) including creating new recognition [19–26], improving protein stability [27–29], and protein-protein [30–34] or protein-DNA interactions [35, 36]. We can find procedures to engineer a protein that binds a specific cofactor [37] or a calcium-binding site [38, 39], redesign an enzyme by stabilizing the transition state [40], or create new activity from scratch [41].

A special mention involves the design of new proteins from scratch, commonly known as de novo design, and literature displays many truly interesting examples of new proteins [42–45]. Currently one of the most common strategies to design new enzymes is based on encountering complementary active sites for the transition states of interest [46, 47]. Despite the success of de novo design in providing novel structures and activity, its difficulties in achieving fast kinetics make it still preferable to modify templates available in nature for the desired chemistry. Indeed, a recent computational study pointed out how target reactivity can be one mutation away from a nonenzymatic protein (if well picked) [48]. Due to the scope of this book, we refer the reader interested in de novo design to recent studies on this topic [49].

Despite many promising studies, rational computational protein redesign has its limitations: it requires a reliable 3D structure of the system of interest and an in-depth comprehension of the catalytic mechanism; understanding the relationships between a protein's primary sequence, its three-dimensional structure, and its function is therefore a fundamental goal. Regrettably, our knowledge of enzyme activity is still incomplete which makes our attempts to modifying them often limited. Detailed understanding of the enzymatic structure/function relation is, however, not necessary in directed evolution, an alternative engineering technique based on massive mutations and selective evolution.

Directed evolution (DE) has proven to be a powerful tool for adapting enzymes to wider applications [50–53]. Briefly, in DE diversity is first created through mutagenesis or recombination, followed by screening for improvements in desired properties. One of the main advantages of DE is most certainly that it does not require a thorough understanding of structure/function relationships, unlike rational or de novo design. The introduction of random mutations throughout the gene allows the discovery of mutations that could be difficult to predict with studies based on structure-function knowledge (mostly focused at the active site region). However, the low frequency of improved mutations, some experimental bias, and the combinatorial explosion of possibilities limit this technique. Furthermore, DE requires the development of high-throughput screening and not all processes can be adapted. The methodologies and achievements of directed evolution were already discussed in other sections of this book and will not be included here. Also we refer the reader to interesting reviews [54–59].

A remarkable observation of many DE experiments is that the location of the beneficial mutations varies considerably. For example, most modifications in enantioselectivity or substrate specificity are located in the vicinity of the active site or in the access/exit of reactants/products [58, 60, 61]. Stability and activity however can be affected by mutations in any part of the protein, close or far from the active site [62], increasing significantly the number of possible mutations. To avoid screening massive number of mutations, one can reduce the region to explore by using functional information (from point mutations, random mutagenesis, or deduction from sequence alignments), or when structural information exists (by visual inspection, analysis, etc.), it would be advantageous to exploit this by concentrating mutations where they might be the most effective [62]. Methods such as saturation mutagenesis (where all other 19 amino acids are tested) on specific positions, generally near the active site, can increase the probability of finding beneficial mutations [63–65]. This approach is particularly advantageous when a high-throughput screening method is not available. Generally known as **semi-rational approaches**, these are based on "smart" libraries that, in principle, should have a higher success rate and try to overcome the limitations of the directed evolution and rational design [66–69].

Although it is true that many computational approaches exist to complement DE experiments [69–71], the scope of this chapter is to center on how physics-based molecular modeling can aid in laboratory DE. For this reason, sequence-based strategies that use evolutionary information or statistical data from previous DE rounds will not be explored. These often use phylogenetic analyses and multiple sequence alignments for exploring the amino acid conservation and relationships between homologues of protein sequences [67, 72–78]. Instead we will center our attention on how computations and structural information may aid and focus mainly in physics-based methods to assist in the improvement of three major aspects in enzyme design: catalytic rate constants, protein stability, and protein-ligand binding processes.

The atomic/molecular detailed computational exploration of a protein's amino acid sequence space is a complex problem. As in most simulation fields, a compromise between sampling quality and quantity is necessary. Sampling quality involves construction of the models together with the energy and scoring functions necessary to rank them and evaluate molecular interactions, topics extensively reviewed previously [63, 79–92]. An energy function describes the internal energy of the protein and its interactions with the environment such as other proteins, substrates, and solvent, aiming at reproducing the features of the folded protein [34, 84, 93]. The level of theory used in these and their parameters vary considerably, but most implementations include bonded (bonds, angles, and torsions) and nonbonded terms (van der Waals and electrostatics) and solvent components. Associated to the energy functions is the ability to efficiently score a large number of protein structures and protein-ligand interactions. Scoring functions are used to assess the quality of the designed protein and help select the preferred sequence and the lowest-energy protein-substrate complex. Just as the energy functions, also these vary considerably and can be statistics-based or empirically based methods such as DMutant or PopMuSiC or physics-based and rely on the derivation of energy terms from basic principles to calculate free energy changes [94–97].

The second key aspect involves the system (model) sampling. Given the large number of degrees of freedom, including possible mutations and structural changes associated with them, sampling near-native protein conformations is difficult. Moreover, in situations where protein-ligand interactions exist, sampling must also extend to all (relevant) protein-ligand conformations. And if we don't consider these issues to be enough of a headache, scoring and sampling are not independent! So, to overcome these limitations, it is essential to introduce many approximations. Strategies to limit the sampling include restricting the backbone and side-chain degrees of freedom [34, 82]. In most protein design strategies, sampling is simplified by using a fixed backbone which is normally obtained from an experimentally determined protein structure [98] or a high-quality homology model. Although controversy exists on the importance of dynamics in catalysis [99–101], currently we see more and more cases where backbone flexibility is being taken into account [18, 102–106]. As shown below, development in molecular dynamics (such as high-throughput molecular dynamics (HTMD), steered MD, etc.) and Monte Carlo techniques is gaining importance in enzyme engineering.

As mentioned, this book chapter will center on physics-based methods to assist in the improvement of catalytic rate constants, protein stability, and protein-ligand binding processes. Before entering in these three topics, we refer to Fig. 10.1 and Table 10.1 for a quick guide describing the main computational methods and models being used for these purposes (a guide for nonexperts in theoretical modeling). Finally, we conclude this book chapter by introducing our perspective on how we believe these techniques will aid in future enzymatic directed evolution.

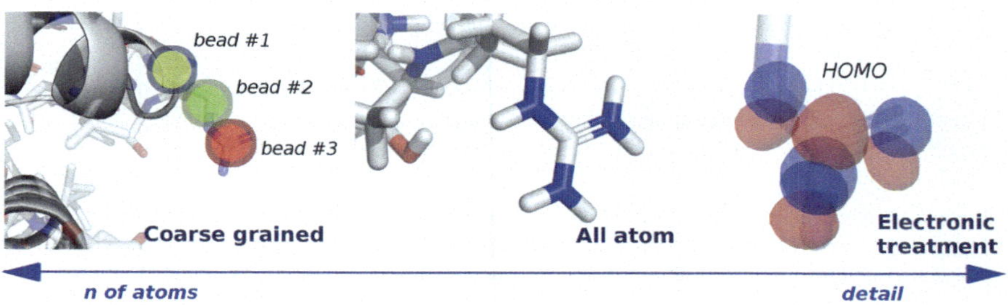

Fig. 10.1 Scheme showing three different levels of granularity used in molecular modeling: coarse grained, all atom, and electronic. In the coarse-grained model, the smallest particle is a bead that includes condensed information on a set of atoms. All atom, as indicated by the name, uses the atom as the smallest unit, while in an electronic treatment, electrons and nuclei are explicitly included. Here we show the highest-energy molecular orbital (HOMO), only possible in an electronic treatment of the system

Table 10.1 A quick guide to simulation methods and theoretical concepts

Methodology guide
Length-size-based models (see also Fig. 10.1)
Coarse-grained (*CG*) *model*. A group of atoms is described by a bead enclosing the properties of the aggregation. For example, according to the MARTINI model [107], amino acids are represented with one to four beads, classified as charged, polar, nonpolar, or apolar, and also subdivided depending on their hydrogen bonding capacity. Reduction in the number of beads decreases the number of pairwise interactions thus increasing the speed of the simulation
All-atom model. All the atoms of the system are included in the model, where the energy function used (see below) must describe their interaction. Electrons and the nuclei information are condensed to a single particle that must contain an averaged description of those properties
Electronic treatment model. Each atom is described as a nucleus with its electrons, requiring, for its description, approximate solution of the Schrödinger equation
Physical theoretical methods
Molecular mechanics (MM) [108]. These methods perform a classical description where atoms (or beads in a coarse-grained model) are represented as spheres (or spheroids) connected by bonds, behaving as springs. Based on MM there arise several computing simulations such as molecular dynamics, Monte Carlo, and docking methods
Force field. It is a set of parameters that define the property of atoms or beads (predefined with a partial charge and radius) and the energy function describing their interactions in MM methods. Typically they include bonding (bond, angle and torsion) and nonbonding (electrostatic and van de Waals) terms
Elastic network model (ENM) [109]. It describes the collective dynamics of proteins by an elastic network, typically using a reduced set of nodes, such as alpha carbons
Molecular dynamics (MD) [110]. It simulates the motion of a model accordingly to the classical Newton's equation. Most MD software uses force fields to describe the properties of atoms and its interactions. With current high-performance computing (HPC), simulations can be expanded up to the millisecond timescale [111] and few millions of atoms; typical values, however, involve thousands of atoms and hundreds of nanoseconds
Monte Carlo (MC) *simulations*. The dynamics of the system are obtained by random (stochastic) motion of the system to assemble a non-time-dependent trajectory [112]. As in MD, it is mostly based on a force field description of the model
PELE [113]. The protein energy landscape exploration (PELE) software is a Monte Carlo-based technique including protein structure prediction techniques (such as ENM) capable of quickly modeling protein dynamics and protein/DNA-ligand interactions [114, 115]
Docking simulations. These propose the preferred relative bound orientation between molecules, mostly used in protein-ligand (substrate) or protein-protein interactions. Usually, docking methods first provide several conformations which are then classified by scoring functions
Scoring functions. These are mathematical functions that predict the strength of intermolecular interactions [116]. Scoring functions are mostly parameterized from MM force fields, empirical data, or knowledge-based functions

Table 10.1 (continued)

Methodology guide
Rotamer library. It contains a restricted number of the most probable conformations (torsion angle values) for a molecule, mostly applied to amino acid side chains, protein backbone, and ligands. They are built from experimental structural data or from accurate quantum simulations (e.g., in ligands). When used with sampling methods, they accelerate the exploration by adopting discrete states instead of continuous values
Quantum mechanics (QM). These methods are based on solving the Schrödinger equation (normally using approximations) under an electronic model description of the system. The solution provides the wave function which fully describes the system: the electronic distribution, the energy, and the gradients to describe the motion of the system. The main limitation of QM methods is their high computational cost, limiting the system's size and simulation speed
Ab initio methods. These are quantum mechanics methods which parameters are obtained exclusively from first principles solution of the equations (still under approximations) but without any usage of parameterized data (see semiempirical methods)
Semiempirical methods. These referred to quantum mechanics methods that use parameters derived from experimental data (or ab initio calculation), typically for the parameterization of the electron-electron interaction terms (the most expensive to compute). Thus, they are less computationally expensive and faster than ab initio ones, capable of dealing with large systems. Their lack of accuracy, especially when fragments are not in the parameterized data set, is their main limitation
QM/MM. This methodology is a combination of QM and MM methods to handle (large) biological all-atom systems [117]. One part of the system where we require an electronic description, such as the active site in an enzyme, is treated at the QM level, and the rest of the model (remainder of the protein, solvent, etc.) is treated at the MM level.

10.2 State of the Art of Molecular Modeling in Protein Design

We provide here a general view of recent computational work on protein design. We do not aim to review all studies produced in the field but to underline several ones which we believe to be important for future developments of *in silico* DE approaches.

10.2.1 Protein Stability Improvement

Understanding and quantifying the effect of mutations on the thermodynamical stability of a protein is of paramount importance for industrial applications. Two of the most popular tools to prepare and score mutated proteins are Rosetta [118, 119] and FoldX [27]. After introducing a mutation, the protein's torsional degrees of freedom (usually side-chain rotamers) are optimized using an energy function that estimates the folding free energy for the created variant. Such energy functions depend on (i) physics-based terms, which account for van der Waals, hydrogen bond, and solvation and electrostatic energies, and (ii) knowledge-based contributions, which determine the probability of a given rotamer according to the Protein Data Bank (PDB) statistics [120]. Apart from these common energy terms, these functions have unique features. For example, Rosetta approximates the free energy change in the unfolded state due to a mutation with context-independent reference energies for each residue [121]. On the other hand, FoldX explicitly estimates the entropic cost to restrict a rotamer in the native state [27]. The relatively low computational cost of these protocols permits to generate and score a large number of mutations in a short time. As shown by Potapov et al., the accuracy/cost trade-off is such that these tools can reproduce overall trends and therefore suggest stabilizing mutations with acceptable probabilities, but they are not good enough to provide detailed results [122].

Following Potapov's accuracy assessment, Kellogg et al. tested the ability of Rosetta to score mutations combining several energy functions and sampling methods with variable resolution [119]. As a main result, the authors concluded that the choice of the sampling algorithm should be tuned with the resolution of the energy function adopted. In other words, an accurate energy function performs better on a finer sampling; likewise, roughly sampled structures should be scored by smoother functions which can tolerate steric clashes better. Still, flexible backbone protocols improved small to large residue mutations, where significant structural changes can occur. In addition, the authors found that conformational sampling was still insufficient to recover the crystal conformation when a large to small hydrophobic residue mutation was introduced, due to poor packing. Larger errors were found when the polarity of the residue drastically changed upon mutation, which suggests poor trade-off between polar desolvation and buried polar interactions. The lack of explicit water molecules and ligand contacts was another factor in some failed predictions. Finally, the lack of a context-dependent unfolded state modeling (a given mutation was considered to have the same effect on the unfolded state independently of the environment [121]) was considered as a source of error, although not a major one. In fact, a free energy variation of the unfolded state upon mutation might change protein stability as well as a variation in the folded state. However, a recent paper shows that an accurate conformational and energetic characterization of the unfolded protein is not trivial, and its inclusion in protein stability scoring significantly worsened the prediction [123].

The entropic scoring in FoldX [27, 124] only takes into account the change in conformational entropy, which depends on the number of accessible conformers in the unfolded state and their probabilities [124]. Although this entropic variation dominates folding [125], large discrepancies in vibrational entropy (the intrinsic entropy of a given protein conformer [124]) have been calculated between thermophilic and mesophilic proteins [126]. Therefore, the thoughtful inclusion of a vibrational entropy contribution in protein design free energy functions might pay off. Najmanovich and coworkers implemented this strategy in the ENcoM server [126], where they combine FoldX [27] with their ENcoM protocol to rapidly estimate vibrational entropy. ENcoM combines ENM techniques with a pairwise atom-type nonbonded interaction term to include the specific nature of amino acids [127].

In an attempt to quantify free energies more rigorously, de Groot and coworkers employed alchemical free energy MD simulations to score 109 mutants of ribonuclease barnase [128]. In this technique, sampling a convenient number of unphysical ("alchemical") intermediates renders a rigorous evaluation of the free energy difference (ΔG) between two states (e.g., wild-type and mutant protein). Unfolded state's free energy differences were calculated using a generic Gly-XXX-Gly peptide with capped termini. This choice provides a universal, albeit less accurate and context-independent, reference state whose values need to be calculated only once and then are stored as a database. The overall Pearson's correlation coefficient with experimental values was 0.86, providing ~72% of the predicted values within 1 kcal/mol of the experimental one when using 30 ns of simulation time. Notably, most of this accuracy (65%) is retained with only 5 ns. The generality of this accuracy/cost ratio will need to be tested against a wider benchmark of mutations. Larger errors were detected for mutations that introduced changes in the electrostatics of buried residues or large structure fluctuation: mutations to glycine, involving bulky and/or well-packed residues, etc.

Due to the impossibility of scoring the entire sequence (mutation) space, several strategies have been developed to focus the search in smaller regions. These include (i) the identification of flexible backbone sites which can be rigidified [129, 130] introducing salt bridges [131] and/or disulfide bonds [132], (ii) the optimization of surface charge-charge interactions [133–135], (iii) the optimization of core packing [136], (iv) the removal of unsatisfied buried polar groups [137], and (v) the localization of critical residues in the active site entry tunnels, especially for co-solute tolerance, with MD [138] or other algorithms like our in-house software PELE [113].

Recently, Wijma and coworkers developed, and applied with success, a mixed approach which aims to obtain highly thermostable protein variants in a short time with minimum experimental screening [139]. In their computational workflow, potentially stabilizing mutations were firstly produced and scored with Rosetta [119] and FoldX [27]. To minimize the risk of affecting catalysis, only residues beyond 10 Å of the substrate were mutated. Mutations were considered potentially stabilizing if $\Delta\Delta G_{fold} \leq -5$ kJ/mol or if $|\Delta\Delta G_{fold}| < 5$ kJ/mol, and the mutation type was contained in the set XXX→Arg, XXX→Pro, and Gly→XXX. These were then filtered to avoid undesired, typically destabilizing features such as increased unsatisfied hydrogen bond donors and acceptors or hydrophobic surface exposure to water. Then, multiple short MD simulations were used to discard variants with increased backbone flexibility. Finally, variants with experimentally confirmed higher thermostability and preserved activity were combined in the lab. This computational hierarchical workflow helps to unmask false positives (~50% of the potentially stabilizing mutations), aiding to focus on reliable mutations; it is, therefore, a plausible strategy for future computer-aided directed evolution of thermostable proteins. The main drawback is the exclusion of mildly damaging mutations that could be coupled synergically to others to improve thermostability.

As reported in a recent review [140], there is still substantial room for improvement of structure- and physics-based (thermo)stability design. This will likely pass through a strong synergy of computational and experimental efforts to improve our understanding of protein stability. In addition, significant work will have to center on developing more accurate energy functions, including polarization, solvation, and vibrational entropy terms. These methodological developments will necessarily have to couple with improvement of sampling algorithms, including a more effective modeling of unfolded state changes.

10.2.2 Protein-Ligand Binding Redesign

Whether we are talking about enzymes or receptors, they all share a common feature: at some stage a protein-ligand recognition process must occur. These are, however, notoriously slow and complex processes that require extensive sampling of the protein-ligand dynamics which in many cases includes induced-fit protein conformational changes. The accurate *in silico* design of protein-ligand interactions is thus a challenging step [141] toward the engineering of proteins for therapeutic [85] and enzymatic purposes [140]. Its difficulty roots in the low tolerance to error due to the reduced number of protein-ligand interactions. In addition, these are largely dominated by polar interactions, which are very sensitive to small changes in geometry [142]. It is worth noting that, despite the small size of the protein-ligand interface, we still face a huge number of possible combinations in sequence space (for 10 positions there are ~10^{13} sequences).

In a recent attempt to benchmark the state of the art of computational protein-ligand interaction design, Allison and coworkers tested Rosetta's [12] sequence recovery (with respect to the wild type) capability over a set of 43 protein-ligand complexes [142]. The Rosetta protocol involved simultaneous ligand motion and side-chain rotamer discrete optimization. Overall, sequence recovery was more successful when (i) a near-optimal pose was inputted and subjected to limited sampling instead of blindly searched; (ii) the ligand was small, nonpolar, and rigid; and (iii) the binding pocket packing was neither overcrowded nor poor. Another interesting result was the significantly higher recovery for nonpolar residues. The authors suggested that new terms should be added to the energy function to correct this bias toward nonpolar interactions [142]. However, this bias could be an artifact of poor sampling, which might limit the accuracy of polar interaction estimation (see above). In fact, other suggested areas of improvement were the use of continuous, instead of discrete, sampling of backbone and side-chain rotamers [143], the concerted rotation of the backbone of two adjacent residues allowing larger side-chain motion (the so-called backrub motion) [144, 145], and the calculations of partition functions providing a link between molecular behavior and bulk thermodynamic quantities over structural ensembles [102, 146, 147]. All these features are grasped by OSPREY [143], a recent open-source solution to protein design which includes graphic processing units (GPU) acceleration [148], dead-end elimination algorithm [149, 150], and the K* method [150]. K* aims to approximate the partition function of the bound and unbound states over an ensemble of structures; their ratio provides an estimation of the binding constant. The conceptual advantage of this methodology is a mathematically rigorous, albeit approximate, free energy difference calculation that explicitly simulates the free ligand and protein. Consequently, ligand and binding site pre-organizations are, in principle, included in the calculation. On the other hand, this absolute free energy calculation is neither accurate nor efficient for systems with a large number of energy minima [151], requiring extensive sampling to reduce errors. However, the error of the method most likely compensates between complex and free monomer calculations, making this strategy a valuable tool for fast free binding energy simulations. Regardless of the methodology chosen, an effort to produce new experimental data will be fundamental to benchmark these high-throughput computational protocols and improve their predictive power [86].

An inaccurate description of the binding site is yet another possible source of error. Indeed, some mutations could shift the pK_a of ligand's and protein's titratable sites or introduce a new titratable residue. Therefore, the system should be prepared thoroughly before computational mutagenesis, and quick pK_a predictors [152] should be used to treat critical mutations. On the other hand, for situations where pK_a is close to the experimental pH conditions, simulation of all the possible combinations for the ambiguous titratable sites is required. For instance, a recent laccase design effort required the simulation of sinapic acid in both protonation states [153]: if one of the two protonation would have been picked, activity changes would have been missed since they mostly involved one of the two accessible protonation states. Finally, waters in the binding site might have an important role in binding, and their neglect could affect the quality of the calculation [154].

A way to filter and correct designs is based on MD simulations [155]. Many features of designed protein-ligand complexes can be inspected with this technique, including hydrogen bond geometries, binding site structural integrity, solvent exposure, and binding site pre-organization. In particular HTMD [156] was used by Baker and coworkers to filter computationally designed candidates according to the fraction of near-attack conformations (NAC), structures that resemble the transition state (TS) [157]. Moreover, MD can help in the future to discern long-range effects. In fact, it has been recently used to highlight the impact of distant mutations on active site pre-organization in evolved enzymes [158, 159]. Furthermore, proteins are dynamical entities organized in a network of correlated fluctuations whose changes can significantly affect binding at large distances [160]. Importantly, such network can be identified through a correlation matrix (which quantifies the correlation degree of a pair of amino acids) and partitioned in communities of highly correlated residues, giving insights on allosteric interactions [161]. These analyses, with contact and hydrogen bond maps, might be used in the future to identify regions whose motion influences the binding site's dynamics (e.g., making the side chain of a catalytic residue too flexible), which can then be subjected to mutagenesis in the lab.

A possible error when studying protein-ligand association arises when focusing mostly on the binding site, as some mutations along a possible entrance channel could hinder the ligand entrance/exit process. Sampling algorithms like PELE

[113, 162] can help to recognize such mutations. Its combination of ENM, side-chain prediction, ligand perturbation (translations and rotations), all-atom minimization, and implicit solvation makes it a suitable tool to quickly map the whole ligand migration process with good accuracy, taking protein flexibility into account [163–165]. Analogous MD-based techniques, such as HTMD [156], RAMD [166], and steered MD [167], have addressed this problem. These provide more accurate simulations, as it explicitly models water molecules, but also significantly more expensive ones, difficult to apply to massive mutation studies. Additional tools such as Fpocket [168] or CAVER [169] are widely used to quickly identify tunnels and cavities. These techniques, however, do not explicitly simulate ligand or protein dynamics.

Finally, quantum chemical calculations can be used to validate promising mutations, especially when charge transfer and polarization have an important role in the binding process. Mixed QM/MM schemes [170], widely used to model large systems, can significantly improve protein-ligand binding prediction directly, through explicit energy calculations [171], or indirectly [172] by recalculating ligand's atomic charges in an attempt to model ligand polarization effects. An alternative, more accurate but slower approach to large systems is the fragment molecular orbital (FMO) method [173]. FMO divides a system in N nonoverlapping fragments (e.g., one for each protein residue and ligand) and calculates the total energy as the sum of one-body fragment energies and two-body interaction energy corrections, providing a ~N^2 scalable fully parallelizable QM calculation. Jensen and coworkers used this methodology to energetically score the cleavability of peptides by HIV-1 protease [174] by looking at the protein-peptide interaction energy.

10.2.3 Catalytic Rate Constant Enhancement

The improvement of catalytic activity of bond breaking/formation passes through the modeling of the TS of the slowest chemical step of the targeted reaction; see below for electron transfer (ET) processes. The problems with the design of optimal activation energies are multiple: (i) the energy function should be sensitive enough to effectively discriminate between the reactant (substrate) and the TS, whose charges and geometries might be similar; (ii) the nature of the TS can change upon mutation; and (iii) activation energies are very sensitive to molecular geometry changes.

In OptZyme the TS is approximated by a transition state analogue (TSA), a stable molecule which electronically and sterically resembles the TS [175]. Once the TSA and the substrate are docked in the active site, two parameters drive mutant selection: the enzyme-substrate (ES) and the enzyme-TSA interaction energies. These last two quantities are obtained using classical force fields. Through a number of conceptual and mathematical simplifications, the authors show that the former energy correlates with the Michaelis constant (K_M) while the latter with the specificity constant (defined as the ration between k_{cat} and K_M). Although it yielded satisfactory correlations for their specific case, it is worth noting that these have no general validity. If the rate constant is comparable or much higher than the ES dissociation constant, the pre-equilibrium approximation is no longer valid. Then, the Michaelis constant cannot be approximated by the ES dissociation equilibrium constant (K_D) [176]. Notwithstanding, ES and enzyme-TSA interaction energies are still valuable tools for a fast semiquantitative evaluation of enzyme variants.

Khersonsky et al. combined computational design and directed evolution to optimize a previously designed Kemp eliminase [177]. As in the previous example, the classical (force field) interaction energy between the enzyme and an explicit TS model was the parameter to be optimized during the sequence exploration. The TS model, however, was derived from QM calculations in solution including key catalytic residues. The authors individuated three main factors for the improved activity: a more favorable electrostatic environment, a better packed active site, and a higher degree of active site pre-organization.

Although the last two methods, based on classical interaction energies, allow to test a big number of mutants, they both present a conceptual limitation: the use of the enzyme-TS (model) interaction energy, which is size-dependent (extensive property), to score the activation energy. To correct for this approximation, the ES interaction energy should be taken into account, providing a relative value. However, poor sampling and inaccurate energy functions might introduce uncertainties that could make its introduction useless (as it often happens in molecular mechanics/generalized born surface area (MM/GBSA) free energy calculations [178]). Still, they are currently the best approach to test a large number of mutations and find promising protein variants which can then be filtered with MD and quantum chemical methods [155].

The only way to properly calculate the activation energy barriers is the introduction of a QM methodology, capable of describing the electronic effects associated with TS formation. QM/MM schemes, which have been widely applied to elucidate enzymatic reaction mechanisms [179], have been employed to rescore promising candidates [155, 179]. A remarkable example is the hierarchical approach of Zheng et al. used to design a cocaine hydrolase [180]. Firstly, the reaction coordinate and the TS of the rate-determining limiting step are determined in the wild type. Then, many mutations are scored according to their protein-TS interaction energy; if this is lower than the wild type, a QM/MM calculation along the reaction coordinate is used to estimate the energy barrier. To allow fast computation, the authors use a reaction coordinate approach, freezing at each step the reactive coordinate and minimizing all the other degrees of freedom. A main drawback is still the need of extensive sampling, which makes the presented methodology too expensive for a general use.

A cheaper alternative to QM/MM calculations is empirical valence bond (EVB) [181]. EVB is based on a semiempirical Hamiltonian which describes reactants and products with their resonance structure (explicitly defining atom connectivity).

Although EVB energies are less accurate than ab initio and DFT QM/MM methods, free energy calculations are orders of magnitude quicker and still can achieve accurate results [182, 183], making EVB a suitable tool to score a bigger number of mutants [184].

In the attempt to describe the entire enzyme or a large part of it with QM calculations, Jensen and coworkers approximated the reaction coordinate with the linear interpolation between reactant and product optimized geometries and calculated each point with semiempirical methods [185–187]. These fast electronic calculations, united with algorithms for large-scale systems such as FMO [173, 188] or the much faster effective fragment molecular orbital (EFMO) [189], make "ab initio biochemistry" [190] closer, albeit still far away for design purposes.

In the particular case of oxidoreductases, where charge transfer processes dominate, additional complexity is added to the protein design problem. Electrons must move from a donor to an acceptor, sometimes through a long-range electron transfer. According to Marcus' theory, the ET rate constant [191] depends on three parameters: (i) electronic coupling, the probability to jump from the reactant to the product's diabatic state, which exponentially depends on the donor-acceptor distance; (ii) reorganization energy, which is the energy penalty that accompanies electron transfer; and (iii) the free energy difference between product and reactant (driving force). The ET rate constant has a maximum when the sum of reorganization energy and driving force equals zero. Although accurate QM/MM methodologies have been developed to study electron transfer rate in proteins [192, 193], their use in enzyme design is limited by their computational cost. To overcome this barrier, we have recently developed a new methodology to approximately evaluate ET rates, which combines fast conformational sampling [162] and quick QM/MM spin density calculations and has been used to evaluate the activity of laccase variants [153, 194]. While PELE provides a thorough and quick mapping of enzyme's and substrate's dynamics, substrate's spin density permits to promptly score the relative changes in driving force upon mutation (the higher the spin density, in principle, the higher the driving force). This protocol was successfully applied to the rational design of a laccase toward the production of conductive polyaniline at low pH [195]. In the same spirit, we rationally improved the oxidation rate of 2,2′-azino-bis(3-ethylbenzothiazoline-6-sulfonic acid) (ABTS) by a highly stable manganese peroxidase (Fig. 10.2) relying on electron coupling calculations, after the entire protein-ligand migration studies were performed [196]. In this case, it was assumed that the driving force and the reorganization energies did not change upon mutation, which can be a reasonable approximation in surface ET.

Since long-range ET is often the rate-limiting step in catalysis, engineering efforts have also centered on mutating residues along the ET pathway. By using the QM/MM e-pathway method [197], Vidal-Limon et al. studied P450BM3's suicide inactivation [198], a common process in heme peroxidases. From the QM-MM calculations, they identified key residues in the second heme coordination sphere, aiming at reducing electron delocalization and obtaining a more stable enzyme against H_2O_2. After mass spectrometry assays confirmed the oxidized sites predicted by QM/MM, they generated a variant 260 times more stable against H_2O_2 inactivation.

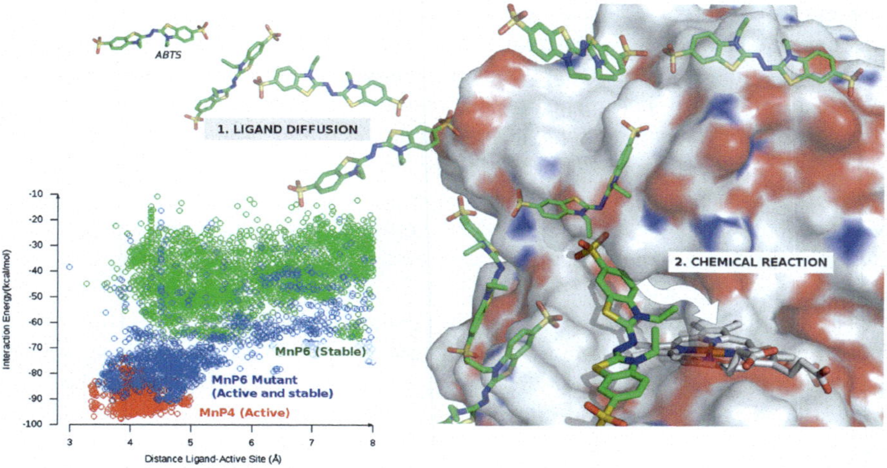

Fig. 10.2 General scheme for the rational MnP6 engineering [196]. The study was divided into two steps: (1) ligand diffusion and (2) electronic transfer process. First, we compare the ligand diffusion in the active and inactive enzyme by computing the interaction energy (*red* and *green dots*, respectively) and the distance between ligand and receptor. From the active enzyme, we extract information about the active site environment that help us to redesign the inactive enzyme by introducing two specific surface mutations (*blue dots*). Importantly, these mutations involved solvent-exposed residues with low conservation and mutability score provided by bioinformatics techniques. The activation was confirmed *in silico* by electronic coupling calculations in the second step. Site-directed experimental mutagenesis validated the success of the new mutant, which combines both stability and activity

10.3 Computer-Aided Directed Evolution, a Perspective

In the previous sections, we have seen multiple examples of computationally driven enzyme engineering. While they use, to more or less degree, structure/function knowledge for the modeling, we observe a tendency toward more random massive sequence sampling; we expect to see in the near future full *in silico* directed evolution studies. By full we obviously do not refer to a complete study of the sequence space (all residues and all possible mutations), but to an exhaustive random mutagenesis combination on a large subset of selected mutants. For 100 residues, for example, we have a sequence space of ~10^{130}, which would take several lifetimes to be evaluated even if using current supercomputers. If we restrict the exploration to single, double, and triple mutants, we have now ~10^9 possible variants to model. While this is still a huge number, one can think in a hierarchical scheme where this sequence space can be explored in days. This is particularly true with the current (and future) developments in lower-cost multicore servers based on mobile technology (see, e.g., the MontBlanc project at https://www.montblanc-project.eu/).

We find a promising example in the work by Wijma et al. where a hierarchical protocol is used to increase thermostability [139]. In this line, we expect the development of additional techniques combining quick bioinformatics (or knowledge-based)

screening of a large sequence space, with a molecular modeling refinement of selected mutants. This last step could be further hierarchically broken down into a first classical molecular modeling screening followed by selected quantum chemical reevaluations, in a similar manner to the previously described study by Zheng et al. [180]. Even though computational techniques are becoming more precise and easy to implement, a synergic effort between *in silico* predictions and experimental validation will be, in our opinion, the preferred solution. Figure 10.3 shows a possible workflow combining these ideas. In order to apply such a combined effort, we should keep in mind that molecular modeling will require an accurate 3D structural model, a possible limiting factor.

The workflow must start with a careful preparation of the wild-type structure, a fundamental step as it determines the outcome of the computational design. This preparation should include some sampling, aimed at generating conformational diversity and providing useful information for design (such as the protein regions where to look for improved variants). We should emphasize that most molecular modeling predictions are based on relative values (rather than absolute ones), in which case a wild-type reference value is needed. In the next step, high-throughput screening of mutations is carried out with quick methods, such as bioinformatics and knowledge-based methods. This step will have to rank the initial sequence space, similar to a high-throughput screening in drug design. Taking into account that each bioinformatics score can be accomplished in less than a second, we can aim for several millions of mutants in a "doable" time; we are still looking at several days of hundreds/thousands of core dedication, a feasible effort, however, in near future multicore and accelerated computers (or cloud computing). Bioinformatics screening of millions (billions) of mutants will benefit from new sequencing and storage of mutational data in the years to come and, in particular, from its processing using machine learning techniques [199]. At the present stage, such techniques are mostly used to restrict mutagenesis to relevant protein regions [72–75, 77, 78]

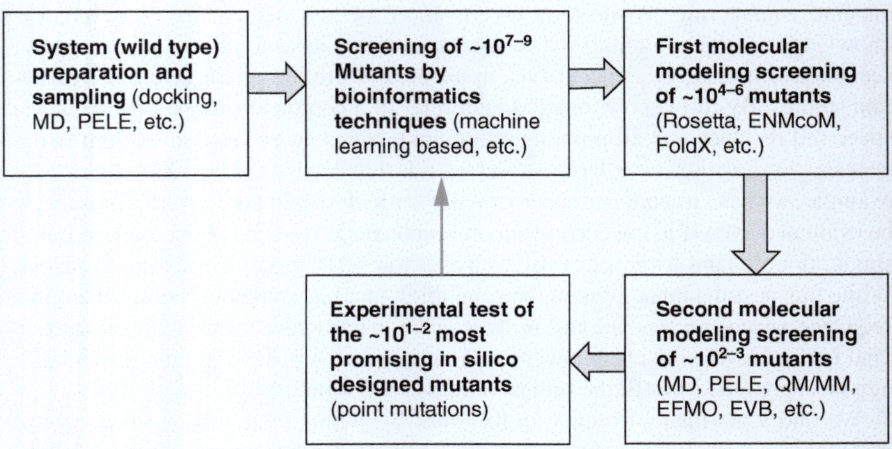

Fig. 10.3 Proposed computer-aided directed evolution workflow

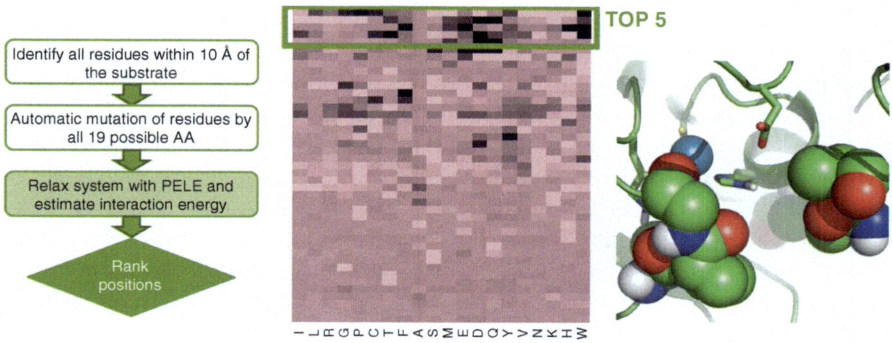

Fig. 10.4 *Left panel*: scheme of the used computational protocol. It includes identification of all residues close to the bound substrate, mutation of these amino acids (AA), system relaxation, and energy scoring. The image in the center is a heat map of the tested 43 mutations. The *pink color* corresponds to an equal or worse value than the wild type, while increasing *darker color* corresponds to improved classical scoring (protein-substrate interaction energy). From this heat map, the best positions (with the largest number of improved variants relative to the wild type) are identified, in this example the top five. On the *right panel*, we find the five amino acid positions (in van der Waals representation) suggested to be tested experimentally

or to guide directed evolution "on the fly" [200]. A second filtering, for example, could then be performed with fast molecular modeling techniques such as FoldX or Rosetta. These techniques could be applied to (the best ranked) several tens of thousands of compounds. The final goal is to provide a reduced set of candidates, few hundreds/thousands, where we can apply a more accurate molecular modeling refinement. The simulation time required for this last step will highly depend on two factors: (i) the exhaustiveness of conformational sampling and (ii) the nature of the property to improve. Conformational sampling aims to determine possible changes in the structure produced by the mutation. Quick assessments, in the order of minutes to hours, can be currently performed by Monte Carlo techniques [153]; using MD will require significantly more computational time, limiting the study to only few hundred mutants. Another important aspect is how to reevaluate the desired property to engineer. Substrate-binding energies and thermal stability could be quickly evaluated, for example, with alchemical MM free energy calculations (with respect to the wild type) [128]. Catalytic design, on the other hand, will require expensive QM calculations. Due to their very high cost (hours to days), future work will have to center in designing cheaper methods [201–203] and/or property evaluations. For example, in our current efforts in oxidoreductases' engineering, the driving force is approximated with the amount of spin density, calculated after five steps of QM/MM geometry optimization, localized on the substrate (with respect to the wild type) [194].

The proposed workflow includes an iterative computational-experimental approach, where several schemes could be imagined. Currently we find very few studies following such an approach, where we can underline, for example, Privett et al. [204]. When thinking of future implementation, an initial less accurate *in silico* evolution could be tested in the lab and a more accurate second one performed only on those regions that show more promising experimental results. Similarly, more accurate simulations could be performed only in single mutants, followed by experimental site-directed mutagenesis, and expanded then to a second *in silico* round involving double mutants and so on. In this way, synergic mutations can be partially recovered. An alternative strategy to retrieve cooperative mutations, while computing single mutants only, is to rank sequence positions instead of point mutations. Positions are ordered according to their frequency of beneficial mutations, following a fast computational saturated mutagenesis protocol, and the most promising are communicated to the experimental laboratory for (iterative) combinatorial saturated mutagenesis. Contrary to the previous strategy, false positives are not filtered out since a position, instead of a given mutation, is chosen. On the other hand, true positives can be recovered. We are currently employing this strategy to improve oxidoreductases' activity. In our initial test on a high redox potential fungal laccase, experimental and theoretical evolutions were run in parallel. In the first DE experimental generation, one improved variant was identified. In the *in silico* round, over 40 positions were screened with PELE, using the protein-substrate interaction energy after an induced-fit procedure, and the best five were identified (Fig. 10.4, central panel). Interestingly the improved variant found experimentally was within these five top positions. Then through combinatorial saturation mutagenesis using NDT degenerated codons, three new variants were identified recovering synergetic effect of two of the suggested *in silico* positions. This approach allows quickly guiding experimental mutagenesis: using 100 CPUs ~200 positions can be scored in 1 day. Although this protocol requires testing more mutants in the lab, it permits to go from the computer to the lab in 24 h with a focused library of mutants, an appealing feature for industrial purposes where large number of mutants can be assayed.

Conclusion

Biotechnology needs accurate enzyme evolution techniques, capable of designing new catalysts able to work in environmentally friendly conditions and, importantly, under industrial requirements. In this line, we find great efforts in developing (and improving) site-directed mutagenesis and directed evolution techniques, with computer simulations increasingly being used for this purpose. Different methodologies, from a quick bioinformatics sequence analysis to a robust solution of the Schrödinger equation, seek to aid the experimental efforts. In this book chapter, we overviewed recent developments in molecular modeling for three different engineering tasks: protein stability, protein-substrate binding, and catalytic rate, with the goal of illustrating how these techniques can influence directed evolution in the near future. We underline two key factors in future implementations: (i) hierarchical combination of different computational solutions (with increasing accuracy but also computational cost), and (ii) close iterative efforts between *in silico* and *in vitro* approaches.

References

1. Schmid A, Dordick JS, Hauer B, Kiener A, Wubbolts M, Witholt B (2001) Industrial biocatalysis today and tomorrow. Nature 409(6817):258–268
2. Patel RN (2008) Synthesis of chiral pharmaceutical intermediates by biocatalysis. Coord Chem Rev 252(5–7):659–701
3. Sukumaran J, Hanefeld U (2005) Enantioselective C-C bond synthesis catalysed by enzymes. Chem Soc Rev 34(6):530–542
4. Koenig SH, Brown RD (1972) H(2)CO(3) as substrate for carbonic anhydrase in the dehydration of HCO(3)(−). Proc Natl Acad Sci U S A 69(9):2422–2425
5. Bar-Even A, Noor E, Savir Y, Liebermeister W, Davidi D, Tawfik DS, Milo R (2011) The moderately efficient enzyme: evolutionary and physicochemical trends shaping enzyme parameters. Biochemistry 50(21):4402–4410
6. Milo R, Last RL (2012) Achieving diversity in the face of constraints: lessons from metabolism. Science 336(6089):1663–1667
7. Currin A, Swainston N, Day PJ, Kell DB (2015) Synthetic biology for the directed evolution of protein biocatalysts: navigating sequence space intelligently. Chem Soc Rev 44(5):1172–1239
8. Gutte B, Däumigen M, Wittschieber E (1979) Design, synthesis and characterisation of a 34-residue polypeptide that interacts with nucleic acids. Nature 281(5733):650–655
9. Russell AJ, Fersht AR (1987) Rational modification of enzyme catalysis by engineering surface charge. Nature 328(6130):496–500
10. Hellinga HW, Caradonna JP, Richards FM (1991) Construction of new ligand binding sites in proteins of known structure: II. Grafting of a buried transition metal binding site into *Escherichia coli* thioredoxin. J Mol Biol 222(3):787–803
11. Jemli S, Ayadi-Zouari D, Hlima HB, Bejar S (2016) Biocatalysts: application and engineering for industrial purposes. Crit Rev Biotechnol 36(2):246–258
12. Schueler-Furman O, Wang C, Bradley P, Misura K, Baker D (2005) Progress in modeling of protein structures and interactions. Science 310(5748):638–642
13. Steiner K, Schwab H (2012) Recent advances in rational approaches for enzyme engineering. Comput Struct Biotechnol J 2:e201209010
14. Richardson JS, Richardson DC (1989) The de novo design of protein structures. Trends Biochem Sci 14(7):304–309
15. Ponder JW, Richards FM (1987) Tertiary templates for proteins: use of packing criteria in the enumeration of allowed sequences for different structural classes. J Mol Biol 193(4):775–791
16. Bolon DN, Marcus JS, Ross SA, Mayo SL (2003) Prudent modeling of core polar residues in computational protein design. J Mol Biol 329(3):611–622
17. Dahiyat BI, Mayo SL (1997) Probing the role of packing specificity in protein design. Proc Natl Acad Sci U S A 94(19):10172–10177
18. Desjarlais JR, Handel TM (1999) Side-chain and backbone flexibility in protein core design1. J Mol Biol 290(1):305–318
19. Scrutton NS, Berry A, Perham RN (1990) Redesign of the coenzyme specificity of a dehydrogenase by protein engineering. Nature 343(6253):38–43
20. Carter P, Nilsson B, Burnier JP, Burdick D, Wells JA (1989) Engineering subtilisin BPN' for site-specific proteolysis. Proteins Struct Funct Bioinforma 6(3):240–248
21. Wells JA, Powers DB, Bott RR, Graycar TP, Estell DA (1987) Designing substrate specificity by protein engineering of electrostatic interactions. Proc Natl Acad Sci U S A 84(5):1219–1223
22. Cedrone F, Ménez A, Quéméneur E (2000) Tailoring new enzyme functions by rational redesign. Curr Opin Struct Biol 10(4):405–410
23. Looger LL, Dwyer MA, Smith JJ, Hellinga HW (2003) Computational design of receptor and sensor proteins with novel functions. Nature 423(6936):185–190

24. Craik C, Largman C, Fletcher T, Roczniak S, Barr P, Fletterick R, Rutter W (1985) Redesigning trypsin: alteration of substrate specificity. Science 228(4697):291–297
25. Bastianelli G, Bouillon A, Nguyen C, Crublet E, Pêtres S, Gorgette O, Le-Nguyen D, Barale J-C, Nilges M (2011) Computational reverse-engineering of a spider-venom derived peptide active against *Plasmodium falciparum* SUB1. PLoS ONE 6(7):e21812
26. Oelschlaeger P, Mayo SL (2005) Hydroxyl groups in the ββ sandwich of metallo-β-lactamases favor enzyme activity: a computational protein design study. J Mol Biol 350(3):395–401
27. Guerois R, Nielsen JE, Serrano L (2002) Predicting changes in the stability of proteins and protein complexes: a study of more than 1000 mutations. J Mol Biol 320(2):369–387
28. Yu H, Huang H (2014) Engineering proteins for thermostability through rigidifying flexible sites. Biotechnol Adv 32(2):308–315
29. Kuhlman B, Baker D (2004) Exploring folding free energy landscapes using computational protein design. Curr Opin Struct Biol 14(1):89–95
30. Kortemme T, Baker D (2004) Computational design of protein–protein interactions. Curr Opin Chem Biol 8(1):91–97
31. Kortemme T, Joachimiak LA, Bullock AN, Schuler AD, Stoddard BL, Baker D (2004) Computational redesign of protein-protein interaction specificity. Nat Struct Mol Biol 11(4):371–379
32. Reina J, Lacroix E, Hobson SD, Fernandez-Ballester G, Rybin V, Schwab MS, Serrano L, Gonzalez C (2002) Computer-aided design of a PDZ domain to recognize new target sequences. Nat Struct Mol Biol 9(8):621–627
33. Shifman JM, Mayo SL (2002) Modulating calmodulin binding specificity through computational protein design. J Mol Biol 323(3):417–423
34. Lippow SM, Tidor B (2007) Progress in computational protein design. Curr Opin Biotechnol 18(4):305–311
35. Ashworth J, Havranek JJ, Duarte CM, Sussman D, Monnat RJ, Stoddard BL, Baker D (2006) Computational redesign of endonuclease DNA binding and cleavage specificity. Nature 441(7093):656–659
36. Chevalier BS, Kortemme T, Chadsey MS, Baker D, Monnat RJ Jr, Stoddard BL (2002) Design, activity, and structure of a highly specific artificial endonuclease. Mol Cell 10(4):895–905
37. Cochran FV, Wu SP, Wang W, Nanda V, Saven JG, Therien MJ, DeGrado WF (2005) Computational de novo design and characterization of a four-helix bundle protein that selectively binds a nonbiological cofactor. J Am Chem Soc 127(5):1346–1347
38. Yang W, Wilkins AL, Ye Y, Z-r L, S-y L, Urbauer JL, Hellinga HW, Kearney A, van der Merwe PA, Yang JJ (2005) Design of a calcium-binding protein with desired structure in a cell adhesion molecule. J Am Chem Soc 127(7):2085–2093
39. Palmer AE, Giacomello M, Kortemme T, Hires SA, Lev-Ram V, Baker D, Tsien RY (2006) Ca2+ indicators based on computationally redesigned calmodulin-peptide pairs. Chem Biol 13(5):521–530
40. Lassila JK, Keeffe JR, Oelschlaeger P, Mayo SL (2005) Computationally designed variants of *Escherichia coli* chorismate mutase show altered catalytic activity. Protein Eng Des Sel 18(4):161–163
41. Bornscheuer UT, Pohl M (2001) Improved biocatalysts by directed evolution and rational protein design. Curr Opin Chem Biol 5(2):137–143
42. Faiella M, Andreozzi C, de Rosales RTM, Pavone V, Maglio O, Nastri F, DeGrado WF, Lombardi A (2009) An artificial di-iron oxo-protein with phenol oxidase activity. Nat Chem Biol 5(12):882–884
43. Tinberg CE, Khare SD, Dou J, Doyle L, Nelson JW, Schena A, Jankowski W, Kalodimos CG, Johnsson K, Stoddard BL, Baker D (2013) Computational design of ligand binding proteins with high affinity and selectivity. Nature 501(7466):212–216
44. Kaplan J, DeGrado WF (2004) De novo design of catalytic proteins. Proc Natl Acad Sci U S A 101(32):11566–11570

45. Dahiyat BI, Mayo SL (1997) De novo protein design: fully automated sequence selection. Science 278(5335):82–87
46. Jiang L, Althoff EA, Clemente FR, Doyle L, Röthlisberger D, Zanghellini A, Gallaher JL, Betker JL, Tanaka F, Barbas CF, Hilvert D, Houk KN, Stoddard BL, Baker D (2008) De novo computational design of retro-aldol enzymes. Science (New York, NY) 319(5868):1387–1391
47. Rothlisberger D, Khersonsky O, Wollacott AM, Jiang L, DeChancie J, Betker J, Gallaher JL, Althoff EA, Zanghellini A, Dym O, Albeck S, Houk KN, Tawfik DS, Baker D (2008) Kemp elimination catalysts by computational enzyme design. Nature 453(7192):190–195
48. Moroz YS, Dunston TT, Makhlynets OV, Moroz OV, Wu Y, Yoon JH, Olsen AB, McLaughlin JM, Mack KL, Gosavi PM, van Nuland NAJ, Korendovych IV (2015) New tricks for old proteins: single mutations in a nonenzymatic protein give rise to various enzymatic activities. J Am Chem Soc 137(47):14905–14911
49. Zanghellini A (2014) De novo computational enzyme design. Curr Opin Biotechnol 29:132–138
50. Petrounia IP, Arnold FH (2000) Designed evolution of enzymatic properties. Curr Opin Biotechnol 11(4):325–330
51. Arnold FH (2001) Combinatorial and computational challenges for biocatalyst design. Nature 409(6817):253–257
52. Minshull J, Willem Stemmer PC (1999) Protein evolution by molecular breeding. Curr Opin Chem Biol 3(3):284–290
53. Packer MS, Liu DR (2015) Methods for the directed evolution of proteins. Nat Rev Genet 16(7):379–394
54. Jaeger K-E, Eggert T (2004) Enantioselective biocatalysis optimized by directed evolution. Curr Opin Biotechnol 15(4):305–313
55. Jestin J-L, Kaminski PA (2004) Directed enzyme evolution and selections for catalysis based on product formation. J Biotechnol 113(103):85
56. Tao H, Cornish VW (2002) Milestones in directed enzyme evolution. Curr Opin Chem Biol 6(6):858–864
57. Williams GJ, Nelson AS, Berry A (2004) Directed evolution of enzymes for biocatalysis and the life sciences. Cell Mol Life Sci CMLS 61(24):3034–3046
58. Dalby PA (2003) Optimising enzyme function by directed evolution. Curr Opin Struct Biol 13(4):500–505
59. Bershtein S, Tawfik DS (2008) Advances in laboratory evolution of enzymes. Curr Opin Chem Biol 12(2):151–158
60. Park S, Morley KL, Horsman GP, Holmquist M, Hult K, Kazlauskas RJ (2005) Focusing mutations into the P. fluorescens esterase binding site increases enantioselectivity more effectively than distant mutations. Chem Biol 12 (1):45–54
61. Strausberg SL, Ruan B, Fisher KE, Alexander PA, Bryan PN (2005) Directed coevolution of stability and catalytic activity in calcium-free subtilisin. Biochemistry 44(9):3272–3279
62. Chockalingam K, Chen Z, Katzenellenbogen JA, Zhao H (2005) Directed evolution of specific receptor–ligand pairs for use in the creation of gene switches. Proc Natl Acad Sci U S A 102(16):5691–5696
63. Chica RA, Doucet N, Pelletier JN (2005) Semi-rational approaches to engineering enzyme activity: combining the benefits of directed evolution and rational design. Curr Opin Biotechnol 16(4):378–384
64. Hill CM, Li W-S, Thoden JB, Holden HM, Raushel FM (2003) Enhanced degradation of chemical warfare agents through molecular engineering of the phosphotriesterase active site. J Am Chem Soc 125(30):8990–8991
65. Reetz MT, Carballeira JD (2007) Iterative saturation mutagenesis (ISM) for rapid directed evolution of functional enzymes. Nat Protocol 2(4):891–903
66. Lutz S, Patrick WM (2004) Novel methods for directed evolution of enzymes: quality, not quantity. Curr Opin Biotechnol 15(4):291–297

67. Lutz S (2010) Beyond directed evolution–semi-rational protein engineering and design. Curr Opin Biotechnol 21(6):734–743
68. Lippow SM, Moon TS, Basu S, Yoon S-H, Li X, Chapman BA, Robison K, Lipovšek D, Prather KLJ (2010) Engineering enzyme specificity using computational design of a defined-sequence library. Chem Biol 17(12):1306–1315
69. Sebestova E, Bendl J, Brezovsky J, Damborský J (2014) Computational tools for designing smart libraries. Methods Mol Biol 1179:291–314
70. Voigt CA, Mayo SL, Arnold FH, Wang Z-G (2001) Computationally focusing the directed evolution of proteins. J Cell Biochem 84(S37):58–63
71. Zaugg J, Gumulya Y, Gillam EM, Boden M (2014) Computational tools for directed evolution: a comparison of prospective and retrospective strategies. Methods Mol Biol 1179:315–333
72. Damborsky J, Brezovsky J (2009) Computational tools for designing and engineering biocatalysts. Curr Opin Chem Biol 13(1):26–34
73. Pei J (2008) Multiple protein sequence alignment. Curr Opin Struct Biol 18(3):382–386
74. Pavelka A, Chovancova E, Damborsky J (2009) HotSpot Wizard: a web server for identification of hot spots in protein engineering. Nucleic Acids Res 37(suppl 2):W376–W383
75. Kuipers RK, Joosten H-J, van Berkel WJH, Leferink NGH, Rooijen E, Ittmann E, van Zimmeren F, Jochens H, Bornscheuer U, Vriend G, Martins dos Santos VAP, Schaap PJ (2010) 3DM: systematic analysis of heterogeneous superfamily data to discover protein functionalities. Proteins Struct Funct Bioinforma 78(9):2101–2113
76. Jochens H, Bornscheuer UT (2010) Natural diversity to guide focused directed evolution. ChemBioChem 11(13):1861–1866
77. Goldsmith M, Tawfik DS (2012) Directed enzyme evolution: beyond the low-hanging fruit. Curr Opin Struct Biol 22(4):406–412
78. Barak Y, Nov Y, Ackerley DF, Matin A (2007) Enzyme improvement in the absence of structural knowledge: a novel statistical approach. ISME J 2(2):171–179
79. Rosenberg M, Goldblum A (2006) Computational protein design: a novel path to future protein drugs. Curr Pharm Des 12(31):3973–3997
80. Poole AM, Ranganathan R (2006) Knowledge-based potentials in protein design. Curr Opin Struct Biol 16(4):508–513
81. Koder RL, Dutton PL (2006) Intelligent design: the de novo engineering of proteins with specified functions. Dalton Trans 25:3045–3051
82. Butterfoss GL, Kuhlman B (2006) Computer-based design of novel protein structures. Annu Rev Biophys Biomol Struct 35(1):49–65
83. Ambroggio XI, Kuhlman B (2006) Design of protein conformational switches. Curr Opin Struct Biol 16(4):525–530
84. Vizcarra CL, Mayo SL (2005) Electrostatics in computational protein design. Curr Opin Chem Biol 9(6):622–626
85. Morin A, Meiler J, Mizoue LS (2011) Computational design of protein-ligand interfaces: potential in therapeutic development. Trends Biotechnol 29(4):159–166
86. Malisi C, Schumann M, Toussaint NC, Kageyama J, Kohlbacher O, Höcker B (2012) Binding pocket optimization by computational protein design. PLoS ONE 7(12):e52505
87. Saven JG (2011) Computational protein design: engineering molecular diversity, nonnatural enzymes, nonbiological cofactor complexes, and membrane proteins. Curr Opin Chem Biol 15(3):452–457
88. Ollikainen N, Smith CA, Fraser JS, Kortemme T (2013) Methods in enzymology: "Flexible backbone sampling methods to model and design protein alternative conformations". Methods Enzymol 523:61–85
89. Park S, Yang X, Saven JG (2004) Advances in computational protein design. Curr Opin Struct Biol 14(4):487–494
90. Samish I, MacDermaid CM, Perez-Aguilar JM, Saven JG (2011) Theoretical and computational protein design. Annu Rev Phys Chem 62(1):129–149

91. Smith RD, Damm-Ganamet KL, Dunbar JB, Ahmed A, Chinnaswamy K, Delproposto JE, Kubish GM, Tinberg CE, Khare SD, Dou J, Doyle L, Stuckey JA, Baker D, Carlson HA (2015) CSAR benchmark exercise 2013: evaluation of results from a combined computational protein design, docking, and scoring/ranking challenge. J Chem Inf Model ASAP 56:1022
92. Wijma HJ, Janssen DB (2013) Computational design gains momentum in enzyme catalysis engineering. FEBS J 280(13):2948–2960
93. Boas FE, Harbury PB (2007) Potential energy functions for protein design. Curr Opin Struct Biol 17(2):199–204
94. Boas FE, Harbury PB (2008) Design of protein-ligand binding based on the molecular-mechanics energy model. J Mol Biol 380(2):415–424
95. Sirin S, Pearlman DA, Sherman W (2014) Physics-based enzyme design: predicting binding affinity and catalytic activity. Proteins Struct Funct Bioinforma 82(12):3397–3409
96. Wickstrom L, Gallicchio E, Levy RM (2012) The linear interaction energy method for the prediction of protein stability changes upon mutation. Proteins 80(1):111–125
97. Mendes J, Guerois R, Serrano L (2002) Energy estimation in protein design. Curr Opin Struct Biol 12(4):441–446
98. Schneider M, Fu X, Keating AE (2009) X-ray vs. NMR structures as templates for computational protein design. Proteins 77(1):97–110
99. Adamczyk AJ, Cao J, Kamerlin SCL, Warshel A (2011) Catalysis by dihydrofolate reductase and other enzymes arises from electrostatic preorganization, not conformational motions. Proc Natl Acad Sci U S A 108(34):14115–14120
100. Gagné D, French Rachel L, Narayanan C, Simonović M, Agarwal Pratul K, Doucet N (2015) Perturbation of the conformational dynamics of an active-site loop alters enzyme activity. Structure 23(12):2256–2266
101. Bhabha G, Lee J, Ekiert DC, Gam J, Wilson IA, Dyson HJ, Benkovic SJ, Wright PE (2011) A dynamic knockout reveals that conformational fluctuations influence the chemical step of enzyme catalysis. Science 332(6026):234–238
102. Allen BD, Nisthal A, Mayo SL (2010) Experimental library screening demonstrates the successful application of computational protein design to large structural ensembles. Proc Natl Acad Sci 107(46):19838–19843
103. Fu X, Apgar JR, Keating AE (2007) Modeling backbone flexibility to achieve sequence diversity: the design of novel α-helical ligands for Bcl-xL. J Mol Biol 371(4):1099–1117
104. Smith CA, Kortemme T (2008) Backrub-like backbone simulation recapitulates natural protein conformational variability and improves mutant side-chain prediction. J Mol Biol 380(4):742–756
105. Lassila JK (2010) Conformational diversity and computational enzyme design. Curr Opin Chem Biol 14(5):676–682
106. Mandell DJ, Kortemme T (2009) Backbone flexibility in computational protein design. Curr Opin Biotechnol 20(4):420–428
107. Marrink SJ, Risselada HJ, Yefimov S, Tieleman DP, De Vries AH (2007) The MARTINI force field: coarse grained model for biomolecular simulations. J Phys Chem B 111(27):7812–7824
108. Bowen JP, Allinger NL (2007) Molecular mechanics: the art and science of parameterization. Rev Comput Chem 2:81–97
109. Doruker P, Atilgan AR, Bahar I (2000) Dynamics of proteins predicted by molecular dynamics simulations and analytical approaches: application to α-amylase inhibitor. Proteins Struct Funct Bioinforma 40(3):512–524
110. Berendsen H (1988) Dynamic simulation as an essential tool in molecular modeling. J Comput Aided Mol Des 2(3):217–221
111. Grossman J, Towles B, Greskamp B, Shaw DE (2015) Filtering, reductions and synchronization in the anton 2 network. In: Parallel and Distributed Processing Symposium (IPDPS), 2015 IEEE International. IEEE, pp 860–870

112. Rathore N, de Pablo JJ (2002) Monte Carlo simulation of proteins through a random walk in energy space. J Chem Phys 116(16):7225–7230
113. Borrelli KW, Vitalis A, Alcantara R, Guallar V (2005) PELE: protein energy landscape exploration. A novel Monte Carlo based technique. J Chem Theory Comput 1(6):1304–1311
114. Cabeza de Vaca I, Lucas MF, Guallar V (2015) New Monte Carlo based technique to study DNA–ligand interactions. J Chem Theory Comput 11(12):5598–5605
115. Borrelli KW, Cossins B, Guallar V (2010) Exploring hierarchical refinement techniques for induced fit docking with protein and ligand flexibility. J Comput Chem 31(6):1224–1235
116. Halperin I, Ma B, Wolfson H, Nussinov R (2002) Principles of docking: an overview of search algorithms and a guide to scoring functions. Proteins Struct Funct Bioinforma 47(4):409–443
117. Gao J, Truhlar DG (2002) Quantum mechanical methods for enzyme kinetics. Annu Rev Phys Chem 53(1):467–505
118. Korkegian A (2005) Computational thermostabilization of an enzyme. Science 308(5723):857–860
119. Kellogg EH, Leaver-Fay A, Baker D (2011) Role of conformational sampling in computing mutation-induced changes in protein structure and stability. Proteins 79(3):830–838
120. Dunbrack RL Jr (2002) Rotamer libraries in the 21st century. Curr Opin Struct Biol 12(4):431–440
121. Kuhlman B, Baker D (2000) Native protein sequences are close to optimal for their structures. Proc Natl Acad Sci 97(19):10383–10388
122. Potapov V, Cohen M, Schreiber G (2009) Assessing computational methods for predicting protein stability upon mutation: good on average but not in the details. Protein Eng Des Sel 22(9):553–560
123. Estrada J, Echenique P, Sancho J (2015) Predicting stabilizing mutations in proteins using Poisson-Boltzmann based models: study of unfolded state ensemble models and development of a successful binary classifier based on residue interaction energies. Phys Chem Chem Phys 17(46):31044–31054
124. Karplus M, Ichiye T, Pettitt BM (1987) Configurational entropy of native proteins. Biophys J 52(6):1083–1085
125. Chong S-H, Ham S (2015) Dissecting protein configurational entropy into conformational and vibrational contributions. J Phys Chem B 119(39):12623–12631
126. Frappier V, Chartier M, Najmanovich RJ (2015) ENCoM server: exploring protein conformational space and the effect of mutations on protein function and stability. Nucleic Acids Res 43(W1):W395–W400
127. Frappier V, Najmanovich RJ (2014) A coarse-grained elastic network atom contact model and its use in the simulation of protein dynamics and the prediction of the effect of mutations. PLoS Comput Biol 10(4):e1003569
128. Seeliger D, Daniel S, de Groot BL (2010) Protein thermostability calculations using alchemical free energy simulations. Biophys J 98(10):2309–2316
129. Huang X, Gao D, Zhan C-G (2011) Computational design of a thermostable mutant of cocaine esterase via molecular dynamics simulations. Org Biomol Chem 9(11):4138–4143
130. Joo JC, Pack SP, Kim YH, Yoo YJ (2011) Thermostabilization of Bacillus circulans xylanase: computational optimization of unstable residues based on thermal fluctuation analysis. J Biotechnol 151(1):56–65
131. Lee C-W, Wang H-J, Hwang J-K, Tseng C-P (2014) Protein thermal stability enhancement by designing salt bridges: a combined computational and experimental study. PLoS ONE 9(11):e112751
132. Pikkemaat MG, Linssen ABM, Berendsen HJC, Janssen DB (2002) Molecular dynamics simulations as a tool for improving protein stability. Protein Eng 15(3):185–192
133. Gribenko AV, Patel MM, Liu J, McCallum SA, Wang C, Makhatadze GI (2009) Rational stabilization of enzymes by computational redesign of surface charge-charge interactions. Proc Natl Acad Sci 106(8):2601–2606
134. Spector S, Wang M, Carp SA, Robblee J, Hendsch ZS, Fairman R, Tidor B, Raleigh DP (2000) Rational modification of protein stability by the mutation of charged surface residues. Biochemistry 39(5):872–879

135. Schweiker KL, Arash Z-A, Davidson AR, Makhatadze GI (2007) Computational design of the Fyn SH3 domain with increased stability through optimization of surface charge-charge interactions. Protein Sci 16(12):2694–2702
136. Borgo B, Havranek JJ (2012) Automated selection of stabilizing mutations in designed and natural proteins. Proc Natl Acad Sci U S A 109(5):1494–1499
137. Hendsch ZS, Thorlakur J, Sauer RT, Bruce T (1996) Protein stabilization by removal of unsatisfied polar groups: computational approaches and experimental tests. Biochemistry 35(24):7621–7625
138. Koudelakova T, Chaloupkova R, Brezovsky J, Prokop Z, Sebestova E, Hesseler M, Khabiri M, Plevaka M, Kulik D, Kuta Smatanova I, Rezacova P, Ettrich R, Bornscheuer UT, Damborsky J (2013) Engineering enzyme stability and resistance to an organic cosolvent by modification of residues in the access tunnel. Angew Chem Int Ed Engl 52(7):1959–1963
139. Wijma HJ, Floor RJ, Jekel PA, Baker D, Marrink SJ, Janssen DB (2014) Computationally designed libraries for rapid enzyme stabilization. Protein Eng Des Sel 27(2):49–58
140. Wijma HJ, Floor RJ, Janssen DB (2013) Structure- and sequence-analysis inspired engineering of proteins for enhanced thermostability. Curr Opin Struct Biol 23(4):588–594
141. Schreier B, Stumpp C, Wiesner S, Hocker B (2009) Computational design of ligand binding is not a solved problem. Proc Natl Acad Sci 106(44):18491–18496
142. Allison B, Combs S, DeLuca S, Lemmon G, Mizoue L, Meiler J (2014) Computational design of protein-small molecule interfaces. J Struct Biol 185(2):193–202
143. Gainza P, Roberts KE, Georgiev I, Lilien RH, Keedy DA, Chen C-Y, Reza F, Anderson AC, Richardson DC, Richardson JS, Donald BR (2013) OSPREY: protein design with ensembles, flexibility, and provable algorithms. Methods Enzymol 523:87–107
144. Keedy DA, Georgiev I, Triplett EB, Donald BR, Richardson DC, Richardson JS (2012) The role of local backrub motions in evolved and designed mutations. PLoS Comput Biol 8(8):e1002629
145. Davis IW, Bryan Arendall W, Richardson DC, Richardson JS (2006) The backrub motion: how protein backbone shrugs when a sidechain dances. Structure 14(2):265–274
146. Chen C-Y, Georgiev I, Anderson AC, Donald BR (2009) Computational structure-based redesign of enzyme activity. Proc Natl Acad Sci U S A 106(10):3764–3769
147. Frey KM, Georgiev I, Donald BR, Anderson AC (2010) Predicting resistance mutations using protein design algorithms. Proc Natl Acad Sci U S A 107(31):13707–13712
148. Zhou Y, Xu W, Donald BR, Zeng J (2014) An efficient parallel algorithm for accelerating computational protein design. Bioinformatics 30(12):i255–i263
149. Hallen MA, Keedy DA, Donald BR (2013) Dead-end elimination with perturbations (DEEPer): a provable protein design algorithm with continuous sidechain and backbone flexibility. Proteins 81(1):18–39
150. Lilien RH, Stevens BW, Anderson AC, Donald BR (2005) A novel ensemble-based scoring and search algorithm for protein redesign and its application to modify the substrate specificity of the gramicidin synthetase a phenylalanine adenylation enzyme. J Comput Biol 12(6):740–761
151. Leach AR (2001) Molecular modelling: principles and applications. Pearson Education, New York
152. Shields GC, Seybold PG (2013) Computational approaches for the prediction of pKa values. CRC Press, Boca Raton
153. Pardo I, Santiago G, Gentili P, Lucas F, Monza E, Medrano F, Galli C, Martínez A, Guallar V, Camarero S (2016) Re-designing the substrate binding pocket of laccase for enhanced oxidation of sinapic acid. Catal Sci Technol ASAP 6:3900
154. Young T, Abel R, Kim B, Berne BJ, Friesner RA (2007) Motifs for molecular recognition exploiting hydrophobic enclosure in protein–ligand binding. Proc Natl Acad Sci 104(3):808–813
155. Kiss G, Çelebi-Ölçüm N, Moretti R, Baker D, Houk KN (2013) Computational enzyme design. Angew Chem Int Ed 52(22):5700–5725
156. Doerr S, De Fabritiis G (2014) On-the-fly learning and sampling of ligand binding by high-throughput molecular simulations. J Chem Theory Comput 10(5):2064–2069

157. Wijma HJ, Floor RJ, Bjelic S, Marrink SJ, Baker D, Janssen DB (2015) Enantioselective enzymes by computational design and in silico screening. Angew Chem Int Ed Engl 54(12):3726–3730
158. Jiménez-Osés G, Osuna S, Gao X, Sawaya MR, Gilson L, Collier SJ, Huisman GW, Yeates TO, Tang Y, Houk KN (2014) The role of distant mutations and allosteric regulation on LovD active site dynamics. Nat Chem Biol 10(6):431–436
159. Osuna S, Jiménez-Osés G, Noey EL, Houk KN (2015) Molecular dynamics explorations of active site structure in designed and evolved enzymes. Acc Chem Res 48(4):1080–1089
160. DuBay KH, Bowman GR, Geissler PL (2015) Fluctuations within folded proteins: implications for thermodynamic and allosteric regulation. Acc Chem Res 48(4):1098–1105
161. Sethi A, Eargle J, Black AA, Luthey-Schulten Z (2009) Dynamical networks in tRNA:protein complexes. Proc Natl Acad Sci U S A 106(16):6620–6625
162. Madadkar-Sobhani A, Guallar V (2013) PELE web server: atomistic study of biomolecular systems at your fingertips. Nucleic Acids Res 41(Web Server issue):W322–W328
163. Lucas MF, Guallar V (2012) An atomistic view on human hemoglobin carbon monoxide migration processes. Biophys J 102(4):887–896
164. Takahashi R, Gil VA, Guallar V (2014) Monte Carlo free ligand diffusion with Markov state model analysis and absolute binding free energy calculations. J Chem Theory Comput 10(1):282–288
165. Hosseini A, Brouk M, Lucas MF, Glaser F, Fishman A, Guallar V (2015) Atomic picture of ligand migration in toluene 4-monooxygenase. J Phys Chem B 119(3):671–678
166. Lüdemann SK, Lounnas V, Wade RC (2000) How do substrates enter and products exit the buried active site of cytochrome P450cam? 1. Random expulsion molecular dynamics investigation of ligand access channels and mechanisms. J Mol Biol 303(5):797–811
167. Grubmüller H, Heymann B, Tavan P (1996) Ligand binding: molecular mechanics calculation of the streptavidin-biotin rupture force. Science 271(5251):997–999
168. Le Guilloux V, Schmidtke P, Tuffery P (2009) Fpocket: an open source platform for ligand pocket detection. BMC Bioinforma 10(1):168
169. Chovancova E, Eva C, Antonin P, Petr B, Ondrej S, Jan B, Barbora K, Artur G, Vilem S, Martin K, Petr M, Lada B, Jiri S, Jiri D (2012) CAVER 3.0: a tool for the analysis of transport pathways in dynamic protein structures. PLoS Comput Biol 8(10):e1002708
170. Senn HM, Walter T (2009) QM/MM methods for biomolecular systems. Angew Chem Int Ed 48(7):1198–1229
171. Chaskar P, Prasad C, Vincent Z, Röhrig UF (2014) Toward on-the-fly quantum mechanical/molecular mechanical (QM/MM) docking: development and benchmark of a scoring function. J Chem Inf Model 54(11):3137–3152
172. Cho AE, Victor G, Berne BJ, Richard F (2005) Importance of accurate charges in molecular docking: quantum mechanical/molecular mechanical (QM/MM) approach. J Comput Chem 26(9):915–931
173. Fedorov DG, Nagata T, Kitaura K (2012) Exploring chemistry with the fragment molecular orbital method. Phys Chem Chem Phys 14(21):7562–7577
174. Jensen JH, Willemoës M, Winther JR, De Vico L (2014) In silico prediction of mutant HIV-1 proteases cleaving a target sequence. PLoS ONE 9(5):e95833
175. Grisewood MJ, Gifford NP, Pantazes RJ, Li Y, Cirino PC, Janik MJ, Maranas CD (2013) OptZyme: computational enzyme redesign using transition state analogues. PLoS ONE 8(10):e75358
176. Atkins PW (1998) Physical chemistry. W H Freeman & Company, New York
177. Khersonsky O, Rothlisberge D, Wollacott AM, Dym O, Baker D, Tawfik DS (2011) Optimization of the in silico designed Kemp eliminase KE70 by computational design and directed evolution. J Mol Biol 407(3):391–412
178. Genheden S, Samuel G, Ulf R (2015) The MM/PBSA and MM/GBSA methods to estimate ligand-binding affinities. Expert Opin Drug Discov 10(5):449–461
179. van der Kamp MW, Mulholland AJ (2013) Combined quantum mechanics/molecular mechanics (QM/MM) methods in computational enzymology. Biochemistry 52(16):2708–2728

180. Zheng F, Fang Z, Wenchao Y, Mei-Chuan K, Junjun L, Hoon C, Daquan G, Min T, Hsin-Hsiung T, Woods JH, Chang-Guo Z (2008) Most efficient cocaine hydrolase designed by virtual screening of transition states. J Am Chem Soc 130(36):12148–12155
181. Kamerlin SCL, Arieh W (2011) The empirical valence bond model: theory and applications. Wiley Interdiscip Rev Comput Mol Sci 1(1):30–45
182. Frushicheva MP, Cao J, Chu ZT, Warshel A (2010) Exploring challenges in rational enzyme design by simulating the catalysis in artificial kemp eliminase. Proc Natl Acad Sci U S A 107(39):16869–16874
183. Frushicheva MP, Cao J, Warshel A (2011) Challenges and advances in validating enzyme design proposals: the case of kemp eliminase catalysis. Biochemistry 50(18):3849–3858
184. Amrein BA, Ireneusz Szeler F, Purg M, Kulkarni Y, Kamerlin SCL (2017) CADEE: Computer-aided directed evolution of enzymes. IUCrJ 4:50–64.
185. Hediger MR, De Vico L, Svendsen A, Besenmatter W, Jensen JH (2012) A computational methodology to screen activities of enzyme variants. PLoS ONE 7(12):e49849
186. Hediger MR, Casper S, De Vico L, Jensen JH (2013) A computational method for the systematic screening of reaction barriers in enzymes: searching for Bacillus circulans xylanase mutants with greater activity towards a synthetic substrate. PeerJ 1:e111
187. Hediger MR, De Vico L, Rannes JB, Christian J, Werner B, Allan S, Jensen JH (2013) In silico screening of 393 mutants facilitates enzyme engineering of amidase activity in CalB. PeerJ 1:e145
188. Ito M, Mika I, Tore B (2014) Novel approach for identifying key residues in enzymatic reactions: proton abstraction in ketosteroid isomerase. J Phys Chem B 118(46):13050–13058
189. Steinmann C, Fedorov DG, Jensen JH (2012) The effective fragment molecular orbital method for fragments connected by covalent bonds. PLoS ONE 7(7):e41117
190. Steinmann C, Casper S, Fedorov DG, Jensen JH (2013) Mapping enzymatic catalysis using the effective fragment molecular orbital method: towards all ab initio biochemistry. PLoS ONE 8(4):e60602
191. Marcus RA (1993) Electron transfer reactions in chemistry. Theory and experiment. Rev Mod Phys 65(3):599–610
192. Blumberger J, Jochen B (2008) Free energies for biological electron transfer from QM/MM calculation: method, application and critical assessment. Phys Chem Chem Phys 10(37):5651
193. Wallrapp FH, Voityuk AA, Guallar V (2013) In-silico assessment of protein-protein electron transfer. A case study: cytochrome c peroxidase–cytochrome c. PLoS Comput Biol 9(3):e1002990
194. Monza E, Lucas MF, Camarero S, Alejaldre LC, Martínez AT, Guallar V (2015) Insights into laccase engineering from molecular simulations: toward a binding-focused strategy. J Phys Chem Lett 6(8):1447–1453
195. Gerard S, Felipe de S, Fátima Lucas M, Emanuele M, Sandra A, Ángel TM, Susana Camarero, VG (2016) Computer-aided laccase engineering: toward biological oxidation of arylamines. ACS Catalysis, 6:5415–5423
196. Acebes S, Fernandez-Fueyo E, Monza E, Lucas M, Almendral D, Ruiz-Dueñas FJ, Lund H, Martinez AT, Guallar V (2016) Rational enzyme engineering through biophysical and biochemical modeling. ACS Catal ACS Catalysis 6(3):1624–1629
197. Guallar V, Wallrapp F (2008) Mapping protein electron transfer pathways with QM/MM methods. J R Soc Interface 5(0):S233
198. Vidal-Limón A, Águila S, Ayala M, Batista CV, Vazquez-Duhalt R (2013) Peroxidase activity stabilization of cytochrome P450 BM3 by rational analysis of intramolecular electron transfer. J Inorg Biochem 122:18–26
199. Fox RJ, Huisman GW (2008) Enzyme optimization: moving from blind evolution to statistical exploration of sequence–function space. Trends Biotechnol 26(3):132–138
200. Feng X, Sanchis J, Reetz MT, Rabitz H (2012) Enhancing the efficiency of directed evolution in focused enzyme libraries by the adaptive substituent reordering algorithm. Chem Eur J 18(18):5646–5654

201. Cui Q, Elstner M (2014) Density functional tight binding: values of semi-empirical methods in an ab initio era. Phys Chem Chem Phys 16(28):14368–14377
202. Christensen AS, Elstner M, Cui Q (2015) Improving intermolecular interactions in DFTB3 using extended polarization from chemical-potential equalization. J Chem Phys 143(8):084123
203. Yilmazer ND, Korth M (2015) Enhanced semiempirical QM methods for biomolecular interactions. Comput Struct Biotechnol J 13:169–175
204. Privett HK, Kiss G, Lee TM, Blomberg R, Chica RA, Thomas LM, Hilvert D, Houk KN, Mayo SL (2012) Iterative approach to computational enzyme design. Proc Natl Acad Sci 109(10):3790–3795

The manufacturer's authorised representative in the EU is Springer Nature Customer Service Centre GmbH, Europaplatz 3, 69115 Heidelberg, Germany. If you have any concerns regarding our products, please contact ProductSafety@springernature.com

Printed and bound by CPI Group (UK) Ltd, Croydon, CR0 4YY
05/10/2026
02228129-0006